Social Foraging Theory

MONOGRAPHS IN BEHAVIOR AND ECOLOGY

Edited by John R. Krebs and Tim Clutton-Brock

Five New World Primates: A Study in Comparative Ecology, by *John Terborgh*

Reproductive Decisions: An Economic Analysis of Gelada Baboon Social Strategies, *by R.I.M. Dunbar*

Honeybee Ecology: A Study of Adaptation in Social Life, by *Thomas D. Seeley*

Foraging Theory, *by David W. Stephens and John R. Krebs*

Helping and Communal Breeding in Birds: Ecology and Evolution, *by Jerram L. Brown*

Dynamic Modeling in Behavioral Ecology, *by Marc Mangel and Colin W. Clark*

The Biology of the Naked Mole-Rat, *edited by Paul W. Sherman, Jennifer U. M. Jarvis, and Richard D. Alexander*

The Evolution of Parental Care, *by T. H. Clutton-Brock*

The Ostrich Communal Nesting System, *by Brian C. R. Bertram*

Parasitoids: Behavioral and Evolutionary Ecology, *by H.C.J. Godfray*

Sexual Selection, *by Malte Andersson*

Polygyny and Sexual Selection in Red-Winged Blackbirds, *by William A. Searcy and Ken Yasukawa*

Leks, *by Jacob Höglund and Rauno V. Alatalo*

Social Evolution in Ants, *by Andrew F. G. Bourke and Nigel R. Franks*

Female Control: Sexual Selection by Cryptic Female Choice, *by William G. Eberhard*

Sex, Color, and Mate Choice in Guppies, *by Anne E. Houde*

Foundations of Social Evolution, *by Steven A. Frank*

Parasites in Social Insects, *by Paul Schmid-Hempel*

Levels of Selection in Evolution, *edited by Laurent Keller*

Social Foraging Theory, *by Luc-Alain Giraldeau and Thomas Caraco*

Social Foraging Theory

LUC-ALAIN GIRALDEAU
AND THOMAS CARACO

Princeton University Press
Princeton, New Jersey

Published by Princeton University Press, 41 William Street, Princeton, New Jersey 08540
In the United Kingdom: Princeton University Press, Chichester, West Sussex

Library of Congress Cataloging-in-Publication Data

Giraldeau, Luc-Alain, 1955–
Social foraging theory / Luc-Alain Giraldeau and Thomas Caraco.
p. cm. — (Monographs in behavior and ecology)
Includes bibliographical references and index.
ISBN 0-691-04876-2 (cl : alk. paper) —
ISBN 0-691-04877-0 (pb : alk. paper)
1. Animals—Food—Mathematical models.
2. Social behavior in animals—Mathematical models. I. Caraco, Thomas, 1946–
II. Title. III. Series.
QL756.5.G57 2000
591.56'5118—dc21 99-052838

This book has been composed in Times Roman and Univers Light

The paper used in this publication meets the minimum requirements of ANSI/NISO Z39.48-1992 (R1997) (*Permanence of Paper*)

http://pup.princeton.edu

Printed in the United States of America

1 3 5 7 9 10 8 6 4 2

1 3 5 7 9 10 8 6 4 2
(Pbk.)

To my parents, Germaine Levac and Aimé Giraldeau, who so often told me, *"Reste à l'école."* I did. And to my children, Félix and Ophélie, to whom I will say the same.—L.-A.G.

Contents

Preface

The sciences do not try to explain, they hardly even try to interpret, they mainly make models.
(John von Neumann)

When we decided to write this book, we wanted to offer readers a set of ideas they could apply to research on the foraging economies of social organisms. The result is a book that consists mainly of new models or reviews of older models; together they constitute a *Social Foraging Theory*. Our focus on modeling, rather than an emphasis on empirical results, reflects the somewhat disorganized way social foraging has been studied. Compared to conventional foraging theory, which concerns the behavior of independent individuals, the analysis of social foraging has lacked unifying themes, clear recognition of the problems that define social foraging, and consistent interaction between theory and experiments. So, we developed models that provide a thematic framework for the behavioral ecology of social foraging. The models collectively delineate a series of problems that span the theory as well as suggest quantitative methods for further development and application of the theory.

For most models discussed in the book, we include a "Summary Box" listing the particular model's distinguishing assumptions and main predictions. These boxes provide essential elements of the models for readers interested in concise statements of predictions to test. At the end of some chapters, we add a "Math Box," which either explores a general concept or provides a model's derivation for those readers interested in formal details. We hope these two devices will provide individual readers some options that make the book more useful.

Hassell and May (1985; see Ives 1995) observe that population ecologists cannot ignore behavior completely in the formulation of population-growth models, but behavioral ecologists can remain unconcerned about the population-level consequences of the phenomena they study. An original impetus in North America for studying behavioral ecology was an interest in exploring mechanisms underlying community structure. "Scaling up" from individual behavior to population dynamics remains a significant but elusive objective (e.g., Murdoch and Oaten 1975; Parker 1985; Schoener 1986; Green 1989; Goss-Custard et al. 1995; Sutherland 1996). We suspect that social foraging can provide some initial steps toward linking behavior to population-level patterns. So, where appropriate, we indicate implications of our models for population ecology.

The models we discuss in this book deal with groups of interacting foragers. At a general level, each model can be categorized as asking one of the two central questions of social foraging theory:

1. Given a particular set of ecological conditions, what can we predict about the size of foraging groups?
2. Given a foraging group of G members, what can we predict about their exploitation of particular resources, as individuals and as a group?

The first chapter defines social foraging and specifies the methods we use. We organize the rest of the book into five parts. The parts are linked by unifying themes, but each part develops a different traditional approach to research on social foraging. Part One emphasizes the first central question. Chapter 2 envisions two individuals attempting to avert starvation and avoid predation, and models the economics of foraging solitarily versus foraging as a group. Chapter 3 assumes that the two foragers belong to the same group and models the decision whether or not to share food when the individual discovers a resource patch.

Chapter 4 models foraging group size when each member's direct fitness increases with the size of the group, at least when groups are small. We refer to this as an *aggregation economy*. We analyze equilibrium group size for different "rules of entry" and show how genetic relatedness among group members may increase or decrease the equilibrium group size, depending on the rule of entry.

Chapter 5 reviews models for the size of foraging groups when each member's fitness declines as group size increases. We refer to this case as a *dispersion economy*; the models include the well-known Ideal Free Distribution (IFD).

The next three parts elaborate models following from the second central question. Part Two (chapters 6 and 7) considers the interaction between producers and scroungers in a group of G foragers. We define producers as group members that search for food clumps, and feed only when they discover food. Scroungers do not find food but attempt to feed whenever a producer locates a clump. The amount of food consumed by any group member and, hence, any group member's probability of starving depend on the frequency of producer and scrounger among the G individuals. The focal problem is predicting the equilibrium frequencies of the two resource-exploitation roles.

Chapter 6 selectively reviews empirical studies of the producer-scrounger interaction and organizes the extensive terminology that has become associated with social parasitism. Chapter 6 also presents a deterministic model for the equilibrium frequency of producer and scrounger in a foraging group.

Chapter 7 analyzes two stochastic models of the producer-scrounger game and suggests that group cohesion and cooperation may constrain the incidence of scrounging.

Part Three contains only a single chapter, but the topics considered merit recognition as a separate set of social foraging models. Chapter 8 reviews models that parallel the most prominent models that conventional theory has developed for solitaries: patch residence time and dietary choice. Models for group members' patch residence times introduced some of the basic concepts for a social foraging theory, but those ideas have not yet been afforded the empirical attention they deserve.

In Part 4 we focus on phenotypic variation among members of a foraging group. We fix group size and treat phenotypic diversity from different perspectives. Chapter 9 reviews an ecological method for partitioning phenotypic diversity between its within-individual and among-individual components. We indicate how patterns in resource quality or abundance may govern the components of phenotypic diversity.

Chapter 10 analyzes the spread of a learned foraging trait among members of a group. When an individual acquires the trait, its rate of finding food clumps increases. When clumps are shared among group members, opportunities to learn the trait through individual experience decline as more group members acquire the trait. However, a naive individual has more opportunity to acquire a trait by observing others as a trait becomes more common. Hence the advance of a learned trait, and phenotypic variation based on the presence or absence of the trait, may exhibit complex, frequency-dependent dynamics.

Chapter 11 models a group's phenotypic variation economically. We compare three types of groups: specialists on a preferred resource, generalists, and specialists that each search for a different resource (the skill pool). We show how the interaction of within-individual and between-individual constraints influences which group's members have the greatest chance of survival.

In Part Five, we conclude with a brief chapter, chapter 12, that synthesizes the book's recurring themes. We hope the book helps readers discover, or rediscover, some attractive directions for research.

Acknowledgments

We thank John Krebs for inviting us to work on this book and for providing punctual, greatly appreciated encouragment. Over the course of writing we have benefited from the comments and criticisms of a large number of people. These include the numerous participants of the Montréal Inter-university Behavioral Ecology Discussion Group. We are especially grateful to Zoltán Barta, Guy Beauchamp, Mats Bjöklund, Wolf Blanckenhorn, Jim Grant, Lucy Jacobs, Alejandro Kacelnik, Donald Kramer, Kevin Laland, Jonathan Newman, Kinya Nishimura, H. R. Pulliam, Derek Roff, Jennifer Templeton, Tom Valone, Bill Vickery, Larry L. Wolf, and Norio Yamamura for discussions and or comments on various sections of the manuscript. We are also grateful to David W. Stephens and two anonymous reviewers for their valuable and detailed comments on an earlier version of the manuscript.

Parts of this manuscript were written while L.-A.G. was on sabbatical leave at Université Paris-XIII and McGill University. L.-A.G. thanks these institutions, and especially Jaap Kalff, for their hospitality. L.-A.G. and T.C. acknowledge financial support from Canadian and American taxpayers through research grants awarded by the Natural Sciences and Engineering Research Council (Canada), le Fonds pour la Formation des Chercheurs et l'Aide à la Recherche (Québec), and the National Science Foundation (USA). The various drafts of each chapter were transported across the Canada-USA border many times by our countries' valiant postal services. Given the number of delayed deliveries and occasional lost chapters, we are certain that a good number of postal workers of both countries have become assiduous readers of our work. We thank them for their interest. We thank the Montréal artist Michel Pimparé for allowing us to place one of his works (entitled *Jeux* [Games]) on the cover of our book. We also thank Alice Calaprice and Sam Elworthy of PUP for their excellent work. Finally, L.-.A.G. wishes to thank Dominique Proulx most warmly for her constant, often needed encouragement over the extended period of book writing and editing.

Montréal, Québec, Canada
Albany, New York, USA
July 1999

Social Foraging Theory

1 Social Foraging Theory: Definitions, Concepts, and Methods

1.1 What Is Social Foraging?

Social Foraging Requires Economic Interdependence of Payoffs

In this book we develop models where the functional consequence of an individual's foraging behavior depends on both the individual's own actions and the behavior of other foragers. We refer to the set of questions we analyze as "social foraging," defined by the concurrent economic interdependence among different individuals' payoffs or penalties. For the most part we restrict attention to effects of conspecifics' foraging decisions on individual survival, but several chapters will conclude with an examination of how predators' social foraging can have community-level consequences. We differentiate our subject from conventional (individual) foraging theory, reviewed by Pyke (1984), Stephens and Krebs (1986), and Schoener (1987). We emphasize that discriminating solitary versus social foraging involves more than extending previously analyzed questions to groups of foragers; economic interdependence implies a more fundamental biological difference (Maynard Smith 1984).

To make the point that social and conventional foraging are distinct, we can select a problem that has been well studied by conventional foraging theory and then identify the added complexity required by a social perspective. The classic prey-choice model specifies a predator's attack probabilities for different prey types upon sequential encounters (see Stephens and Krebs 1986). Typically, one associates a specific net rate of energy intake with the choices available to the solitary forager. That is, we could find the model predator's net rate of intake for a series of rational prey-selection strategies, e.g., take all prey as encountered versus take only the most profitable prey.

Now consider the same problem but allow two or more predators to exploit the same clump of prey. What now is the strategy that maximizes net intake rate? The answer is: it depends (see chapter 8 for details). The payoff obtained from any prey-selection strategy depends on the strategy adopted by the individual's competitors. So, taking only the best prey type will generate one intake rate if the competitors also do the same, but a different rate if the competitors take all prey types encountered instead. That is, economic interdependence means that the reward for a foraging policy depends simultaneously on all competitors' behavior. Furthermore, interdependent payoffs require that game theory replace the simple optimality models of conventional foraging theory, so that predictions follow from evolutionarily stable strategies (see below).

We recognize that the word "social" is semantically ambiguous. Indeed, its connotations can be rather diverse. For example, some researchers restrict use of the term "social" to organisms exhibiting a certain amount of familial dependence, those bearing elaborate behavioral displays, or those living within demographically structured groups. Others, however, use the term more liberally to include any animal that spends a good part of its life in groups, even if these groups are open, unstructured, and temporary. We use the word "social" in its broadest sense to mean any set of individuals that can be linked by identifiable, mutual relationships. Our criterion for social foraging simply requires that two or more individuals concurrently influence each other's energetic gains and losses. However, our simple definition merits a brief comment to clarify its range of application.

Social Foraging Does Not Require "Social" Organisms

Traditionally, some animal groups have been denied the status of social, often being referred to as mere aggregations. We contend that social foraging theory may apply whether animals are recognized as "social" or not. Both proximate and ultimate (or functional) distinctions have been proposed to separate true social groups (i.e., collections of social animals) from aggregations (collections of nonsocial animals). Ethologists tended to distinguish social groups from aggregations on the basis of the proximate causes of group formation. Social groups were viewed as the result of a genuine attraction between individuals, while aggregations were merely statistical coincidences of animals, often around a common resource: "Not all aggregations of animals however are social. When, on a summer night, hundreds of insects gather round our lamp, these insects need not be social. They may have arrived one by one, and their gathering just here may be clearly accidental; they gather because each of them is attracted by the lamp" (Tinbergen 1964, 1). Ecologists, for their part, have proposed a distinction at the ultimate, or functional, level. They discriminate social groups from aggregations because social groups are composed of individuals that derive specific evolutionary advantages from the presence of others while members of aggregations do not: "Intrinsic gregariousness has evolved to provide such concrete advantages . . . as the procurement of food. . . . In practice it may not be easy to distinguish social groups (in which individuals derive benefits by virtue of their presence with others) from aggregations" (Morse 1980, 271–272).

The ecologist's functional distinction between social groups and aggregations can be important when one is interested in the evolutionary origin of sociality. However, neither proximate nor functional distinction limits the application of social foraging theory. The essential property of social foraging is the interdependence of individuals' benefits and costs, whether foragers are attracted to one another or to the same food resource, and whether they mainly derive antipredatory or foraging benefits from their group membership.

The generality of our approach to social foraging can be illustrated with an example. Lions (*Panthera leo*) lead an apparently more complex social existence than do, say, pigeons (*Columba livia*). Lionesses live in permanent, structured social units composed of genetically related individuals that together raise and defend young as well as hunt prey (Schaller 1972; Bertram 1978; Packer 1986; Heinsohn and Packer 1995). Pigeons, on the other hand, form loose breeding colonies of probably unrelated individuals (Goodwin 1954) that do not forage in permanent, structured groups (Lefebvre and Giraldeau 1984). Instead, pigeons have individualized itineraries over a number of foraging stations, forming at each an open flock characterized by changing membership over the course of a day (Lefebvre and Giraldeau 1984; Lefebvre 1986). Despite the extensive differences between lion and pigeon sociality, we hold that the same economic approach of social foraging theory can help predict and explain the functional significance of both species' feeding-group sizes (Clark and Mangel 1984, 1986; Pulliam and Caraco 1984). For both lions (Caraco and Wolf 1975; Giraldeau and Gillis 1988; Packer 1986; Clark 1987) and pigeons (Lefebvre 1983), the amount of food available to an individual likely depends on the number of foragers within the group. It is this interdependence, coupled with the hypothesis that both organisms must forage efficiently to survive and reproduce, that allows the different foragers' equilibrium group size to be predicted by a single social foraging model. Therefore, even though the social life of lions and pigeons differ markedly, similar functional analyses unite them conceptually as social foragers. Emphasis on the distinctions, whether proximate or functional, between social groups and aggregations has obscured the apparent similarity of foraging decisions made by both types of organisms.

Social Foraging Is Not the Study of the Origin of Foraging Groups

It is often argued that animals such as small fish (Pitcher 1986), birds (Lazarus 1972), or primates (Wrangham 1986) may have become gregarious in response to predation. For certain other animals, such as lions, some experts argue convincingly that foraging benefits do not account for the origin of sociality (Packer et al. 1990). We can reasonably conclude that the evolutionary origins of some animals' group-feeding habits lie outside of their foraging economies. Should we exclude such species from studies of social foraging? We think not. The economic interdependence of group members' foraging behavior that defines the issues of social foraging neither requires nor precludes particular origins of sociality. Failing to discriminate between the evolutionary origins of sociality and contemporary functions of foraging groups as objects of study can confuse the issues of social foraging (as we define it). One may argue how effectively food, predation, or both explain sociality's origins. However, social foraging analyses, unlike those pertaining to the origin of sociality, do not treat antipredatory and energetic benefits as competing hypotheses. A model for social foraging in a contemporary environment will more likely emphasize how the two types of benefits may interact to govern some currency of fitness (e.g., Houston et al. 1993).

Several of our models take the probability of surviving as the currency of fitness. In general, we can write the probability of surviving some specified time interval as the product of the probability of avoiding starvation and the independent probability of avoiding predation. This approach combines the two sources of mortality into a single currency (e.g., Pulliam et al. 1982; Mangel and Clark 1986; McNamara and Houston 1986; Newman 1991). When decreases in starvation imply increases in predation, we can profitably use this approach to investigate social foragers' compromises between food and safety (e.g., Caraco et al. 1980a; Elgar 1986a; Newman and Caraco 1989; Abrahams and Dill 1989; Rayor and Uetz 1990; Houston et al. 1993).

Not every social foraging model will require analysis of a trade-off between foraging gains and avoiding predators. Suppose that all strategic options yield exactly the same hazard of predation; then it may be safe to ignore effects of predators (see Lindström 1989). When appropriate or convenient, one might choose to incorporate antipredatory requirements as constraints in a model for social foraging. Concern about antipredatory behavior seems to appear more frequently when group foraging is considered, possibly because predation is so often cited in discussion of sociality's origins. It is important to emphasize that predation is not more of a problem for group foragers (Martindale 1982; Edwards 1983; Lima et al. 1985; Mangel and Clark 1986; Stephens and Krebs 1986). In fact, one could turn the standard view around and suggest that since groups may provide reduced predation hazard, the foraging of animals in groups is less likely to be constrained by predators than is solitary foraging. Social foraging theory, then, focuses on contemporary adaptive function and says little about origins of sociality.

1.2 Concepts and Methods of Social Foraging Theory

There are at least two currently popular methods used to generate hypotheses about the functional significance of behavior. One, the comparative method, involves accumulating information about several populations or several species and then correlating ecological conditions with the populations' or species' attributes (e.g., Crook 1965; Altmann 1974; Jarman 1982; Clutton-Brock and Harvey 1984; Harvey and Pagel 1991). For instance, in their study of avian kleptoparasitism, Brockmann and Barnard (1979) review all published instances of the behavior to draw conclusions concerning the ecological circumstances that promote its evolution. The alternative method employs optimization techniques to formulate specific, often quantitative hypotheses about behavior (e.g., Pyke 1984; Parker 1984a; Mangel and Clark 1988). For instance, Vickery et al. (1991) and Caraco and Giraldeau (1991) both develop foraging games designed to analyze the economics of kleptoparasitic behavior, to predict the ecological circumstances under which the behavior is maintained. We favor the latter method because it has been particularly successful in predicting the behavior of solitary foragers (reviewed by Stephens and Krebs 1986), and has guided research programs across a broad spectrum of questions in behavioral ecology (Grafen 1991). The use of optimization methods in evolutionary ecology has been criticized at times (e.g., Cohen 1976; Gould and Lewontin 1979; Lande 1982; Gray 1987; Hines 1987). Most, if not all, of these criticisms have been answered reasonably and rigorously; rather than repeat the general arguments, we refer the reader to an appropriate literature (Maynard Smith 1978; Oster and Wilson 1978; Mayr 1983; Krebs and Davies 1984; Stephens and Krebs 1986; Schoener 1987). We shall, however, specify the game-theoretical concepts we apply to model the behavior of social foragers.

Optimization Models in Social Foraging Theory

The mathematical methods used in conventional models of adaptive foraging seldom can serve social foraging theory, where the efficiency of a particular behavioral strategy depends, by definition, on the frequencies of feasible strategies among an individual's competitors. Questions in social foraging theory still concern the adaptive significance of individual behavior, but the models rely on game-theoretical equilibria and the concept of an evolutionarily stable strategy (ESS) (Maynard Smith 1982; Parker 1984a; Hines 1987).

An evolutionarily stable behavior possesses both optimality and stability properties (Parker 1984a; Vincent and Brown 1984). An ESS combines these attributes because an ESS must qualify as a Nash equilibrium (e.g., Hines 1980; Thomas 1985). A Nash equilibrium is a set of strategies, one for each player, such that no player can improve its payoff by changing strategy when the other players continue using their Nash equilibrium strategies (e.g., Riley 1979; Caraco and Pulliam 1984; Parker and Sutherland 1986). So, an ESS maximizes a player's expected payoff in the sense that no other feasible strategy does better against an ESS than the ESS itself. Readers not inclined toward an optimality-based interpretation of an ESS may find comfort in Hines's (1987) discussion of a polymorphic evolutionarily stable strategy.

An ESS also has, of course, a stability property following from the Nash equilibrium concept. Depending on one's detailed characterization of the game (see below), stability may imply dynamics resistant to invasion by a rare alternative strategy or combination of strategies, or stability may refer to the consequences an individual incurs when deviating unilaterally from an ESS (Parker 1984a; Vincent and Brown 1984; Brown and Vincent 1987; Hines 1987).

It is worth noting that game theory may also be applied to problems where conspecifics differ in their ability to sequester resources (i.e., an asymmetric game), as well as to questions concerning interactions between species (e.g., Parker and Sutherland 1986). So we adopt a game theory approach, because with or without the full analysis of an ESS, it can help us investigate asymmetric competitive interactions, including circumstances where a social forager makes "the best of a bad job" (e.g., Parker 1984a; Caraco et al. 1989).

Phenotypic, Rather than Genetic, Focus

As mentioned in the preceding paragraph, the term "ESS" currently is used in several senses; Hines (1987) summarizes the theory. The definition and existence of an ESS, as well as the criteria for stability, can vary according to several properties of the model population or group. Strategies may be equivalent to alleles or assumed to be modified by the environment. The population may have a finite or infinite size (e.g., Riley 1979; Vickery 1987; M. Schaffer 1988; Nishimura and Stephens 1997). Interactions may or may not involve repeated play (e.g., Axelrod and Hamilton 1981). Only one or several different strategies deviating from a candidate ESS may occur simultaneously (e.g., Boyd and Lorberbaum 1987; Brown and Vincent 1987; Farrell and Ware 1989; Lorberbaum 1994). Each of these differences, and several more, has theoretical significance, but we cannot consider them all. In applying game theory to social foraging we adopt Hines's (1987) suggestion and look for the "practical relevance of the (ESS) concept to actual biological phenomena."

Some approaches to ESS theory use frequency-dependent payoffs to generate the dynamics of a system of competing alleles (e.g., Taylor and Jonker 1978; Hines 1980). Change in gene frequencies is the essence of evolution, but reducing complex social behavior to simple genotypes, as for instance, in Brockmann et al.'s (1979) field study of digging and entry in golden digger wasps (*Sphex icheumeneus*) we find generally too restrictive. We want to appreciate not only behavioral diversity due to attributes such as size, age, and sex, but also the important strategic variation due to learning. Our predictive models therefore focus on phenotypes. Like most behavioral ecologists, we rely on what Grafen (1991) calls the "phenotypic gambit." Most of our models invoke the original definition of an ESS (Maynard Smith 1982; see Parker 1984a; Houston and McNamara 1988), which essentially ignores genetic constraints once the set of possible phenotypes has been established (see below). Genetic and phenotypic models of a particular behavior should be viewed as complementary analyses. They often lead to identical conclusions (e.g., Aoki 1984; Michod 1984; Thomas 1985; Maynard Smith 1988; Moran 1992; Weissing 1996), and each approach has its advantages. Essentially, we assume in most cases that selection has enhanced learning capacities and decision-making rules that allow an individual to vary its behavior efficiently across a range of environmental and social conditions (Dawkins 1980; Barnard 1984a; Pulliam and Caraco 1984; Cosmides and Tooby 1987; León 1993).

Although we do not consider the intergenerational dynamics of gene frequencies, we do examine how learning can govern the frequencies of various phenotypes within a cohesive group. Social foragers must often experience a great deal of spatial and temporal variation in both resource characteristics and competitive interactions. If natural selection favors some form of phenotypic plasticity in such an environment (Via and Lande 1985; Fagan 1987a; West-Eberhard 1989; Lessells 1991; Houston and McNamara 1992; Moran 1992), a significant fraction of the variability in foraging behavior might be acquired through learning (e.g., Norton-Griffiths 1967; Krebs 1973; Werner and Sherry 1987). The acquisition of behavioral traits is usually the domain of learning psychologists, but behavioral ecologists recently have appreciated the importance of the process in terms of the problems faced by solitary foragers (e.g., Krebs et al. 1978; Pulliam 1980; Kamil 1983; Shettleworth 1984; Stephens and Krebs 1986). But learning from experience necessarily proceeds differently in a social context, where learning itself can generate polymorphisms in a manner analogous to frequency-dependent selection (Giraldeau 1984, 1997).

Terminological Clarifications

In this subsection we associate some behavioral terms with the game-theoretic concepts we use to solve our models. The same behavioral term can, of course, mean different things to different authors. For example, cooperation can mean a genetically determined attribute (e.g., Nowak and May 1992), a small set of phenotypically plastic behaviors (e.g., Pulliam et al. 1982; Noë 1990), or a broad class of phenomena ranging from mutualism to traits favored under group selection (e.g., Mesterton-Gibbons and Dugatkin 1997; see Brown 1983; Dugatkin 1997). Therefore, we want to specify the meaning of the behavioral terms we use in modeling social foraging. We use the definitions we assign consistently, and intend no criticism of alternative terminologies.

To begin, we view the difference between observable behavior and a foraging strategy as similar to the distinction between "territory and map"; the latter is an idealized guide to the former. As stated above, a strategic model for social foraging usually takes the form of a game, where each competitor pursues its own objective (e.g., increasing its consumption of discovered food). The game's solution predicts how the different players' often conflicting, but sometimes coincident, objectives can be resolved. However, models in two of the book's chapters are simpler; they assume a single strategic objective can predict social foraging behavior. Taking an example from chapter 4, suppose an individual increases its inclusive fitness by economically joining or avoiding a group of relatives. That is, we assume a single decision-maker takes an action rendering the sum of fitness effects on self and effects on relatives (weighted by degree of relatedness) positive. Then three possibilities can be favored under Hamilton's Rule (Hamilton 1964; Grafen 1984). If effects on both self and relatives are positive, we term the behavior a *mutualism*, without the need for special reference to kinship (see below). If the effect on self is negative but the effect on relatives is positive (and of greater absolute value), we term the actor's behavior *kin-directed altruism*. If the effect on self is positive and the effect on relatives is nonpositive (and, if negative, of lesser absolute value), we term the behavior *selfish*, again requiring no special reference to kin. We consider both mutualism and kin-directed altruism special cases of what we call *unconditional cooperation*. That is, the decision-maker's action promotes the fitness of other individuals, and natural selection can maintain the trait without it being conditioned on either an immediate or delayed response-in-kind by those other individuals (below we define *conditional cooperation*). The behavior defined as selfish does not, of course, qualify as cooperation, although it can be favored by Hamilton's Rule.

The other type of model where we require only a single strategic objective assumes that members of the same foraging group work essentially as a team (see Oster and Wilson 1978, 302). The models of chapter 11 suppose that group members begin the day in the same physiological condition, and then divide each food clump they discover equally. Consequently, a single currency assesses each group member's expected benefit obtained from using feasible patch-exploitation strategies. Our treatment of a group as a team resembles several other dynamic-optimization models of social foraging (Clark 1987; Mangel 1990; Székely et al. 1991). The terminology we use in chapter 11 differs little from conventional foraging theory (see Mangel and Clark 1988) and needs no explanation here.

Some of our models assume a game between two players, and others assume an N-player game. All of our applications of game theory assume uninformed play (see Bram and Mattli 1993). This means that each player chooses an action once per play of the game without knowledge of (or communication concerning) the action any other player is about to take. This is a common assumption in behavioral ecology; for alternatives, see Maynard Smith and Parker (1976), Pulliam and Caraco (1984), or Noë (1990).

Most of our models take the form of a symmetric game. Symmetry implies competitive equivalence of the players, so that each has the same payoff (or penalty) function. So, symmetry means that an individual's payoff or penalty is specified completely by the combination of interacting strategies, without reference to any other phenotypic attribute of this individual. In a few instances we consider a competitive asymmetry due to aggressive dominance. Nearly all of our game-theory models are discrete. That means each player's action on any single play of a game belongs to a finite set (as opposed, for example, to a War of Attrition that affords players a continuous set of alternative strategies); the number of elements in the set equals the number of possible pure strategies (see below). Consequently, the payoffs or penalties for two-player games can be arrayed in matrix form. For clarity and simplicity, the majority of our matrix games have only two pure strategies. Following Parker (1984a), we first identify a class of behaviors associated with social foraging. Then the model lets one action represent the presence of the class of behaviors, and lets the alternative action represent its absence.

Some of our models consider only a single round of play. But certain models' payoffs or penalties may conform to a Prisoner's Dilemma (e.g., Axelrod and Hamilton 1981; Nowak and Sigmund 1992; Mowbray 1997; Nishimura and Stephens 1997); when this occurs we consider the consequences of probabilistically repeated play. As pointed out by Mesterton-Gibbons and Dugatkin (1997), the distinction between single and repeated interaction of the same individuals, and between behaviors associated with single versus repeated play, should depend on a logical temporal scaling. Most of our models define the duration of a round of play as a foraging period τ time units long (see Newman and Caraco 1989). At the end of a foraging period, each player aquires some benefit or pays some cost, and then a new round of play may commence. Hence our temporal scaling of play mimics physiological and environmental constraints on the timing of foraging. More importantly, assuming repeated play of the same individuals helps focus attention on the fundamental significance of population spatial structure for the economics of individual interactions (Houston 1993; Ferriere and Michod 1996; Caraco et al. 1997; Levin et al. 1997).

In our applications of game theory, a strategy is a rule for using feasible actions; see discussion in Vincent and Grantham (1981) or Weissing (1996). In general, an individual's strategy may assign positive probability to two (or more) different actions and may be conditioned on environmental variables (e.g., food density) and/or the behavior of another player. Most of our models predict "pure" strategies, a solution with a single action. As mentioned above, we invoke the well-known Tit-for-Tat strategy (TFT) (Axelrod and Hamilton 1981) when the game qualifies as a probabilistically iterated Prisoner's Dilemma.

Next consider some concepts we apply in solving our game-theory models. The familiar notion of an evolutionarily stable strategy (ESS) envisions introduction of a single, rare strategy (which deviates from the ESS) into a population where all other individuals use the ESS. The stability property of an ESS requires that the rare, deviating strategy be disfavored. Recall that we require that an ESS qualify as a Nash equilibrium to the specified game (Parker and Sutherland 1986; Recer et al. 1987; Weissing 1996). A Nash equilibrium for a two-player game implies that neither player is tempted to change its strategy (hence behavior) as long as the other player continues to use its same strategy. For N players, no individual is tempted to change strategy as long as all others continue with their Nash-equilibrium strategies. If a player is not tempted to change strategy, it is because the individual cannot increase its payoff (or decrease its penalty) by altering its behavior (Vincent and Grantham 1981). As a convenience, we describe a Nash equilibrium as stable if unilateral deviation reduces that individual's payoff. We may describe a Nash equilibrium as neutrally stable if unilateral deviation has no effect on the individual's payoff. Not every Nash equilibrium qualifies as an ESS, but any ESS in our models will have the Nash-equilibrium property. Math Box 1.1 at the end of this chapter illustrates the application of the Nash equilibrium concept to solve first a discrete game, and then a continuous game.

Above we defined unconditional cooperation to include both kin-directed altruism (a cost to the actor benefits relatives) and mutualism (both individuals, whether related or not, benefit). A number of studies focus on nonrelated individuals, seeking only to increase their own payoff, that may be penalized if they fail to cooperate mutualistically (Caraco and Brown 1986; Mesterton-Gibbons 1991; Mesterton-Gibbons and Dugatkin 1992; see Clements and Stephens 1995). So, mutualism may occur with appreciable frequency among social foragers, when a Nash solution to a foraging game produces a mutualistic interaction. To discriminate between unconditional cooperation and our applications of the iterated Prisoner's Dilemma, we use the term *conditional cooperation* for the latter, where cooperation arises conditionally on simultaneous (conditional mutualism) or delayed (reciprocal altruism) reciprocation of behavior (see Summary Box 1.1).

While a Nash solution requires that an individual's unilateral deviation in strategy cannot be rewarded, a Pareto optimal solution allows that an individual might deviate and so increase its payoff, but only by reducing at least one other player's payoff (e.g., Vincent and Grantham 1981). So, Pareto

SUMMARY BOX 1.1 A SIMPLE TERMINOLOGY FOR CLASSES OF BEHAVIORS THAT MAY OCCUR AMONG SOCIAL FORAGERS

I. Selfish behavior
II. Cooperative behavior
    A. Group-selected altruism
    B. Unconditional cooperation
        1. Kin-directed altruism
        2. Mutualism
    C. Conditional cooperation
        1. Conditional mutualism
        2. Reciprocal altruism

Any behavior favored by natural selection results from a fundamentally competitive process. But we use the term "selfish behavior" when the actor increases its payoff and, as a consequence, decreases the payoff to one or more individuals (related or unrelated to the actor).

Cooperation implies that an action increases the payoff to one or more other individuals; the actor's payoff may increase or decrease. The actor cooperates unconditionally if the benefit to relatives is sufficiently large, or if the actor benefits via mutualism. The actor cooperates conditionally when the economics of cooperative behavior require a response in-kind; some form of social organization fostering repeated interactions of the same individuals is assumed. The cooperator may benefit from this response immediately (conditional mutualism; the standard Prisoner's Dilemma) or later (reciprocal altruism).

optimal solutions to a game assume that players mutually coordinate their strategies to advance each individual's payoff; hence Pareto optimality helps us identify consequences (but not always the ecological causes) of mutual cooperation. Of course, an advantage to cooperation between nonrelatives requires that mutual cooperators acquire a greater payoff than individuals lacking such cooperation (Axelrod and Hamilton 1981; Pulliam et al. 1982; Caraco and Brown 1986; Yamamura and Tsuji 1987; Mesterton-Gibbons 1991; Mowbray 1997; Nishimura and Stephens 1997).

Suppose two players have the choice between a cooperative and a noncooperative behavior, as in the Prisoner's Dilemma. Ordinarily, Pareto optimality identifies a continuous solution set, strategy pairs where the two players' payoffs satisfy the Pareto optimal condition (see Math Box 1.2). The Pareto-optimal solution set has the interesting property that increasing one player's payoff necessarily decreases the other player's payoff (Oster and Wilson 1978; Vincent and Grantham 1981). Hence, mutually cooperative players might have to "negotiate" a particular Pareto solution (Caraco and Pulliam 1984; Noë 1990). For a symmetric game, a Nash solution qualifying as an ESS implies that the competitively equivalent players receive the same expected payoff. Hence equality of expected payoffs, over some biologically relevant timescale (Nowak and Sigmund 1994; Mesterton-Gibbons and Dugatkin 1997), should also apply to models of mutual cooperation. Following Axelrod and Hamilton (1981), the most common application of the Prisoner's Dilemma in evolutionary biology assumes equivalent players and focuses on equality of expected payoffs at each round of repeated play.

Consider a single round of play when two equivalent individuals' payoff matrices conform to the Prisoner's Dilemma (defined in Math Box 1.2). Each player chooses either to cooperate or defect during that round of play. Defect is the more rewarding response to both cooperate and itself. Then pure defect is an ESS, and a player will always be tempted to defect against cooperation. But mutual cooperation rewards each player more than does mutual defection; hence the dilemma. The game serves as a reasonable metaphor for asking questions about possible cooperative interactions between nonrelatives (see Dugatkin 1997). Pure cooperation is a Pareto optimal solution for a single round of play; cooperation yields a greater reward than does the Nash equilibrium strategy, but cooperation is unstable against defection.

One escape from the dilemma assumes probabilistically repeated play and the Tit-for-Tat (TFT) strategy (Axelrod and Hamilton 1981; Mesterton-Gibbons 1991). TFT cooperates in the initial round of play and thereafter behaves as its opponent last did, until play is terminated. If the expected duration of the interaction between the same two players is sufficiently large, TFT is a Nash solution to the iterated game and an ESS (see Maynard Smith 1984 or Math Box 1.2). TFT is a reactive strategy; continued cooperation by a TFT strategist is conditional upon cooperation of the other player. Hence we refer to any behavior associated with the TFT strategy as *conditional cooperation*. That is, the decision-maker's action promotes the fitness of other individuals, but selection can maintain the trait only when it is conditioned on an immediate or delayed response-in-kind by the other individuals. For convenience, we use the term *conditional mutualism* when mutual cooperators benefit simultaneously, as in the (standard) iterated Prisoner's Dilemma. We use the term *reciprocal altruism* when nonrelated players alternate between altruist and recipient roles, so that the response-in-kind is delayed (see Noë 1990). For completeness we note that reciprocal altruism, as we define it, is not a solution to the standard iterated Prisoner's Dilemma (cf. Mesterton-Gibbons and Dugatkin 1997).

Conditional cooperation may or may not occur commonly among social foragers (see Dugatkin et al. 1992; Clements and Stephens 1995). But conditional cooperation can be quite interesting because its predictions often differ from behavior predicted in its absence. As we stated previously, perhaps the most important insight from TFT's stability is the significance of a social environment and a spatial grain in a population that together permit or promote repeated interaction between the same individuals. Recent evolutionary theory suggests that natural selection should relax genetic constraints on strategic behavior (e.g., Hammerstein 1996; Weissing 1996). In that light, our emphasis on phenotypic plasticity and economic decision-making invokes the iterated Prisoner's Dilemma as a general metaphor for ecological conditions that may favor repeated interactions and conditional cooperation. Summary Box 1.1 collects our behavioral terminology.

1.3 Interactions among Social Foragers

Having defined social foraging in terms of economic interdependence, and introduced both concepts and methods of social foraging theory, we now point out the general properties of questions analyzed in this book. Problems in social foraging require detailed assumptions about the process of searching for food. At least some group members must attempt to find food, but certain individuals may try to avoid the cost of searching (Barnard and Sibly 1981). Among the active searchers, different foragers may focus their effort on different resources (Giraldeau 1984). Furthermore, the probability density of the time taken to locate a patch of a particular resource can depend on both the number of individuals searching for that resource and the way their efficiencies interact (Ekman and Rosander 1987; Caraco et al. 1995). Most models of foraging processes, social or not, either neglect or deemphasize the often inherently random nature of food discovery. In contrast, we treat searching for (or capturing) food as stochastic, hypothesizing that the economics of social foraging commonly depend on the dynamics of food discovery. Math Box 1.3 provides examples of the tools we use to characterize the probabilistic processes of food discovery.

The next set of important assumptions in social foraging theory concerns the allocation of discovered food among group members. Food may be actively shared, divided equally among competitively equivalent foragers, or allocated asymmetrically according to dominance status (Caraco and Giraldeau 1991; Ranta et al. 1993; Ruxton et al. 1995). Differences in the way social foragers search for food, and patterns in the way they divide discovered food, together suggest a simple introduction to the economic interactions we model.

We can categorize any social forager as one of three types according to its food-searching behavior (following Vickery et al. 1991): producers, scroungers, and opportunists. Producers search their environment for food clumps (or divisible prey). When a producer discovers food, it can prevent other producers from usurping any of the resource but may elect to let them feed. Scroungers do not search for food directly. Instead, they attend to other foragers' clump discoveries and aggressively or stealthfully sequester some food at each clump found by either a producer or an opportunist. Opportunists both search for food directly and attempt to obtain food at clumps located by producers or other opportunists. An opportunist then simultaneously searches both as a producer and a scrounger but may be less efficient than either more specialized forager (Vickery et al. 1991). It is important to note that the same categories can be applied to individuals that switch among strategies. Hence, at any one time an individual may play one of the following: producer, scrounger, or opportunist. During the duration of that play of the game, it behaves according to the rules of the alternative it chose.

Producer versus Producer

Suppose two producers forage in close proximity. When one finds a clump, it might choose to share the food with the other producer, perhaps giving a "food call" to attract the other forager (chapter 3). Food-sharing might arise as a consequence of kinship (McNamara et al. 1994; Emlen 1995), conditional cooperation (Caraco and Brown 1986; Mesterton-Gibbons 1991), the danger of predation when feeding alone (Newman and Caraco 1989), or a mutualism (Stephens et al. 1995). More generally, consider a group of $(G - 1)$ producers that share food mutualistically. Their choice of admitting or repelling another producer presents the problem of equilibrium group size under a group-controlled entry rule (chapter 4; Giraldeau and Caraco 1993; Higashi and Yamamura 1993).

Producer versus Scrounger

Social foraging theory includes the producer-scrounger interaction (Part 2; Barnard and Sibly 1981; Giraldeau et al. 1990), which has been applied to a diversity of questions in population biology (Parker 1984b). In a large enough group, a rare scrounger can have an economic advantage over producers since the scrounger feeds at each clump found. However, scroungers require at least some producers to feed. Hence, we may anticipate an equilibrium mix of producers and scroungers (Vickery et al. 1991) unless the "cost of scrounging" is large enough to eliminate the latter (Caraco and Giraldeau 1991).

Producer versus Opportunist

For simplicity, suppose a group of two contains one producer and one opportunist (chapter 2). Both foragers search for food, but only the opportunist feeds at each clump discovered. Greater dominance status of the opportunist might produce this situation; the same result might arise when the producer finds resources much faster than does the opportunist. Symmetric competition will not likely maintain this interaction (Pulliam and Caraco 1984).

Opportunist versus Opportunist

Now each forager searches for food, and each acquires some food at every clump discovered. Groups of opportunists are sometimes termed an "information-sharing system" (chapter 6; Clark and Mangel 1984; but see Vickery et al. 1991). The problem of comparing solitary versus social foraging is often framed as an interaction of opportunists; two foragers, dividing each clump found, may survive better than each of two solitaries. Similarly, more sophisticated questions about equilibrium group size under a free-entry rule (Clark and Mangel 1986; Giraldeau and Caraco 1993; Higashi and Yamamura 1993) can be viewed as an interaction among opportunists (chapter 4).

Suppressing the importance of the searching process, while extending the spatio-temporal scale of the analysis, leads to questions about the ideal free distribution of competitors across resource patches (chapter 5; Fretwell 1972; Rosenzweig 1981; Sutherland and Parker 1985; Krivan 1996) and its various modifications. If different opportunists search for different resources, each individual searches as a specialist and feeds as a generalist. Hence the problem of the skill pool economy (Giraldeau 1984) is an interaction among opportunists (chapter 11).

Scrounger versus Opportunist

Only the opportunists discover food, and all group members feed at each clump. Each individual should achieve a greater foraging rate as the frequency of opportunists in the group increases. But the scrounger-opportunist group can persist due to knowledge limitation (chapter 10; Giraldeau et al. 1994b). That is, when some group members have acquired a skill needed to make a given resource available, other foragers may have few opportunities to learn the skill (and become a producer). In this case, social foraging constrains the group members' collective capacity to increase their economic efficiency.

Some other problems in social foraging are drawn directly from questions in conventional foraging theory. These include diet choice as a group member and patch residence time when resources deplete (chapter 9).

1.4 Concluding Remarks

Readers may have noticed recurring themes in our view of social foraging theory. It may be useful to collect these themes here, since we return to them in the chapters that follow. We stress the importance of the mutual dependence of individuals' payoffs and penalties. Economic interdependence may occur during the search for food, during the division of food following its discovery, or during both. These interdependencies define social foraging and set it apart from conventional foraging theory.

Another important theme is our view of foraging as a stochastic rather than deterministic process. Stochastic models take into account effects that reward variance exerts on a forager's survival, and more realistically depict the inherent uncertainty of foraging processes. They easily permit inclusion of predation hazard in our theory. So, survival probabilities and risk sensitivity are important themes that run through the book's chapters.

We also emphasize the importance of distinguishing between cooperative and noncooperative solutions to social foraging games. When appropriate, we investigate whether cooperative solutions, whether conditional or unconditional, occur and explore the biological circumstances under which we expect them to arise.

Math Box 1.1 Concepts from Game Theory

Here we review the idea of an evolutionarily stable strategy (ESS), in a form familiar to behavioral ecologists (Parker 1984a). We also review the related, but more general, concept of a Nash equilibrium (Vincent and Grantham 1981; Weissing 1996). We organize the presentation by first describing the structure and solution of a discrete game, and then doing the same for a continuous game. For convenience, we restrict this discussion to symmetric, two-player games. For further technical development, see Hines (1987) or Mesterton-Gibbons (1992).

A Discrete Game

Suppose an interaction, a contest, between two foragers (1 and 2) alters each individual's survival in a manner depending on the behavioral "choice" each makes. Each player selects an action from a set of S behaviors. S is a positive integer ($S \geq 2$), so the game is discrete rather than continuous.

The S^2 feasible behavioral combinations imply well-defined consequences, arrayed for player i ($i = 1, 2$) in an $S \times S$ matrix. We might express the consequences as survival probabilities, mortality probabilities, or another currency of fitness. For survival we let m_i represent player i's *payoff* matrix. For mortality or an energetic shortfall while foraging, we let M_i represent player i's *penalty* matrix. In a symmetric game the players are competitively equivalent, implying that they have the same payoff (penalty) matrix. So, $m_1 = m_2 = m$, and $M_1 = M_2 = M$. The majority of our models calculate probabilities of insufficent energy intake, so here we discuss only penalty matrices M, with elements M_{rc} ($r = 1,2,\ldots,S; c = 1,2,\ldots,S$). Hence, the inequalities associated with an ESS will have the direction opposite that of the more familiar payoff-matrix formulation. But the distinction should be readily apparent, and this discussion parallels the analysis of our models.

Associated with the penalty matrix M, let π_i represent a strategy of player i ($i = 1, 2$). π_i is a column vector with S elements π_{ik} ($k = 1, 2, \ldots, S$). π_{ik} is the probability that player i chooses action k when the foragers interact. Hence, $0 \leq \pi_{ik} \leq 1$, and

$$\sum_{k=1}^{S} \pi_{ik} = 1.$$

Math Box 1.1 *(cont.)*

π_i is a pure strategy if a single element is unity, so that the other elements are 0. π_i is mixed if two or more $\pi_{ik} > 0$.

To motivate our models, let $w_i(\pi_i, \pi_j)$ represent the probability that player i fails to meet its energetic requirement when i has strategy π_i and player j has strategy π_j. The expected penalty w_i depends bilinearly on the strategies

$$w_i(\pi_i, \pi_j) = \pi_i^T M \pi_j, \qquad (1.1.1)$$

where π_i is transposed. Each model player attempts to decrease its expected penalty and avoid an energetic failure.

An ESS

Suppose a strategy $\tilde{\pi}$ is an ESS for the game defined by penalty matrix M. If members of an infinite population use $\tilde{\pi}$, no single rare alternative can be favored (e.g., Parker 1984a). Then, for any feasible $\tilde{\pi} \neq \pi$, either

$$w(\tilde{\pi}, \tilde{\pi}) < w(\pi, \tilde{\pi}) \qquad (1.1.2)$$

or

$$w(\tilde{\pi}, \tilde{\pi}) = w(\pi, \tilde{\pi}) \text{ and } w(\tilde{\pi}, \pi) < w(\pi, \pi). \qquad (1.1.3)$$

When condition (1.1.2) holds, the stability of $\tilde{\pi}$ is clear. $\tilde{\pi}$ responds better to itself than does any rare alternative, since $\tilde{\pi}$ minimizes a player's expected penalty when the opponent uses $\tilde{\pi}$. When condition (1.1.3) holds, adopting $\pi \neq \tilde{\pi}$ against the ESS does not imply a greater penalty, according to the first part of the condition. But $\tilde{\pi}$ implies a lower expected penalty against π than the alternative π does against itself, according to the second part of condition (1.1.3). Hence, use of π must, on average, result in a greater expected penalty than does use of $\tilde{\pi}$, so $\tilde{\pi}$ remains an ESS. $\tilde{\pi}$ may be a pure or mixed ESS; all members of a population attain the same expected penalty, $w(\tilde{\pi}, \tilde{\pi})$, when all adopt the ESS.

Parker (1984a) discusses the "conditional ESS," a set of environmentally determined strategies. Put simply, variation in environmental circumstances (e.g., food density, predation hazard) leads individuals to adjust their behavior accordingly.

Nash Equilibria

Suppose the strategy pair $(\hat{\pi}_1, \hat{\pi}_2)$ is a Nash equilibrium for the game defined by penalty matrix M. As pointed out in Section 1.2, $(\hat{\pi}_1, \hat{\pi}_2)$

Math Box 1.1 *(cont.)*

specifies a strategy pair where neither player is tempted to deviate (i.e., alter its strategy) unilaterally. In terms of our example M, neither forager can reduce its chance of an energetic shortfall by changing its strategy as long as the other player continues to use its Nash strategy. This property does not require equality of expected penalties (consider an asymmetric interaction). But in symmetric games the players may have identical Nash strategies, implying they have the same expected penalties. In this case the common Nash strategy qualifies as an ESS; $\hat{\pi}_1 = \hat{\pi}_2 = \tilde{\pi}$, and

$$w_1(\hat{\pi}_1, \hat{\pi}_2) = w_2(\hat{\pi}_2, \hat{\pi}_1) = w(\tilde{\pi}, \tilde{\pi}). \qquad (1.1.4)$$

Every symmetric two-player game with $S = 2$ (a 2×2 matrix) has at least one Nash solution satisfying an ESS condition (either 1.1.2 or 1.1.3). Larger matrices ($S > 2$) may not have an ESS, but if they do, it will be a Nash equilibrium (Haigh 1975).

Diagonal Dominance

The simplest way to locate an ESS for a symmetric discrete game is to identify diagonally dominant columns (Haigh 1975). For the penalty matrix M, the M_{kk} ($k = 1, 2, \ldots, S$) are the entries on the main diagonal. Each M_{kk} quantifies how well an action responds to itself. Suppose

$$M_{kk} < M_{rk} \text{ for } 1 \leq r \leq S,\ r \neq k. \qquad (1.1.5)$$

Then the kth column of M is *diagonally dominant*. That is, M_{kk} is the minimal chance of an energetic failure when the other player uses the kth behavior. Since the kth behavior is the best response to itself, a pure strategy employing only behavior k is an ESS by condition (1.1.3.) Furthermore, the kth action cannot be part of another, mixed ESS for the same game. Since M has S columns, there can be as many as S pure ESS's (Haigh 1975).

If $S = 2$ and neither column exhibits diagonal dominance, the ESS (see above) may be pure or mixed. If a 2×2 matrix lacks diagonal dominance, the ESS has the form (e.g., Oster and Wilson 1978; Caraco and Pulliam 1984):

$$(\tilde{\pi}_{i1}, \tilde{\pi}_{i2}) = (M_{12} - M_{22}, M_{21} - M_{11}) / (M_{12} + M_{21} - M_{11} - M_{22}). \qquad (1.1.6)$$

The last expression applies in three cases:

1. If $M_{11} = M_{21}$ and $M_{12} < M_{22}$, then ($\tilde{\pi}_{i1} = 1$, $\tilde{\pi}_{i2} = 0$) solves (1.1.6), and is a pure ESS by condition (1.1.3).

Math Box 1.1 (*cont.*)

2. If $M_{11} > M_{21}$ and $M_{12} = M_{22}$, then $(\tilde{\pi}_{i1} = 0, \tilde{\pi}_{i2} = 1)$ solves (1.1.6), and is a pure ESS by condition (1.1.3).

3. If $M_{kk} > M_{rk}$ for $r = 1$ and 2 (and $M_{12} \neq M_{21}$), then the ESS, given by (1.1.6), is mixed.

EXAMPLES: DISCRETE GAMES

Suppose M is a penalty matrix for a two-player discrete game. Each player has two actions, and play is not repeated. M_{rc} is the probability of an energetic shortfall (e.g., Stephens 1982) when player 1 chooses action (row) r, and player 2 chooses action (column) c. Each forager seeks to minimize its failure probability, and thus reduce its chance of starving.

1. Suppose the penalty matrix has the form

$$M = \begin{bmatrix} 0.05 & 0.12 \\ 0.08 & 0.14 \end{bmatrix}. \tag{1.1.7}$$

Diagonal dominance identifies a pure strategy $(\tilde{\pi}_{i1} = 1, \tilde{\pi}_{i2} = 0)$ as the only ESS.

2. After reversing the entries in the second column

$$M = \begin{bmatrix} 0.05 & 0.14 \\ 0.08 & 0.12 \end{bmatrix}. \tag{1.1.8}$$

Diagonal dominance identifies each pure strategy as an ESS; each pure strategy also meets the definition of a Nash equilibrium.

3. The previous examples invoke ESS condition (1.1.2). After changing M_{21} in matrix (1.1.7), we have

$$M = \begin{bmatrix} 0.05 & 0.12 \\ 0.05 & 0.14 \end{bmatrix}. \tag{1.1.9}$$

Neither column exhibits diagonal dominance. But the first pure strategy remains an ESS, since ESS condition (1.1.3) holds.

4. Now consider

$$M = \begin{bmatrix} 0.08 & 0.12 \\ 0.05 & 0.14 \end{bmatrix}. \tag{1.1.10}$$

Neither column exhibits diagonal dominance. Using expression (1.1.6), the mixed ESS is $(\tilde{\pi}_{i1} = 0.4, \tilde{\pi}_{i2} = 0.6)$ for $i = 1, 2$. Note that the same solution results if we consider survival probabilities. That is, the payoff matrix with entries $m_{rc} = (1 - M_{rc})$ has the same ESS.

Math Box 1.1 (*cont.*)

5. Finally, consider the 3 × 3 penalty matrix:

$$M = \begin{bmatrix} 0.05 & 0.12 & 0.11 \\ 0.08 & 0.14 & 0.08 \\ 0.07 & 0.11 & 0.09 \end{bmatrix}. \qquad (1.1.11)$$

$\widetilde{\pi} = (1, 0, 0)$ is a Nash equilibrium, and an ESS by diagonal dominance. Note that the strategy pair $\hat{\pi}_1 = (0, 0, 1)$ and $\hat{\pi}_2 = (0, 1, 0)$ is a Nash equilibrium; neither player is tempted to deviate unilaterally, since the chance of starvation would increase. However, the energetic-shortfall probabilities differ, and this Nash equilibrium cannot qualify as an ESS.

A Continuous Game

Now, when two players interact, each selects an action from a continuous set. Specifically, individual i ($i = 1, 2$) chooses action u_i ($0 \leq u_i \leq u_{\max}$) at each play of the game. Penalty matrices are replaced by penalty functions $W_i(u_i, u_j)$ in continuous games. We assume W_i is continuous in each decision variable. Symmetry of the game implies $W_1(u_1, u_2) = W_2(u_2, u_1)$.

We assume there is a pure strategy $\tilde{u}$ that is an ESS against any rare alternative $u \neq \tilde{u}$ (Parker 1984a). Then each $u_i = \tilde{u}$ at the ESS, which will satisfy conditions for a Nash equilibrium (e.g., Oster and Wilson 1978):

$$\left(\frac{\partial W_1}{\partial u_1}\right)_{\tilde{u}} = \left(\frac{\partial W_2}{\partial u_2}\right)_{\tilde{u}} = 0 \qquad (1.1.12)$$

and

$$\left(\frac{\partial^2 W_i}{\partial u_i^2}\right)_{\tilde{u}} \geq 0 \text{ for } i = 1, 2. \qquad (1.1.13)$$

The identical players incur the same expected penalty at the ESS. A small, unilateral deviation by either player increases that player's penalty. The ESS for an N-player game has a similar solution (see Parker 1984a).

An Example

Supose two social foragers are subject to mortality from starvation and predation. During the day, individual i ($i = 1, 2$) allocates a proportion u_i of its time to vigilance for predators, and allocates a proportion

Math Box 1.1 (*cont.*)

$(1 - u_i)$ of its time to searching for and consuming food. $W_i(u_i, u_j)$ is the probability individual i fails to survive both mortality hazards. Let the penalty functions for the symmetric game be

$$W_1(u_1, u_2) = 0.8 + u_1^2 + 0.25u_2^2 - 0.85(u_1 + u_2)$$
$$W_2(u_2, u_1) = 0.8 + u_2^2 + 0.25u_1^2 - 0.85(u_2 + u_1).$$

Note that $\partial W_i / \partial u_j < 0$; as one individual allocates more time to vigilance, the other forager's mortality declines. Using expression (1.1.12), we find that $\widetilde{u}_1 = \widetilde{u}_2 = 0.425$; each $W_i = 0.3$ at the ESS. Since each $\partial^2 W_i / \partial W_i^2 > 0$, condition (1.1.13) is fulfilled; mortality probabilities are minimized subject to the stability of a Nash equilibrium.

Math Box 1.2 Conditional Mutualism

We apply the term *conditional mutualism* when cooperators benefit simultaneously and each individual's continued cooperation depends on like behavior by the other player, or (in an N-person game) enough other players. The standard device suggesting a logic for conditional mutualism is the iterated Prisoner's Dilemma (IPD). Here we restrict attention to the two-player IPD and invoke the Tit-for-Tat strategy (TFT; Axelrod and Hamilton 1981) as an economic basis for conditional mutualism; we discuss the N-player IPD in the context of producer-scrounger interactions (see chapter 7). After briefly reviewing TFT as an ESS, we consider the concept of Pareto optimality for both the Prisoner's Dilemma and a continuous game.

The Prisoner's Dilemma

Maynard Smith (1984) clearly summarizes the essential features of the IPD, and we take the same approach. As pointed out in section 1.2, we may assume that each of two players chooses to cooperate or to defect during a single round of play. For repeated play between the same two identical players, each chooses between TFT and pure defect. Following Maynard Smith (1984), suppose the payoffs (rather than penalties) for a single round of play are

$$\begin{array}{ccc} & C & D \\ C & 6 & 0 \\ D & 8 & 2. \end{array} \tag{1.2.1}$$

By diagonal dominance, the sole ESS is pure defect. Of course, mutual cooperation yields a greater payoff to each player; mutual cooperation is a Pareto optimal solution (see below).

Now suppose that the same two individuals play repeatedly. Matrix (1.2.1) applies at each round of play, and each individual's payoffs accumulate additively (see Newman and Caraco 1989). The assumed alternatives are TFT and pure defect. TFT cooperates at the first interaction and thereafter does what its opponent last did (Axelrod and Hamilton 1981). Again following Maynard Smith (1984), the payoff matrix for the repeated game after twenty rounds of play is:

$$\begin{array}{ccc} & \text{TFT} & D \\ \text{TFT} & 120 & 38 \\ D & 46 & 40 \end{array} \tag{1.2.2}$$

Math Box 1.2 (*cont.*)

Both TFT and pure D are Nash equilibria, and each qualifies as an ESS for the game defined by matrix (1.2.2). Hence, the conditional mutualism suggested by TFT may be evolutionarily stable. More generally, if play between the same two individuals repeats with a sufficiently large probability, TFT is an ESS for the game with an uncertain duration (Axelrod and Hamilton 1981).

PARETO OPTIMALITY

Suppose that each player acts to help the other, so that their mutual interests are advanced. Then their behavior should correspond to a Pareto solution to the game describing their interaction (Vincent and Grantham 1981). To apply the concept of Pareto optimality to the discrete games considered above, let Ω_i represent the payoff to player i ($i = 1, 2$). Let π_i represent the probability that player i uses the behavior termed cooperative, i.e., C for matrix (1.2.1) and TFT for matrix (1.2.2).

Suppose that the strategy pair $(\pi_1^\circ, \pi_2^\circ)$ is a Pareto solution to a two-player discrete game. Then, following Vincent and Grantham (1981), there exists a θ $(0 \leq \theta \leq 1)$ such that

$$\left(\frac{\partial L_\theta}{\partial \pi_1}\right)_{\pi_2^\circ} = \left(\frac{\partial L_\theta}{\partial \pi_2}\right)_{\pi_1^\circ} = 0, \qquad (1.2.3)$$

where

$$L_\theta = \theta \Omega_1 + (1 - \theta)\Omega_2. \qquad (1.2.4)$$

The function L_θ additively combines the two players' individual payoffs by assigning weights θ and $(1 - \theta)$ to the respective Ω_i. Given that $(\pi_1^\circ, \pi_2^\circ)$ maximizes L_θ, the Pareto optimal strategies will depend on θ. The resulting set of Pareto optimal solutions $\pi^\circ = \{\pi_1^\circ(\theta), \pi_2^\circ(\theta)\}$ then specifies a corresponding set of payoffs $\Omega = \{\Omega_1(\pi^\circ), \Omega_2(\pi^\circ)\}$. Ω has a defining property that the two individuals' payoffs cannot increase simultaneously. Hence, increasing one player's payoff decreases the payoff to the other player (e.g., Oster and Wilson 1978).

Analyses of evolutionary games ordinarily take $\theta = 1/2$, so that each payoff is weighted equally. In this case $\pi_1^\circ = \pi_2^\circ = \pi^\circ$ when condition (1.2.3) holds.

Using (1.1.1), expected payoffs for the single-play PD defined by matrix (1.2.1) are

$$\Omega_i = 2(1 - \pi_i + 3\pi_j)\ i, j = 1,2;\ i \neq j. \qquad (1.2.5)$$

Math Box 1.2 *(cont.)*

Using condition (1.2.3) to find π° yields the following:

1. If $\theta < 1/4$, the Pareto solution is ($\pi_1^\circ = 1$, $\pi_2^\circ = 0$). Payoffs are ($\Omega_1 = 0$, $\Omega_2 = 8$). Player 1 cooperates and player 2 defects when the latter's payoff is weighted sufficiently.
2. If $1/4 < \theta < 3/4$, the sole Pareto solution is mutual cooperation; ($\pi_1^\circ = 1$, $\pi_2^\circ = 1$). Each player's payoff is 6.
3. If $\theta > 3/4$, the Pareto solution is ($\pi_1^\circ = 0$, $\pi_2^\circ = 1$). Payoffs are ($\Omega_1 = 8$, $\Omega_2 = 0$). Player 2 cooperates and player 1 defects when the latter's payoff is weighted sufficiently.

Note that each increase in the payoff to player 1 is accompanied by a decrease in the payoff to player 2, a defining property of the Pareto optimal solution set. Note also that no element of π° qualifies as an ESS. For the repeated game specified by matrix (1.2.2), TFT is a Nash solution and an ESS; no unilateral or coordinated change in strategy can advance either player's payoff. Hence the ecological circumstances promoting repeated interaction between the same individuals should be keys for conditional mutualism.

Math Box 1.3 Topics from Probability Theory

As indicated in section 1.3, several of our models specify search for food as a random process, and then use the probability distribution of the amount of food discovered to derive the individual's chance of starvation. The tools employed to solve these models include probability generating functions and stochastic processes (continuous time "jump" processes). Here we describe essential properties of generating functions, and then use a generating function to solve a Poisson process. Similar methods can be used to solve more complicated processes mentioned in later chapters (see Giffin 1978). We cannot provide this much detail for each model in the text, but we list appropriate references.

Probability Generating Functions

Suppose an event of interest, such as the discovery of a food clump by a foraging group, occurs randomly in time. Let $X(t)$ represent the cumulative number of these events in the time interval $[0, t]$. Then $X(t)$ is a discrete random variable. Let

$$\Pr[X(t) = x] = p_x(t) \text{ for } x = 0, 1, 2, \ldots .$$

A probability generating function is the expected value of a function of a discrete random variable. For any $| z | < 1$, let $g_x(z,t)$ represent the probability generating function of the random variable $X(t)$. Then

$$g_x(z, t) = \mathrm{E}[z^{x(t)}] = \sum_{x=0}^{\infty} z^x p_x(t). \tag{1.3.1}$$

Differentiating with respect to z, we have

$$\frac{\partial g_x\,(z, t)}{\partial z} = \sum_{x=0}^{\infty} xzu^{x-1} p_x(t). \tag{1.3.2}$$

Evaluating this derivative at $z = 1$ yields

$$\left(\frac{\partial g_x\,(z, t)}{\partial z} \right)_{z=1} = \sum_{x=0}^{\infty} x p_x(t) = \mathrm{E}[X]. \tag{1.3.3}$$

Math Box 1.3 (*cont.*)

The probability generating function can be used to find not only the mean, but also the higher moments of $X(t)$ (see Giffin 1978). For our immediate purposes, the more useful partial deriviative is

$$\frac{\partial g_x(z, t)}{\partial t} = \frac{\partial}{\partial t} \sum_{x=0}^{\infty} z^x p_x(t) = \sum_{x=0}^{\infty} \frac{dp_x(t)}{dt} z^x. \tag{1.3.4}$$

Consider the Poisson probability function:

$$\Pr[X(t) = x] = e^{-\lambda t}(\lambda t)^x/x!. \tag{1.3.5}$$

The mean and variance are the same $E[X(t)] = V[X(t)] = \lambda t$. The probability generating function for the Poisson variate is

$$g_x(z, t) = e^{-\lambda t} \sum_{x=0}^{\infty} \frac{(\lambda t)^x}{x!} z^x = e^{\lambda t(z-1)}. \tag{1.3.6}$$

Expressions (1.3.5) and (1.3.6) are equivalent in that each identifies the same Poisson random variable.

A Poisson Process

Suppose a group of G foragers encounters food clumps at constant probabilistic rate $G\lambda$ while searching (we treat G as a pure number in such applications). We can ask how many clumps are discovered by any searching time t, the first step toward specifying how much energy each individual consumes.

Let $X(t)$ represent the cumulative number of clumps found by time t. As above, $\Pr[X(t) = x] = p_x(t)$. Since the foraging process begins at $t = 0$, $p_0(0) = 1$. Given that $G\lambda$ is constant, i.e., independent of $X(t)$, we assume:

1. The probability that a single clump is discovered during the short time interval Δt is $G\lambda\Delta t$.
2. The probability that more than a single clump is discovered during Δt is $o(\Delta t)$, where

$$\lim_{\Delta t \to 0} [o(\Delta t)/\Delta t] = 0.$$

That is, $o(\Delta t)$ goes to 0 faster than does Δt.

3. From the preceding, the probability that no clump is discovered

Math Box 1.3 (*cont.*)

during Δt must be $1 - G\lambda\Delta t - o(\Delta t)$. Then the $p_x(t + \Delta t)$ depend on the $p_x(t)$ as follows:

$$\begin{aligned} p_O(t + \Delta t) &= p_O(t)[1 - G\lambda\Delta t - o(\Delta t)] \\ p_x(t + \Delta t) &= p_{x-1}(t)\, G\lambda\Delta t + p_x(t) \\ &\quad [1 - G\lambda\Delta t - o(\Delta t)];\ x > 1. \end{aligned} \tag{1.3.7}$$

We multiply terms, rearrange and divide by Δt, so that the left side of each equation has the form $([p_x(t + \Delta t) - p_x(t)]/\Delta t)$. Then we take the limit as $\Delta t \to 0$ and obtain

$$\begin{aligned} \frac{dp_O(t)}{dt} &= -\, G\lambda p_O(t) \\ \frac{dp_x}{dt} &= G\lambda p_{x-1}(t) - G\lambda p_x(t);\ x > 1. \end{aligned} \tag{1.3.8}$$

This is an infinite set of differential-difference equations. Time varies continuously (hence the term differential), and the number of clumps varies discretely (hence the term difference).

Next we multiply each equation for $dp_x(t)/dt$ by z^x:

$$\begin{aligned} \frac{dp_O(t)}{dt} z^0 &= -\, G\lambda p_O(t) z^0 \\ \frac{dp_x(t)}{dt} z^x &= G\lambda p_{x-1}(t) z^x - G\lambda p_x(t) z^x;\ x > 1. \end{aligned} \tag{1.3.9}$$

Adding all the equations leads to a single equation:

$$\sum_{x=0}^{\infty} \frac{dp_x(t)}{dt} z^x = G\lambda z \sum_{x=0}^{\infty} p_x(t) z^x - G\lambda \sum_{x=0}^{\infty} p_x(t) z^x. \tag{1.3.10}$$

Using the definition of $g_x(z, t)$ and expression (1.3.4), the preceding becomes

$$\partial g_x(z, t)/\partial t = G\lambda(z - 1)\, g_x(z, t). \tag{1.3.11}$$

Since $p_O(0) = 1$, the preceding is easily solved:

$$g_x(z, t) = \exp\{G\lambda t(z - 1)\}. \tag{1.3.12}$$

From expression (1.3.6) we recognize that the probability function for the number of food clumps discovered by time t must be Poisson with mean and variance $G\lambda t$. A number of models in this book list assumptions concerning the random process of food discovery, and then give

Math Box 1.3 (*cont.*)

the resulting distribution; most of these models can be solved by the generating-function method of the preceding example. Knowing the probabilities for the amount of food the group will discover, we can consider how food is divided among group members and then calculate each individual's probability of starvation or other currency of fitness.

PART ONE Group Membership Games

2 Two-Person Games: Competitive Solutions

2.1 Introduction

This chapter introduces a distinction between aggregation and dispersion economies. Dispersion economies (chapter 5) assume that any increase in group size decreases each member's fitness, so maximal benefits are obtained when individuals are dispersed and solitary. Aggregation economies, on the other hand, emphasize how group membership can increase the quantity or quality of resources available to the individual (e.g., Clark and Mangel 1986). In aggregation economies, a reasonable proxy for survival or fecundity must increase, at least initially, with foraging group size. In this chapter we selectively review some hypothesized mechanisms of aggregation economies. Our review is not extensive; more comprehensive treatments appear in Pulliam and Caraco (1984) and Barnard and Thompson (1985). Moreover, we reserve treatment of special aggregation economies, arising out of cooperative food sharing, for the next chapter. Here we simply examine solitary versus social foraging by analyzing a pair of two-person games where fitness equates with survival. The first game assumes that competitively equivalent players discover food as a random process. The second game assumes a dominance asymmetry between the competitors; both the discovery of food and its division between foragers are random.

Two-person group membership games apply to a number of social foraging situations. The most well known example, documented by Davies and Houston (1981, 1983), involves territory-sharing in the pied wagtail (*Motacilla alba*). A territory holder sometimes allows a satellite to exploit food located within the territory; the satellite then helps defend the food resources. An owner tolerates a satellite only at high food abundance, when the owner gains more from the satellite's defense than it loses to the satellite's resource consumption (see Davies and Houston 1984). Since this interaction involves only two individuals, a clear and careful analysis of the aggregation economy is possible. Therefore, we shall apply a simple game-theoretical analysis to the stability of feeding solitarily versus foraging in pairs, before we examine more general questions about the stable size of foraging groups (chapters 4 and 5).

2.2 Achieving an Aggregation Economy

Improved Searching for Food

When one group member's discovery of food allows other individuals the opportunity to feed, group foraging may generate an aggregation economy. A spatially heterogeneous dispersion of food can imply that individual foraging success depends on the efficiency of the group members' collective searching. Not surprisingly, groups ordinarily discover food clumps faster than a solitary can (e.g., Krebs et al. 1972; Pitcher et al. 1982; Bergelson et al. 1986; see Caraco et al. 1989). However, an individual's expected food intake in a patchy environment need not always increase with group size (e.g., Hake and Ekman 1988).

Suppose all G group members search simultaneously for food clumps. When one finds a clump, each individual expects to consume the same fraction (G^{-1}) of the food. That is, foragers are not constrained by the incompatibility of producing and scrounging (Vickery et al. 1991; Caraco and Giraldeau 1991; see chapters 6 and 7). $\Lambda(G)$ represents the total rate at which the group locates food (see Math Box 1.1). Since the number of group members G is discrete, $\Lambda(G)$ has finite difference:

$$\Delta\Lambda(G) = \Lambda(G + 1) - \Lambda(G). \tag{2.1}$$

This difference is positive whenever adding a group member increases the group's rate of clump discovery. An aggregation economy presumably requires that the total rate at which the group locates food increases fast enough at small group sizes to increase each individual's benefits. For simplicity, we let $\Lambda(G) = \lambda G^{\alpha}$; any α greater than zero assures that the difference is positive. Each group member's expected rate of food consumption (clumps per unit search time) is

$$\Lambda(G)/G = \lambda G^{\alpha - 1}. \tag{2.2}$$

If $\alpha > 1$, individuals' food-discovery rates interact positively. Each forager's average food-consumption rate consequently increases with group size. This could occur if larger groups of predators flush increasingly greater numbers of spatially aggregated, hidden prey.

If $\alpha = 1$, group members search independently, and each individual's average food-consumption rate does not vary with group size. However, if clumps are discovered as random and independent events, the variance of the time spent searching for each individual's required amount of food declines with group size (Caraco 1981; Pulliam and Millikan 1982; Clark and Mangel 1984; Ranta et al. 1993; Ruxton et al. 1995). This variance reduction (or the reduction in the individual's food-consumption variance when foraging time is fixed) can increase a forager's survival probability when group members expect enough food to satisfy physiological requirements (Clark and Mangel 1986; Caraco 1987; Ekman and Rosander 1987; Mangel 1990). Hence $\alpha = 1$ (equivalently $\Lambda(G)/G$ = a constant) can easily imply an aggregation economy.

If the value of α lies between zero and one, the total rate at which the group locates food increases more slowly than group size. Although the total rate at which the group encounters food increases (i.e., $\Delta\Lambda(G) > 0$), each individual's average rate of food consumption nonetheless decreases with group size. Larger groups of predators sometimes alert their prey (Charnov et al. 1976; Goss-Custard 1976), or the areas that different group members search may overlap (Hake and Ekman 1988). When this sort of inefficiency causes average feeding rates to decline with group size, an aggregation economy will probably require either a reduction in food-intake variance or an increase in protection from predation sufficient to increase individual survival (Mangel 1990; Houston et al. 1993). We return to some of these points in chapter 4.

The "information center" hypothesis suggests that the unpredictable appearance of food clumps, in both space and time, can favor not just group foraging, but a colonial social organization (Hoogland and Sherman 1976; Greene 1987; C. Brown 1988; Seeley and Visscher 1988; Götmark 1990; Barta and Szép 1995). In an avian breeding colony, unsuccessful foragers supposedly watch for individuals returning to the colony with food for their dependent offspring. Unsuccessful individuals later follow the successful foragers when the latter return to temporary concentrations of food. The information center hypothesis proposes a plausible aggregation economy based on a group's food-discovery efficiency, but it can be difficult to evaluate the idea empirically (Waltz 1983).

Applications of the information center hypothesis usually classify foragers as succeeding or failing. But in certain cases, individuals acquire more detailed information about food quality and food abundance by observing other foragers (e.g., Greene 1987; Templeton and Giraldeau 1995a,b, 1996). Acquisition of information concerning locations of patchily distributed food may have been an important advantage of sociality when protohominids occupied the savanna environment (Kurland and Beckerman 1985). Contemporary fishermen often choose to share information on the location of resources; see models developed in Mangel and Clark (1983) or Mangel and Plant (1985). In general, pooling information will allow group members to learn locations of available food faster than a solitary is able to.

Enhanced Capture and Consumption of Food

Larger groups of carnivores often can capture larger prey (see Pulliam and Caraco 1984; Gese et al. 1988). Furthermore, an increase in carnivore group size may increase both the probability of success per capture attempt and the frequency of multiple kills (Schaller 1972; Caraco and Wolf 1975; Nudds

1978; Clark 1987; Giraldeau 1988; Packer and Ruttan 1988; also see Beckerman 1991). Individually or collectively, these mechanisms can imply an aggregation economy.

The carnivore examples just mentioned are well known. But the economic interaction defining social foraging can occur in less behaviorally complex animals. Consider the jack-pine sawfly larva (*Neodiprion pratti banksianae*) studied by Ghent (1960). These organisms feed only on the tissues inside pine needles. Access to their food presents a problem, since a larva finds it difficult to bite through a needle's outer covering. When a single larva does break through the cuticle, other larvae sometimes converge at the entry point. The group collectively enlarges the hole, and each individual's effort enhances its own and its neighbor's access to food. Ghent (1960) reports 80% mortality for isolated larvae, while the mortality among group members was 53%—strong evidence of an aggregation economy.

Groups may often defend food against scavengers more effectively than solitaries can (see Pulliam and Caraco 1984). Reversing perspective, groups sometimes overcome a territory holder's defense of food resources, while solitary intruders are repelled (e.g., Diamond 1987). Foster (1985) documents an interesting example. A territorial dusky damselfish (*Stegastes dorsopunicans*) easily prevents any solitary blue tang surgeonfish (*Acanthurus coeruleus*) from feeding on algal mats the owner defends. But groups of surgeonfish can overcome aggressive defense by the same damselfish and gain access to the food.

Reduced Time and Energy Costs of Foraging

Group foraging may reduce an individual's time or energy expenditure per unit of energy consumed. Riechert et al. (1986) compared solitary and social foraging in the tropical spider *Agelena consociata*. A solitary adult female must spend a relatively large amount of energy maintaining her web trap; in groups, each individual spends less energy maintaining the communal web trap. Solitaries incur energy deficits more often than social foragers, and solitaries suffer higher mortality rates than group members. Groups may occupy sites where webs are less susceptible to damage from the elements, and the different energy expenditures would follow (Riechert et al. 1986). But perhaps group members can allocate less energy to the web because sociality somehow increases the rate of prey capture per unit area of the web (see Uetz 1989; Uetz and Hodge 1990; Caraco et al. 1995).

Improved Feeding of Offspring

In some avian and mammalian species, "helpers" assist parents in procuring food for dependent offspring (Emlen 1984, 1995; J. L. Brown 1987). Helpers may be related to the offspring and may have a low probability of breeding successfully outside of the group. Holding the number of breeders and offspring constant, an increase in group size (i.e., more helpers) may imply an increase in each group member's fitness. Each forager need not expend as much time and energy securing food for the offspring. The offspring should be less likely to starve as group size increases, and their mortality hazard due to predation may decline (Caraco and Brown 1986). Consequently, an increase in helper number may enhance the indirect component of each helper's inclusive fitness. So, in fairly simple ways, foraging for dependent offspring may generate an aggregation economy.

Increased Protection from Predators

Suppose an individual's survival often depends on both averting starvation and avoiding predation (see Houston et al. 1993). The probabilities of these events easily can vary with group size and so produce an aggregation economy. Stacy (1986) suggests that avoiding predation may be the primary day-to-day benefit of group foraging in yellow baboons (*Papio cynocephalus*). Given the relative safety of a large group, members can forage farther from protective cover and allocate more time to feeding (less time to vigilance). Therefore, large groups presumably provide individuals an advantage in terms of foraging energetics during periods of low food availability (Stacy 1986). In this case group foraging depends on sociality's influence on predation hazard. Economically similar circumstances arise when avian social foragers spend less time being vigilant than solitaries and more time feeding (e.g., Caraco et al. 1980a; Pulliam et al. 1982; Elgar et al. 1986; Ekman 1986; but see Godin 1986), and when reduced vigilance in groups allows members to shift their dietary selection to more energetically profitable, but more time-consuming foods (Lima and Dill 1990). If group membership allows an individual some advantage in detecting approaching predators (e.g., Hart and Lendrem 1984), in fleeing from predators (Ydenberg and Dill 1986), in repelling predators (C. Brown and Hoogland 1986), or if group membership simply dilutes the individual's chance of being a predator's victim (Caraco and Pulliam 1984), then group members should obtain their required energy intake more rapidly than highly vigilant solitaries.

A Consideration of the Costs of Social Foraging

Even in an aggregation economy, an individual's fitness-related costs may increase with group size, and these costs will eventually exceed benefits as group membership increases. Resource competition may increase with group size. Larger groups often result in increased local density of consumers, so that competition leads to aggression and kleptoparasitism (e.g., Goss-Custard and Durell 1987a,b,c). Members of larger groups may experience greater exposure to parasites and pathogens; Hoogland (1979) demonstrates an increase in ectoparasite load with group size. Larger groups must sometimes attract more frequent predatory attacks; in some cases this effect may induce an increase in predation hazard with group size (Lindström 1989).

A number of reviews and texts list hypothesized benefits and costs of sociality. It is the dependence on group size that makes analysis of aggregation economies interesting and quantitatively feasible.

2.3 A Symmetric Group Membership Game

A Case with No Predation Hazard

Suppose two foragers exploit an environment where food occurs in clumps of a uniform size. A simple model asks how clump density and group searching efficiency might influence preference for solitary versus group foraging.

The game allows a player no more than τ time units to discover and consume the amount of food contained in a clump. Otherwise, the individual starves or incurs a physiological deficit increasing mortality of the forager or its offspring. For simplicity, we treat handling times as negligible.

A solitary must locate a single food clump by time τ to avoid starving. Each player, while foraging solitarily, discovers food clumps at constant probabilistic rate λ. Then a solitary's probability of starving, $M(G = 1)$, is

$$M(G = 1) = e^{-\lambda\tau}. \tag{2.3}$$

If the two individuals forage as a group, they share any food discovered. The group discovers clumps at constant probabilistic rate $K\lambda$. The parameter K specifies the group's searching advantage. A group expects to encounter clumps faster than a solitary when $K > 1$, but a solitary consumes all the food it finds. Each group member starves unless they collectively discover two food clumps by time τ. Then each group member's starvation probability, $M(G = 2)$, is

$$M(G = 2) = e^{-K\lambda\tau} + K\lambda\tau e^{-K\lambda\tau} = e^{-K\lambda\tau}(1 + K\lambda\tau). \tag{2.4}$$

$M(G = 2)$ sums the first two terms of a Poisson probability function with mean $K\lambda\tau$.

Each player has two pure strategies: preference for solitary foraging ($G = 1$) and preference for group foraging ($G = 2$). Group foraging occurs only if both individuals prefer it. So if π_i is the probability that player i will prefer solitary foraging, then $(1 - \pi_1)(1 - \pi_2)$ is the likelihood of group foraging. The resulting penalty matrix M is simple:

$$\begin{array}{ccc} & G = 1 & G = 2 \\ G = 1 & e^{-\lambda\tau} & e^{-\lambda\tau} \\ G = 2 & e^{-\lambda\tau} & e^{-K\lambda\tau}(1 + K\lambda\tau). \end{array} \tag{2.5}$$

By inspection, no mixed strategy can be an ESS (see Math Box 1.1). Group foraging should be favored when the probability of starving when paired is less than the probability of starving when alone. Group foraging is a stable Nash equilibrium and an ESS when

$$\begin{aligned} e^{-K\lambda\tau}(1 + K\lambda\tau) &< e^{-\lambda\tau} \\ \ln(1 + K\lambda\tau) &< \lambda\tau(K - 1). \end{aligned} \tag{2.6}$$

Reversing this inequality makes solitary foraging a neutrally stable Nash equilibrium and an ESS, since a solitary then has a smaller chance of starving than a group member.

Group foraging can never be an ESS if there is no searching advantage (if $K \leq 1$). That is, the group must locate food clumps faster than a solitary can, or else neither individual should prefer group foraging. It is worth noting that group foraging can be an ESS for searching-advantage values between 1 and 2, where a group member's expected food consumption (as opposed to clumps encountered) is lower in a group than as a solitary. In this situation the greater variance of the time a solitary requires to find a food clump increases its chance of starvation and hence favors groups (e.g., Clark and Mangel 1984; Caraco et al. 1995; see section 2.2). Given that each forager tries to reduce its chance of starving, this introductory model predicts that the incidence of group foraging should increase if

1. The searching advantage implies that groups encounter food clumps at a rate sufficiently exceeding a solitary's discovery rate (i.e., $K > 1$).
2. Food density is high, or if food clumps are easily located (i.e., when λ is large).
3. The time available to search for food is long (i.e., when τ is large).

Reversing these conditions, of course, favors solitary foraging.

Suppose inequality (2.6) holds, so that group foraging is the ESS, and two solitaries begin to search for food together. In this case (i.e., in an aggregation economy) group formation constitutes a mutualism; see Summary Box 1.2. Both individuals' survivorship is increased by grouping, hence it would be fair to associate the notion of "cooperative foraging" with the stable competitive solution to this game. This may seem odd, but will contrast appropriately with the result for dispersion economies, where a stable increase in group size constitutes selfish behavior; see chapter 5.

A Case including Predation Hazard

Starvation and predation may impose simultaneous mortality hazards and so influence the economics of social foraging (e.g., Barnard 1980; Caraco et al. 1980b; Barnard and Thompson 1985; Ydenberg and Dill 1986; Newman and Caraco 1989; Houston et al. 1993). To examine the interaction between starvation and predation in our simple two-person game, let $s(1)$ represent a solitary's probability of surviving predation while foraging, and let $s(2)$ represent the probability that the individual survives predation while foraging in a group. For convenience, let ξ represent the relative predation hazard of a group member; i.e., $s(1)/s(2)$. When an individual's mortality hazard due to predation is lower in a group than when alone, $\xi < 1$. However, if groups attract more predatory attacks, solitary foraging could be safer so that $\xi > 1$.

We treat starvation and predation as independent events. So a solitary forager's probability of *survival* is

$$m(G = 1) = s(1)\,(1 - e^{-\lambda\tau}). \tag{2.7}$$

Each group member's probability of surviving both sources of mortality is

$$m(G = 2) = s(2)\,[1 - e^{-K\lambda\tau}\,(1 + K\lambda\tau)]. \tag{2.8}$$

We assume that each individual should try to maximize its chance of surviving. Then, group foraging will be a stable Nash solution and an ESS when

$$1 - e^{-K\lambda\tau}\,(1 + K\lambda\tau) > (1 - e^{-\lambda\tau})\xi. \tag{2.9}$$

When a group member and a solitary suffer the same predation hazard ($\xi = 1$), predation has no influence on group foraging. When group foraging is safer from predators than solitary foraging ($\xi < 1$), the predictions suggested by considering starvation probabilities alone hold qualitatively. As the group's relative safety increases, groups become stable at increasingly lower expected food levels. Figure 2.1 indicates that when the group's predation hazard is sufficiently small compared to a solitary forager, group foraging can be stable at search advantage values lower than 1, where the group collectively encounters patches at a rate lower than a solitary's.

When a group member's predation hazard is greater than a solitary's, there is a minimal value of the searching advantage exceeding unity below which group foraging cannot be stable. For searching advantage values exceeding this minimum, there is an interval of expected food levels ($\lambda\tau$) where group foraging is stable. The extent of this interval increases with increasing searching advantage (see fig. 2.1). For expected food levels above and below this interval, solitary foraging is the ESS. So, when groups are hazardous and searching advantage is large, the likelihood of group foraging increases with increasing expected food levels when this level is low. However, as the expected food levels get larger, the influence of starvation on mortality declines since foragers discover food so quickly. Predation becomes the dominant mortality hazard, and at high expected food levels individuals should return to solitary foraging (see Mangel 1990; Székely et al. 1991).

A number of more complex analyses are possible. Predation hazard could depend on the time spent foraging, and starvation need not be a step function of food consumption (e.g., Newman 1991; Houston et al. 1993). But the

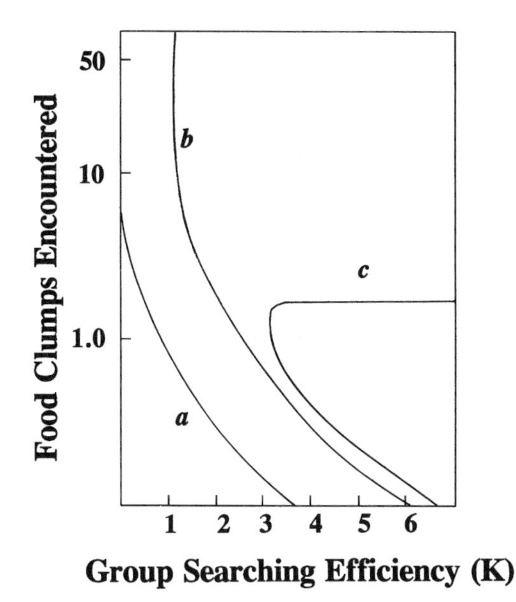

Figure 2.1 Combinations of searching advantage and expected food levels for which the symmetric, two-player group membership game predicts stable group foraging. Crossing a plot from left to right moves from a region of evolutionarily stable solitary foraging to a region of stable group foraging. Each line gives the result for different levels of predation relative to solitary foraging ξ. In *a*, *b* and *c*, $\xi = 0.5$, 1.0, and 1.2, respectively, so that foraging in a group is less, the same as, and more hazardous than solitary foraging.

analysis of this simple two-person game shows that game theory can be applied in a simple way to help understand the conditions under which group foraging can be economically advantageous (see Summary Box 2.1). The next section takes the same approach to a more complex and likely more realistic group-membership game.

2.4 An Asymmetric Group Membership Game

Suppose two foragers, one dominant and one subordinate individual, exploit an environment where food occurs as items rather than as clumps (Caraco 1987). Then there are two immediate differences between this and the preceding analysis. First, the dominant's competitive advantage during group foraging implies an asymmetric game. Second, the term "food items" refers to a resource occurring as indivisible units; a single food item is always consumed in its entirety by one forager (see Recer and Caraco 1989a; Caraco and Giraldeau 1991).

SUMMARY BOX 2.1 THE SYMMETRIC TWO-PLAYER GROUP MEMBERSHIP GAME

ASSUMPTIONS

Currency: Survival

Decision: Forage alone or with another individual

Constraints:

1. Only two players.
2. Players competitively equal.
3. Each play is uninformed.
4. Strategies not genetically based but phenotypically plastic.
5. Food is encountered in clumps which are divided equally between foragers and require no handling time.
6. Habitats do not deplete.

PREDICTIONS

A. No predation hazard

1. Groups stable only when group clump encounter rate exceeds rate when alone.
2. Groups stable when clumps are easily found.
3. Groups stable when foraging period is long relative to energetic requirement.

B. With predation hazard

1. When predation hazard in groups is lower than solitary foraging:
The same as above except that groups can be stable even if the rate of clump encounter within groups is lower than solitary encounter rates.

2. When predation hazard in groups is greater than when foraging alone:
The minimum rate of encounter with clumps that allows group foraging is larger than in the absence of predation.

This model further assumes that a forager learns to locate, or to recognize, food items more quickly as the number of items encountered accumulates. The analysis will suggest how competitive asymmetry and an advantage of group searching might interact to influence preference for solitary versus group foraging.

Each forager has τ time units available to search for food (handling times are negligible). An individual fails to meet its minimal physiological requirement for food if it consumes R or fewer food items by time τ; an energetic failure again imposes starvation or some other fitness cost.

$X(t)$ represents the number of items a solitary forager has consumed by time t; $X(0) = 0$. A solitary discovers food items at probabilistic rate $\lambda(X)$. Learning implies that $\lambda(X)$ increases as $X(t)$ increases. Alternatively, the forager might move toward a greater spatial or temporal concentration of food as its consumption increases. Specifically, we assume that $\lambda(X)$ depends linearly on $X(t)$:

$$\lambda(X) = \lambda[1 + X(t)]. \tag{2.10}$$

The dynamics of the process define how the probabilities $\Pr[X(t) = x]$ change with time and $\lambda(X)$. For a solitary, the total number of items consumed by time τ, $X(\tau)$, follows a geometric probability function with mean

$$\mathrm{E}[X(\tau)] = \mathrm{e}^{\lambda\tau} - 1. \tag{2.11}$$

A solitary's probability of an energetic failure is

$$\Pr[X(\tau) \leq R] = 1 - (1 - \mathrm{e}^{-\lambda\tau})^{R+1}. \tag{2.12}$$

Math Box 2.1 at the end of this chapter provides details of the model.

Now suppose the two indivduals forage as a group. $\chi(\tau)$ represents the total number of items the group finds and consumes during the foraging period. $\chi(\tau) = X_D + X_S$, where X_D is the number of items consumed by the dominant and X_S is the number of items the subordinate individual consumes. The group discovers food items at total probabilistic rate $K\lambda(\chi)$:

$$K\lambda(\chi) = K\lambda[1 + \chi(t)]. \tag{2.13}$$

K is again the group's searching advantage. The form of $K\lambda(\chi)$ implies that each forager learns to recognize prey more quickly as the total number of items consumed accumulates.

When the group discovers an item, the dominant consumes that item with probability θ; $1/2 \leq \theta < 1$. Any increase in θ increases the asymmetry of the game. Each item is treated independently with respect to competition between the dominant and subordinate individual.

The preceding assumptions imply that $\chi(\tau)$ follows a geometric probability function with mean

$$\mathrm{E}[\chi(\tau)] = \mathrm{e}^{K\lambda\tau} - 1. \tag{2.14}$$

In Math Box 2.1 we derive the unconditional distribution for each X_i ($i = D, S$), and use those results to identify each forager's probability of an energetic failure.

For the dominant:

$$\Pr[X_D \leq R] = 1 - \left[\frac{\theta(1 - \mathrm{e}^{-K\lambda\tau})}{\theta(1 - \mathrm{e}^{-K\lambda\tau}) + \mathrm{e}^{-K\lambda\tau}}\right]^{R+1}. \tag{2.15}$$

The subordinate individual's probability of an energetic failure, $\Pr[X_S \leq R]$, is obtained by substituting $(1 - \theta)$ for θ in expression (2.15).

Group foraging is a stable Nash equlibrium if both players have lower failure probabilities as group members. Solitary foraging is a neutrally stable Nash equilibrium, if either player prefers to forage alone. That is, solitary foraging results if both players have lower probabilities of energetic failure as solitaries. Alternatively, the dominant prefers social foraging, but the solitary prefers to forage alone, and, by assumption, both remain solitary. Math Box 2.1 gives details of the game's solutions.

Figure 2.2 shows how the stability of group foraging depends on the search advantage (K) and the degree of asymmetry (θ). It also illustrates how that dependence varies with the expected food level. As the degree of asymmetry increases, the dominant's competitive advantage during group foraging increases. Consequently, the dominant can prefer group foraging at decreasing values of searching advantage as player asymmetry increases. For a given asymmetry, increasing the expected food level decreases the value of the searching advantage where each individual begins to prefer group foraging (fig. 2.2: compare top to bottom panel). In fact, when the asymmetry is small (θ is near ½) and the expected food level grows large, group foraging will be stable even as the searching advantage approaches unity.

Assuming that rates of encountering food items increase with forager experience, the model predicts that the likelihood of group foraging should increase when

1. The competitive asymmetry is not too large (θ is near ½); if the dominant does have a strong advantage, the searching advantage must be large.
2. Expected food levels are high, because items are easily discovered (λ large).
3. Time available for foraging does not strongly constrain the probability of avoiding energetic failure (i.e., τ is large).

Summary Box 2.2 reviews the asymmetric interaction just analyzed.

The games we have considered to this point are similar in that food-discovery rates never decline as more food is located, and increased food levels (specifically $\lambda\tau$) increase the likelihood of group foraging. But Caraco (1987) presents a model where solitary and group foragers encounter food items at rates that decline linearly as food is discovered. The result of this assumption, which mimics a reasonable effect of resource depression (Charnov et al. 1976), is that group foraging grows less likely as the expected food level increases. For a given searching advantage, a sufficiently high initial food density leads a solitary to avoid the greater decline in encounter rate experienced as a group depletes available resources. Hence no simple generality linking expected food consumption to preference for group foraging will always hold.

We note that the physiologically required number of food items (R) does not influence the stability of group foraging in the last model. Since the

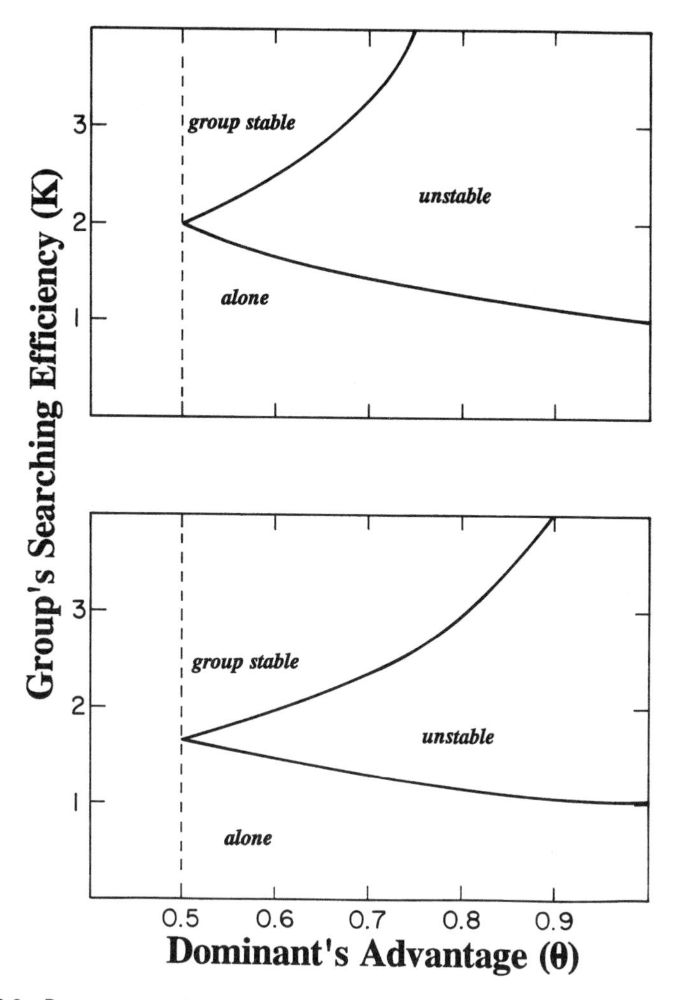

Figure 2.2 Parameter combinations where group searching is stable, where the dominant, but not the subordinate prefers grouping ("unstable" for convenience), and where both prefer searching alone. θ is the probability the dominant acquires any food item discovered, characterizing the asymmetry between players. A solitary's expected food intake ($\lambda\tau$) is increased 50-fold in the lower panel.

game's solution requires comparing geometric distributions, preference for solitary versus group foraging depends only on expected food intake levels. The geometric's variance is a function of its mean only; the individual should always prefer the larger average food intake at any level of its physiological requirement (see Caraco 1987). However, for more complex distributions of resource consumption, the stability of group foraging will depend on a comparison of expected and required food intake levels (see chapter 4).

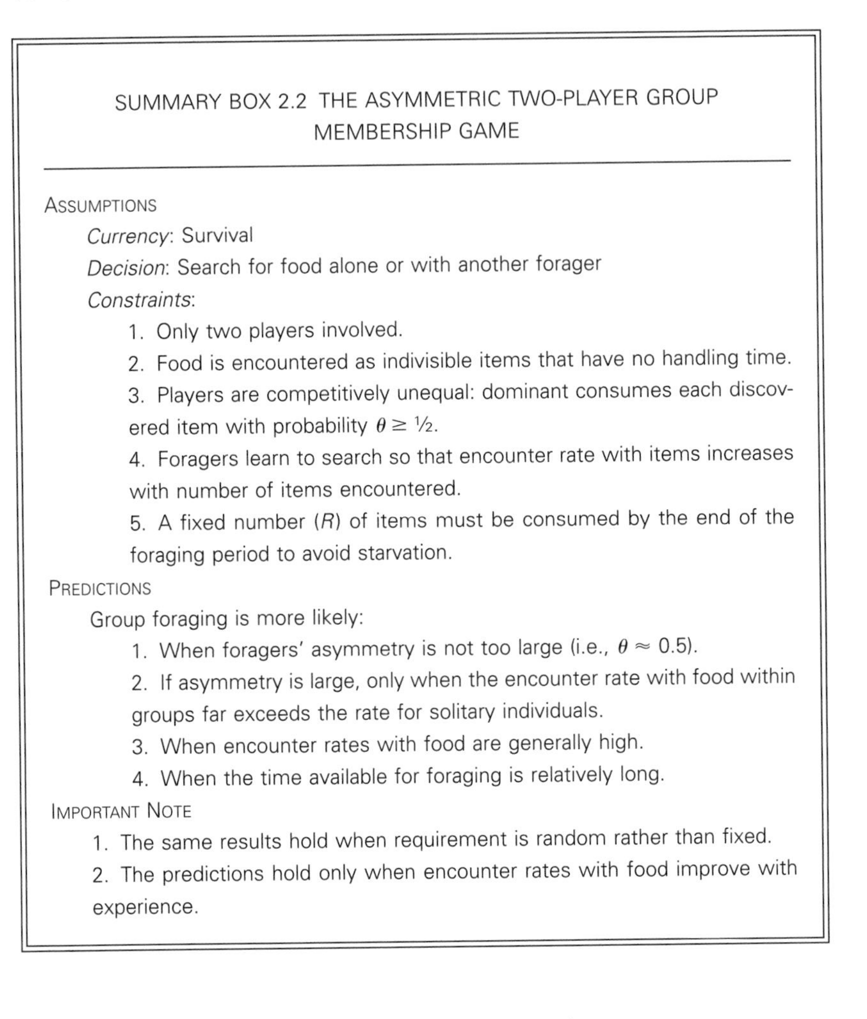

SUMMARY BOX 2.2 THE ASYMMETRIC TWO-PLAYER GROUP MEMBERSHIP GAME

ASSUMPTIONS

Currency: Survival

Decision: Search for food alone or with another forager

Constraints:

1. Only two players involved.
2. Food is encountered as indivisible items that have no handling time.
3. Players are competitively unequal: dominant consumes each discovered item with probability $\theta \geq \frac{1}{2}$.
4. Foragers learn to search so that encounter rate with items increases with number of items encountered.
5. A fixed number (R) of items must be consumed by the end of the foraging period to avoid starvation.

PREDICTIONS

Group foraging is more likely:

1. When foragers' asymmetry is not too large (i.e., $\theta \approx 0.5$).
2. If asymmetry is large, only when the encounter rate with food within groups far exceeds the rate for solitary individuals.
3. When encounter rates with food are generally high.
4. When the time available for foraging is relatively long.

IMPORTANT NOTE

1. The same results hold when requirement is random rather than fixed.
2. The predictions hold only when encounter rates with food improve with experience.

2.5 Concluding Remarks

Implications for Future Work

Testing preference for solitary versus group foraging requires creative experimentation; see Ekman and Hake (1988). Most models of the sort analyzed in this chapter have yet to be tested empirically. One of the most striking predictions of our models concerns the effect of the searching advantage (K). It is surprising to realize that, for sufficiently high food density, group foraging can be evolutionarily stable so long as a pair's combined patch encounter rate exceeds the patch encounter rate of a solitary (i.e., $K > 1$). So, it is the clump encounter rate, not the expected feeding rate per se, that is crucial for the maintenance of stable groups. This means that group foraging can tolerate considerable searching interference between foragers, as long as grouping reduces the variance of the economic outcome.

Decreasing a group member's predation hazard relative to a solitary's hazard can make group foraging evolutionarily stable at even lower expected food levels. That is, safety from predation can make groups stable even for searching advantage values below unity.

Predictions for the asymmetric game assume that foragers encounter food at increasing rates either because of learning or an enrichment of the environment. Under this assumption, group stability depends on the extent of competitive asymmetry and the searching advantage. As the asymmetry (θ) increases, groups will be stable only if the searching advantage increases sufficiently to compensate for the subordinate individual's losses. All these predictions await testing.

Final Comment

This chapter introduces the notion of an aggregation economy. We use the simple device of comparing solitary and paired foragers to establish our approach to foraging games. That is, we let ecological factors governing either the rate of food discovery or the allocation of food following discovery regulate the interaction between or among foragers. Consequently, these factors will influence the outcome of the foraging game and the resulting predictions. The next chapter extends the analysis of two-person foraging games to special cases such as food-sharing and food-calling, classes of behaviors that may require conditional cooperation for economic stability.

Math Box 2.1 Energetic-Failure Probabilities: Asymmetric Game

From the text, a solitary discovers food items as a linear, increasing function of $X(t)$:

$$\lambda(X) = \lambda[1 + X(t)]. \tag{2.1.1}$$

Initially, $X(0) = 0$. The probability a solitary has found no food by time t is $\Pr[X(t) = 0]$, which declines with time:

$$d\Pr[X(t) = 0]/dt = -\lambda \Pr[X(t) = 0]. \tag{2.1.1}$$

For $X(t) > 0$,

$$d\Pr[X(t) = x]/dt = \lambda x \Pr[X(t) = x - 1] - \lambda(1 + x) \Pr[X(t) = x]. \tag{2.1.3}$$

Standard techniques (e.g., Giffin 1978; see Math Box 1.3) identify the state probabilities. At the end of foraging ($t = \tau$), we have

$$\Pr[X(\tau) = x] = e^{-\lambda\tau}(1 - e^{-\lambda\tau})^x. \tag{2.1.4}$$

$X(\tau)$ follows a geometric probability function; the mean and variance, respectively, are

$$E[X(\tau)] = e^{\lambda\tau} - 1;\ V[X(\tau)] = e^{\lambda\tau} E[X(\tau)]. \tag{2.1.5}$$

A solitary's probability of an energetic failure is

$$\Pr[X(\tau) \le R] = e^{-\lambda\tau}\sum_{x=0}^{R}(1 - e^{-\lambda\tau})^x = 1 - (1 - e^{-\lambda\tau})^{R+1}. \tag{2.1.6}$$

From the text, $\chi(\tau)$ is the total number of items a group consumes during a foraging period. $\chi(\tau)$ varies randomly, and

$$\chi(\tau) = X_D + X_S.$$

X_D is the number of items the dominant consumes, and X_S is the subordinate individual's food consumption.

The group's combined rate of food-item discovery increases linearly with $\chi(t)$:

$$K\lambda(\chi) = K\lambda[1 + \chi(t)]. \tag{2.1.7}$$

The dominant consumes any item discovered by the group with probability θ; $1/2 \le \theta < 1$. Hence, the subordinate consumes the item with probability $1 - \theta$. Each item is contested independently.

Math Box 2.1 *(cont.)*

By comparison to (2.1.4), $\chi(\tau)$ has a geometric distribution with mean and variance

$$E[\chi(\tau)] = e^{K\lambda\tau} - 1; \; V[\chi(\tau)] = e^{K\lambda\tau} E[\chi(\tau)]. \tag{2.1.8}$$

Given $\chi(\tau)$, the conditional distribution of (X_D, X_S) is binomial, since each item is an independent trial:

$$\Pr[X_D = d, X_{S=s} | \chi(\tau) = d + s] = \binom{d+s}{d} \theta^d (1 - \theta)^s. \tag{2.1.9}$$

The unconditional probabilities are, by definition,

$$\Pr[X_D = d, X_S = s] = \Pr[X_D = d, X_S = s \,|\, \chi(\tau) = d + s] \bullet \Pr[\chi(\tau) = d + s]. \tag{2.1.10}$$

Since $\chi(\tau)$ has a geometric distribution,

$$\Pr[X_D = d, X_s = s] = \binom{d+s}{d} \theta^d (1 - \theta)^s e^{-K\lambda\tau} (1 - e^{-K\lambda\tau})^{d+s} \tag{2.1.11}$$

$$= \binom{d+s}{d} (\theta[1 - e^{-K\lambda\tau}])^d \, ([1 - \theta][1 - e^{-K\lambda\tau}])^s \, e^{-K\lambda\tau}.$$

Expression (2.1.11) shows that (X_D, X_S) follows a negative multinomial probability function; the moments are

$$E[X_D] = \theta(e^{K\lambda\tau} - 1); \; V[X_D] = E[X_D] \, (E[X_D] + 1) \tag{2.1.12}$$

$$E[X_S] = (1 - \theta)(e^{K\lambda\tau} - 1); \; V[X_S] = E[X_S] \, (E[X_S] + 1) \tag{2.1.13}$$

$$COV[X_D, X_S] = E[X_D] \, E[X_S] > 0. \tag{2.1.14}$$

Expression (2.1.14) shows that, given θ, X_D and X_S covary positively. The probabilities of energetic failure are obtained, in two steps, from the negative multinomial. First, recall that

$$\Pr[X_D = d] = \Sigma_s \Pr[X_D = d, X_S = s]$$

defines the unconditional distribution of X_D; the unconditional distribution of X_S is defined similarly. Unconditionally, each X_i has a geometric distribution (Boswell et al. 1979). For the dominant

$$\Pr[X_D = d] = \rho_D (1 - \rho_D)^d,$$

where

Math Box 2.1 (*cont.*)

$$\rho_D = \mathrm{e}^{-K\lambda\tau}/[\theta(1 - \mathrm{e}^{-K\lambda\tau}) + \mathrm{e}^{-K\lambda\tau}]. \tag{2.1.15}$$

Substituting $(1 - \theta)$ for θ in ρ_D gives ρ_S, and

$$\Pr[X_S = s] = \rho_S(1 - \rho_S)^s.$$

Using (2.1.15), the dominant's chance of an energetic failure is

$$\Pr[X_D \leq R] = 1 - \left[\frac{\theta(1 - \mathrm{e}^{-K\lambda\tau})}{\theta(1 - \mathrm{e}^{-K\lambda\tau}) + \mathrm{e}^{-K\lambda\tau}}\right]^{R+1}. \tag{2.1.16}$$

The subordinate individual's chance of failure, $\Pr[X_S \leq R]$, is obtained by substituting $(1 - \theta)$ for θ in (2.1.16).

Using (2.1.6) and (2.1.16), both players will prefer solitary foraging when

$$[1 - \mathrm{e}^{-\lambda\tau}]^{R+1} > [1 - \rho_i]^{R+1} \tag{2.1.17}$$

for *both* $i = D$ and S. In this case, solitary foraging is a Nash equilibrium. Reversing inequality (2.1.17) for both players implies that group foraging is a stable Nash equilibrium. Since the game is asymmetric, inequality (2.1.17) may hold for the subordinate, but not the dominant, player. In this case, the dominant prefers group foraging, while the subordinate player fares better alone. Our assumption of uninformed play predicts solitary foraging, but over a longer timescale an alternation of solitary and group foraging might occur (see Pulliam and Caraco 1984; Parker 1985).

We simplify these results by substituting for the ρ_i values. Both players prefer solitary foraging when

$$\mathrm{e}^{\lambda\tau} - 1 > \theta(\mathrm{e}^{K\lambda\tau} - 1) \text{ and } \mathrm{e}^{\lambda\tau} - 1 > (1 - \theta)(\mathrm{e}^{K\lambda\tau} - 1). \tag{2.1.18}$$

Equivalently, $\mathrm{E}[X(\tau)] > \mathrm{E}[X_i]$ for $i = D$ and S. Each of the three random variables $X(\tau)$, X_D and X_S has a geometric distribution, and the probability of starvation always declines as the mean increases. Compacting (2.1.18) gives the condition where both individuals prefer solitary foraging as

$$(\mathrm{e}^{K\lambda\tau} - \mathrm{e}^{\lambda\tau})/(\mathrm{e}^{K\lambda\tau} - 1) < \theta < (\mathrm{e}^{\lambda\tau} - 1)/(\mathrm{e}^{K\lambda\tau} - 1). \tag{2.1.19}$$

Math Box 2.1 (*cont.*)

Both competitors prefer group foraging when

$$(e^{K\lambda\tau} - e^{\lambda\tau})/(e^{K\lambda\tau} - 1) > \theta > (e^{\lambda\tau} - 1)/(e^{K\lambda\tau} - 1). \tag{2.1.20}$$

Failure to satisfy either (2.1.19) or (2.1.20) implies that the dominant prefers group foraging, but the subordinate competitor prefers solitary foraging.

3 Two-Person Games: Conditional Cooperation

3.1 Introduction

Conditional cooperation has special significance in evolutionary ecology, probably because an opportunity to behave cooperatively implies an opportunity to defect. That is, interacting cooperators might collectively enhance their benefits. But they risk the chance that an individual will defect, selfishly furthering its own benefits at the cooperators' expense (see Math Box 1.2). This intriguing contrast of "cooperate" and "defect" behaviors has prompted a remarkable series of analyses addressing fundamental issues of sociality (e.g., Axelrod and Hamilton 1981; Maynard Smith 1984; Mesterton-Gibbons 1991; Nowak and Sigmund 1994; Ferièrre and Michod 1996; Mowbray 1997). Most of these studies emphasize that stability of cooperation between nonrelatives, who are selfishly tempted to defect, should be conditioned by a requirement for reciprocity, whether immediate or delayed. Hence we use the term *conditional cooperation* to discriminate forms of reciprocity from unconditional mutualism (see Summary Box 1.2).

Sharing Food

The model in section 2.3 assumes that social foragers share any food they discover. This chapter converts that assumption into questions about the economics of sharing food in a stochastic environment. Food-sharing between unrelated individuals, as we define it, includes a producer's tolerance of a competitor at a *defensible* resource clump. But food-sharing often involves a more readily identifiable action by the individual that shares. Food may be delivered to another individual (e.g., J. Brown and E. Brown 1980), or the individual may actively alert other group members to the location of discovered food (e.g., Elgar 1986a; Chapman and Lefebvre 1990). We begin by describing two examples where individuals clearly act to share food.

The Mexican Jay (*Aphelocoma ultramarina*) is a plural-breeding, communal bird. Individuals ordinarily retain membership in the same demographic unit for long periods of time, and the group defends a communal territory (Brown 1987). Each breeding female lays its eggs in a separate nest. But just before fledging the young, mothers begin feeding offspring of other reproductives in the same communal group (J. Brown and E. Brown 1980). Parents at one nest are not close relatives of fledglings from nearby nests (J. Brown and E. Brown 1981), so that kinship apparently does not maintain this behavior. Furthermore, the degree of mutual parenting cannot be dismissed as a consequence of an adult's inability to recognize its own offspring (Caraco and Brown 1986). Plausibly, reproductives "choose" to share food they find with other reproductives' offspring.

Our second example concerns food-sharing mediated by food-calling (e.g., Stokes and Williams 1972; Wrangham 1977; Mangel 1990). The most extensively studied case is the "chirrup" call of the house sparrow, *Passer domesticus*, which individuals commonly emit when they find a food clump (Elgar 1986a,b).

In a series of experiments, Elgar (1986a,b) has shown that the chirrup attracts other sparrows to the location of food patches; hence the call results in shared food. Both sexes give the call, and more rapid vocalization rates are associated with faster recruitment to the resource. Food-calling has a cost; the caller most often obtains less of the clump it discovered than it would by choosing not to call and exploiting the food as a solitary. Elgar (1986b) attributes food-calling to the likely reduction in predation hazard achieved by foraging in a group, as opposed to feeding alone. Newman and Caraco (1989) model the interaction of starvation and predation and suggest that conditional cooperation may explain the economics of the behavior.

In our models, cooperative food-sharing arises not as an inevitable consequence of food discovery, but as a decision made by the forager finding the resource. In this chapter we model food-sharing and food-calling separately, but treat them as fundamentally similar. Each model asks what environmental conditions promote a conditionally cooperative basis for this sort of behavior, and what conditions render these behaviors unconditional mutualisms (Caraco and Brown 1986; Newman and Caraco 1989; Mesterton-Gibbons 1991; Mesterton-Gibbons and Dugatkin 1992). That is, we discriminate between environments where food-sharing is equivalent to cooperation in the Prisoner's Dilemma, and environments where an unconditional mutualism is sufficient for stability of food-sharing.

As in the preceding chapter, we analyze two-person games. In section 3.2 we model food-sharing of the sort reported by J. Brown and E. Brown (1980, 1981). In our model, an individual can incur a fitness penalty directly or indirectly. Failing to acquire sufficient food imposes a direct penalty. The same individual may acquire sufficient food but incur a penalty as a consequence of the other player's energetic failure. We refer to the latter as an indirect penalty, and term the probability of an indirect penalty the *communal cost*. The assumption of a communal cost might often apply to foragers that live and breed in the same group (Caraco and Brown 1986).

In section 3.3 we model food-calling. The models assume only direct penalties, and so might better apply to ephemeral foraging groups.

3.2 Food-Sharing with a Communal Cost

An individual that discovers or captures a food clump may share the resource simply by tolerating one or more foragers that it could otherwise evict. In more complex social systems, the finder may take food to a central place where other group members share the resource (Isaac 1978; Kurland and Beckerman 1985), or alloparental care by active reproductives may involve sharing food between one's own offspring and the offspring of another, unrelated group member (J. Brown 1987).

We analyze food-sharing as a game between two unrelated foragers searching for clumped food. We treat the consumers as competitively equivalent, so the game is symmetric (with uninformed play). An individual discovering food may consume the entire clump (equivalently, feed only its own family) or may share the clump equally with the other consumer (or with the other player's family). We represent the two, discrete alternatives as N (for not share) and A (for always share).

We assume the two players' interaction involves probabilistically repeated play (Axelrod and Hamilton 1981). Given this type of nonrandom interaction, we can ask when the game conforms to a Prisoner's Dilemma (PD), where each of two cooperators does better than each of two interacting defectors (e.g., Owen 1968; Axelrod and Hamilton 1981; Selten and Hammerstein 1984; Feldman and Thomas 1987). In this case cooperation, conditional upon like behavior, might be anticipated under repeated play (see Math Box 1.2). That is, Tit-for-Tat (Axelrod and Hamilton 1981; Pulliam et al. 1982; Hirshleifer and Rasmussen 1988) may be stable for probabilistically repeated play, assuming that the PD offers a sensible metaphor for the maintenance of conditional cooperation (May 1987; Mesterton-Gibbons 1991; Houston 1993; Crowley and Sargent 1996).

Energy Shortfall and Penalties

The random variable T represents the time elapsing until a designated forager has discovered and consumed a clump or the economic equivalent (two clumps shared equally with the other forager). T_i ($i = 1$ or 2) refers to one of the two arbitrarily designated players.

If $T_1 > \tau$, player 1 suffers a fitness penalty. Exceeding the critical foraging time τ may imply an energy deficit for the forager itself. Similarly, a failure to meet the metabolic requirements of dependent offspring may impair their development or render them more susceptible to disease. In some birds, offspring may begin begging loudly if insufficient food is delivered to them by a critical time, and the noise may attract a predator (e.g., Caraco and Brown 1986; J. Brown 1987). The models suit any of these possibilities.

Suppose $T_1 \leq \tau < T_2$. According to the preceding assumption, player 2 incurs a fitness penalty with certainty. Even though T_1 does not exceed the critical time, player 1 suffers a penalty with probability C; $0 \leq C \leq 1$. C represents the communal cost, the chance of an indirect penalty. For example, when begging by player 2's offspring attracts a predator (since $T_2 > \tau$), the predator might indirectly learn the location of player 1's offspring. In like manner, if player 2's offspring become infective as a consequence of inadequate nourishment, the disease might spread to player 1's offspring. When fitness penalties involve dependent offspring, the communal cost should often depend on the spatial dispersion of the respective players' offspring (Caraco and Brown 1986). If penalties govern only the foragers' individual survivorship, the communal cost may be reduced or may disappear ($C = 0$). In any case, we assume that the players cannot control the communal cost, but can respond to variation in the probability of an indirect penalty.

If both T_1 and T_2 exceed τ, both foragers incur a penalty with certainty. If neither T_1 nor T_2 exceeds τ, neither player is penalized. We assume each player should attempt to minimize its probability of a penalty, given the assumption of probabilistically repeated play.

To offer a contrast to the model in section 2.4, the food-sharing game in this section assumes that the discovery of one food clump decreases the rate at which the next clump is encountered. A model where the individual's clump-discovery rate is a constant, following Caraco and Brown (1986), is summarized in Math Box 3.1 at the end of this chapter, and Math Box 3.2 summarizes a model where the discovery of one food clump increases the rate at which the next clump is encountered.

N VERSUS *N*

When each player chooses not to share, a forager incurs a penalty with certainty if it fails to discover a clump before τ time units elapse. Even if player i locates a clump at $T_i \leq \tau$, that individual may be penalized indirectly if the other forager fails to find food in the time available for searching.

Let X ($X = 0, 1$) represent the number of clumps already discovered during a single round of play. Let $\lambda(X)$ represent an individual's probabilistic rate of locating a food clump as a function of X. We assume that food abundance is limited similarly at the start of each round of play, so that $\lambda(X)$ decreases immediately after the first clump is discovered:

$$\lambda(X) = \lambda(K - X), \tag{3.1}$$

where $K > 1$. Hence one forager's success depletes food enough to affect the other's rate of food discovery (Charnov et al. 1976; Caraco 1979; see Green 1984). Since the two N-strategists search independently and symmetrically, the first clump is encountered at total probabilistic rate $2K\lambda$.

To derive each N-strategist's penalty probability, we note that an individual (player 1) can incur a penalty through three different compound events:

1. Both foragers fail to discover food.
2. Player 1 fails to find food, but player 2 succeeds.
3. Player 1 discovers food, but incurs a penalty indirectly when player 2 fails to locate a food clump.

The first case is the simplest. The probability that both foragers fail to discover food by time τ is $e^{-2K\lambda\tau}$.

Now suppose that player 1 locates a food clump at time $T_1 < \tau$. Food abundance decreased, so that player 2's clump-discovery rate declines to $\lambda(K - 1)$. The probability that player 2 then fails to find food is $\Pr[T_1 \leq \tau < T_2]$:

$$\Pr[T_1 \leq \tau < T_2] = \int_0^\tau e^{-\lambda(K-1)(\tau-t)}\, 2K\lambda e^{-(2K\lambda t)}\, dt$$

$$\Pr[T_1 \leq \tau < T_2] = [2K/(K + 1)]\,[e^{-\lambda(K-1)\tau} - e^{-2K\lambda\tau}]. \tag{3.2}$$

Each player has probability 1/2 of encountering the first clump, given that food is discovered. If player 2 finds the first clump, player 1 may incur a direct penalty. If player 1 finds the first clump, an indirect penalty remains a possibility.

To obtain each individual's penalty probability for N versus N, we sum the probabilities of the three mutually exclusive ways to incur a penalty. The resulting probability of a penalty is

$$M_{NN} = e^{-2K\lambda\tau} + [K(1 + C)/(K + 1)][e^{-\lambda(K - 1)\tau} - e^{-2K\lambda\tau}]. \tag{3.3}$$

Clearly, M_{NN} declines with $K\lambda\tau$ and increases as C increases.

A VERSUS *A*

Since each A-strategist always shares food, $T_1 = T_2 = T_{AA}$. Both incur a penalty if $T_{AA} > \tau$, and no indirect penalty can occur. Given the individual clump-discovery rates in expression (3.1), the first clump is encountered at combined probabilistic rate $2K\lambda$, and the second (given $X = 1$) is encountered at combined rate $2\lambda(K - 1)$. Using standard methods (see Math Box 1.3), Boswell et al. (1979) show that the probability density of T_{AA} can be expressed as

$$f(T_{AA}) = 2\lambda K(K - 1)(1 - e^{-2\lambda T_{AA}})e^{-\lambda T_{AA}}. \qquad (3.4)$$

The mean and variance are

$$E[T_{AA}] = (2K - 1)/[2\lambda K(K - 1)] \qquad (3.5)$$

$$V[T_{AA}] = [K^2 + (K - 1)^2]/[4\lambda^2 K^2 (K - 1)^2]. \qquad (3.6)$$

Our assumptions concerning the searching process imply that T_{AA} is the sum of two exponential waiting times. The first has mean $(2K\lambda)^{-1}$, and the second has mean $([2\lambda(K - 1)]^{-1}$. $E[T_{AA}]$ is the sum of these two quantities. The probability that no food is found is then $e^{-2K\lambda\tau}$. Given that the first clump is found at some time $t < \tau$, the probability that the two foragers fail to discover a second clump in the time available is

$$\int_0^{\tau} e^{-2\lambda(K-1)(\tau-t)} 2K\lambda e^{-2K\lambda t} dt.$$

Summing the result of this integral with the probability that no food is found gives the penalty probability for each of two players that always share food:

$$M_{AA} = \Pr[T_{AA} > \tau] = e^{-2K\lambda\tau} + Ke^{-2(K-1)\lambda\tau}(1 - e^{-2\lambda\tau}). \qquad (3.7)$$

When $K = 2$, the penalty probability is the same as in our model with an increasing rate of clump discovery; see Math Box 3.2.

N VERSUS *A*

When one player chooses not to share and the other always shares, the former has a competitive advantage. If the *N*-strategist first locates a clump, it consumes the resource and stops foraging. In this case the process develops as in the *N* versus *N* interaction. If the *A*-strategist locates the first clump, the food is shared and both players continue searching as in the *A* versus *A* interaction. If the *A*-strategist first discovers food, we assume that the second clump—if found—is always shared, since the *N*-strategist will necessarily have met the time constraint.

The random durations of search time are $T(N, A)$ for the *N*-strategist and $T(A, N)$ for the *A*-strategist. If the *N*-strategist locates the first clump, $T(N, A) < T(A, N)$, and the search times are equal if the *A*-strategist finds the first clump. So, this interaction requires $T(N, A) \leq T(A, N)$; not-sharing may gain an immediate advantage since it cannot have the greater search time.

From the preceding we see that with probability 1/2, the *A*-strategist finds the first clump and the foraging process then develops like an *A* versus *A* interaction, which allows only direct penalties. The contribution of this possibility to each player's unconditional probability of a penalty is

$$[e^{-2K\lambda\tau} + Ke^{-2\lambda(K-1)\tau}(1 - e^{-2\lambda\tau})]/2. \qquad (3.8)$$

The last expression is simply $M_{AA}/2$, since each player must have the same penalty probability if the A-strategist locates the first clump.

With probability 1/2, the N-strategist finds the first clump (if a clump is discovered) and gains an advantage over its opponent. The N-strategist is penalized directly if no food is found and only indirectly if a single clump is found. Under these assumptions, the contribution of this possibility to the N-strategist's unconditional penalty probability is

$$(e^{-2K\lambda\tau}/2) + [KC/(K + 1)][e^{-\lambda(K-1)\tau} - e^{-2K\lambda\tau}]. \tag{3.9}$$

The corresponding contribution to the A-strategist's unconditional penalty probability is obtained simply by letting $C = 1$ in expression (3.9); the A-strategist is penalized directly if either 0 or 1 clump is discovered. The unconditional penalty probabilities for N versus A are as follows. For the N-strategist

$$M_{NA} = e^{-2K\lambda\tau} + \frac{K}{2}[e^{-2(K-1)\lambda\tau} - e^{-2K\lambda\tau}] + \frac{KC}{K+1}[e^{-(K-1)\lambda\tau} - e^{-2K\lambda\tau}]. \tag{3.10}$$

The A-strategist's unconditional probability of a penalty is

$$M_{AN} = e^{-2K\lambda\tau} + \frac{K}{2}[e^{-2(K-1)\lambda\tau} - e^{-2K\lambda\tau}] + \frac{K}{K+1}[e^{-(K-1)\lambda\tau} - e^{-2K\lambda\tau}]. \tag{3.11}$$

We now have the elements of the penalty matrix M and so can analyze the symmetric food-sharing game with decreasing clump-discovery rates.

A Penalty Matrix

Table 3.1 shows the penalty matrix M for the food-sharing game. Table 3.2 shows M evaluated at $K = 2$, where the penalty probability for mutual food-

Table 3.1

Penalty matrix M when the rate of encountering food clumps decreases after an encounter. M_{rc} is the probability of a penalty when the focal player selects the behavior associated with row r, and its opponent chooses the behavior for column c. N is not-sharing; A is always share. For convenience, let $\alpha = e^{-2\lambda\tau}$, and let $\beta = e^{-(K-1)\lambda\tau}$.

	N	A
N	$\alpha^K + [K(1 + C)/(K + 1)][\beta - \alpha^K]$	$\alpha^K + (K/2)[\beta^2 - \alpha^K] + [KC/(K + 1)][\beta - \alpha^K]$
A	$\alpha^K + (K/2)[\beta^2 - \alpha^K] + [K/(K + 1)][\beta - \alpha^K]$	$\alpha^K + K\beta^2[1 - \alpha]$

Table 3.2

Penalty matrix M for the decreasing encounter rate model, with K set equal to 2. Entries obtained by evaluating penalty probabilities in table 3.1 at selected value of K. $\alpha = e^{-2\lambda\tau}$.

	N	A
N	$\alpha^2 + [2(1 + C)/3][\alpha^{1/2} - \alpha^2]$	$\alpha + (2C/3)[\alpha^{1/2} - \alpha^2]$
A	$\alpha + (2/3)[\alpha^{1/2} - \alpha^2]$	$2\alpha - \alpha^2$

sharing, $\Pr[T_{AA} > \tau]$, is the same as in the increasing encounter-rate model (see Math Box 3.2). M has elements M_{rc} ($r,c = N, A$). Since M is a 2×2 matrix, it must have at least one Nash solution qualifying as an ESS (see Math Box 1.1).

M is a decomposable matrix in the sense of Mesterton-Gibbons (1992). That is, the trace of the matrix, $(M_{NN} + M_{AA})$, equals the sum of the off-diagonal elements, $(M_{NA} + M_{AN})$. Math Box 3.3 reviews this property and its application to the Prisoner's Dilemma. Since M is decomposable, the condition for a mixed ESS,

$$M_{AN} < M_{NN} \textit{ and } M_{NA} < M_{AA},$$

cannot hold. Hence diagonal dominance (Math Box 1.1) identifies a stable Nash equilibrium and a pure ESS when

$$M_{NN} < M_{AN} \textit{ or } M_{AA} < M_{NA}. \tag{3.12}$$

The two inequalities in expression (3.12) cannot hold simultaneously since, again, M is decomposable. But one or the other must hold, depending on parameter values. Hence, for a single play of the food-sharing game, we have two possible solutions.

Not-sharing will be the pure ESS if $M_{NN} < M_{AN}$. Using the penalty matrix in table 3.2 (i.e., letting $K = 2$), this condition reduces to

$$C < 3(e^{2\lambda\tau} - 1)/2(e^{3\lambda\tau} - 1). \tag{3.13}$$

Always sharing will be a Nash solution and a pure ESS if $M_{AA} < M_{NA}$. From table 3.2 this condition becomes

$$C > 3(e^{2\lambda\tau} - 1)/2(e^{3\lambda\tau} - 1), \tag{3.14}$$

which simply reverses inequality (3.13). Hence the likelihood that food-sharing is an ESS for a game played once increases as the communal cost increases. That is, a greater chance of an indirect failure when the opponent fails to find enough food increases the likelihood that foragers will share. The right-hand side of inequality (3.14) decreases as $\lambda\tau$ increases, so that the likelihood that food-sharing is an ESS also increases with food-clump density and available foraging time (Caraco and Brown 1986).

Suppose inequality (3.14) holds, so that sharing food is the better response to either alternative. Food-sharing is an ESS by diagonal dominance, but does it constitute cooperation? Constraints on the penalty probabilities, given expression (3.14), imply that $M_{AA} < M_{NN}$ (reversing this inequality requires $2\lambda\tau > \ln[(1 + 2e^{3\lambda\tau})/3]$, which is impossible). Hence food-sharing has a mutualistic property. Food-sharing in this case is stable because neither individual is tempted to change behavior; failing to share increases the chance of a fitness penalty. Then food-sharing, when it is an ESS for a single play of the game, is merely an unconditional mutualism. However, we might speculate that food-sharing could arise as an unconditional mutualism and later be maintained by reciprocity (Mesterton-Gibbons and Dugatkin 1992). Therefore, we examine conditions where food-sharing is a conditional mutualism for repeated play.

Next we use the Prisoner's Dilemma and find a Pareto solution to evaluate conditions that might promote mutually cooperative food-sharing when the same individuals engage in probabilistically repeated play (Axelrod and Hamilton 1981). Above we noted that M_{NN} always exceeds M_{AA}. Therefore, not-sharing can never qualify as the cooperative behavior in a PD. So, assume that not-sharing is the pure ESS for a single round of play. Then, $M_{NN} < M_{AN}$, by the applicable ESS condition. Furthermore, $M_{NA} < M_{AA}$, since M is decomposable. However, M can still take the form of a PD where food-sharing equates with "cooperate" and not-sharing is equivalent to "defect." If M qualifies as a PD, then (see Math Box 1.2)

$$M_{AA} < M_{NN} \textit{ and } M_{AA} < (M_{NA} + M_{AN})/2.$$

The first condition gives an advantage to cooperative food-sharing. The second condition eliminates any advantage to the players alternating behaviors out of phase. Substituting the elements of M, each of these conditions reduces to

$$\frac{3e^{2\lambda\tau} - e^{3\lambda\tau} - 2}{e^{3\lambda\tau} - 1} < C < \frac{3(e^{2\lambda\tau} - 1)}{2(e^{3\lambda\tau} - 1)}. \tag{3.15}$$

When expression (3.15) holds, mutual sharing implies a lower penalty probability than each player incurs at the ESS for a single play of the game. Given (3.15), food-sharing qualifies as conditional cooperation when play is iterated probabilistically.

We can summarize this game by partitioning the $(\lambda\tau, C)$ space into three regions; see fig. 3.1. Not-sharing is the only ESS, and sharing must be disadvantageous when

$$C < (3e^{2\lambda\tau} - e^{3\lambda\tau} - 2)/(e^{3\lambda\tau} - 1). \tag{3.16}$$

With no communal cost ($C = 0$), not-sharing is the only possible ESS; food-sharing cannot be favored.

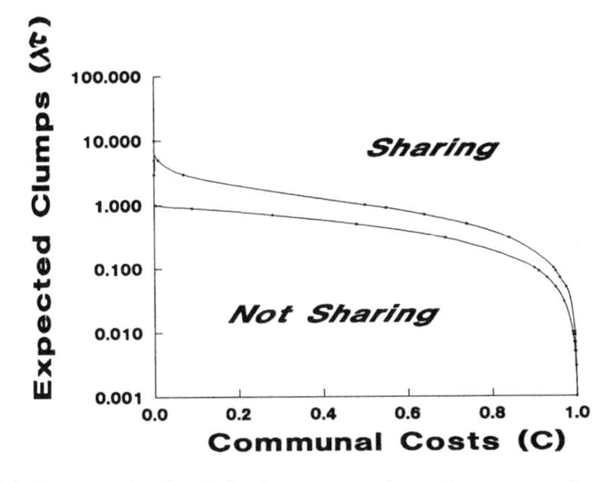

Figure 3.1 Solutions to the food-sharing game when clumps are discovered at a decreasing rate. The model's parameter K equals 2. For the $(\lambda\tau, C)$ combinations in the region nearest the origin, pure N (not sharing) is the Nash solution and ESS; sharing food cannot be favored in this region. In the middle region, pure N remains a Nash solution and an ESS, but the penalty matrix now qualifies as a Prisoner's Dilemma; food-sharing is the cooperative behavior. In the region farthest from the origin, pure A is the Nash solution and ESS; not-sharing is always disadvantageous in this region.

In the second region, expression (3.15) holds. Not-sharing remains an ESS, but mutual food-sharing is advantageous. Conditional mutualism becomes a second ESS for probabilistically repeated play. That is, a Tit-for-Tat strategy (Axelrod and Hamilton 1981), with food-sharing the cooperative behavior, can qualify as an ESS when the chance of continued interaction between the same two individuals is sufficiently large. Related strategies that reward cooperation and retaliate against lack of cooperation (e.g., Pulliam et al. 1982; Mesterton-Gibbons 1992; Nowak and Sigmund 1993) offer plausible bases for maintaining conditional cooperation.

Finally, when inequality (3.16) is reversed, pure sharing is an unconditional mutualism and the only ESS. That is, failing to share is always disadvantageous under this condition. We obtain quantitatively similar results when we assume constant or increasing clump-discovery rates (Math Boxes 3.1 and 3.2); see figures 3.2 and 3.3. For the three models, food-sharing can never be beneficial near the origin of the $(\lambda\tau, C)$ space. In the middle region, food-sharing can be advantageous, but requires conditional cooperation over repeated play. In the third region, food-sharing is the sole ESS.

The major assumptions and results of the food-sharing game with direct and indirect penalties are listed in Summary Box 3.1. Increased density or apparency of food clumps (λ) promotes the likelihood of food-sharing, as does an increase in the time available to fulfill foraging requirements (τ). More interestingly, an increased communal cost, the probability that a suc-

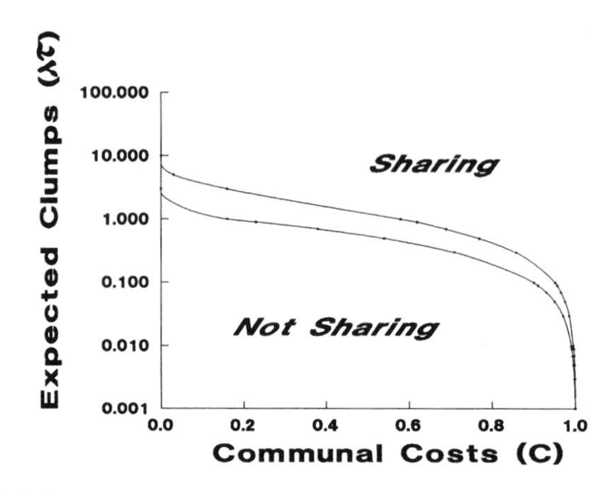

Figure 3.2 Solutions to the food-sharing game when clumps are discovered at a constant rate. The solutions associated with the three regions are the same as in figure 3.1; see Math Box 3.1. The results closely match those for the decreasing-rate model.

cessful forager incurs a penalty as an indirect consequence of a neighbor's failed foraging effort, promotes food-sharing. Different individuals or their offspring may incur fitness costs, particularly predation or parasitism, as a consequence of only one forager's failure. When this occurs, the common ecological challenge may increase the chance of mutualism, conditional or unconditional, among foragers.

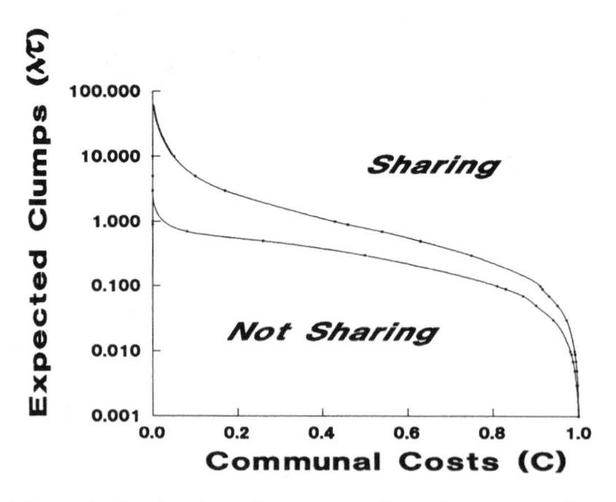

Figure 3.3 Solutions to the food-sharing game when clumps are discovered at an increasing rate; see Math Box 3.2. Solutions associated with the three regions are the same as in figures 3.1 and 3.2. For a given $\lambda\tau$, food-sharing becomes economically feasible at a lower communal cost (C) with an increasing encounter rate than for the constant-rate case.

SUMMARY BOX 3.1 FOOD-SHARING WITH A COMMUNAL COST

ASSUMPTIONS

Currency: Probability of paying a fitness cost

Decision: Allow other forager to share food or not

Constraints:

1. Only two players interact.
2. Handling time is negligible.
3. Consumption of one food clump can affect rate at which next clump is discovered.
4. Penalty incurred directly if individual fails to consume food clump (or two halves) by time τ.
5. Communal cost is chance player i incurs a penalty when only player j fails to meet time constraint.

PREDICTIONS

1. With no communal cost, not-sharing is always the only ESS; communal cost is necessary for cooperative (conditional or unconditional) food sharing.
2. Food sharing can arise as a conditional mutualism, given repeated play, while not-sharing remains an ESS.
3. If communal cost is sufficiently large, food sharing, as an unconditional mutualism, becomes the only ESS.
4. The likelihood that food sharing will be stable increases with food clump density, time available to search for food, and the communal cost.
5. Whether foragers encounter food clumps at decreasing, constant, or increasing rates has little effect on the particular games analyzed.

SOME GENERALIZATIONS

We may suppose that spatial proximity can promote food-sharing when the communal cost can make choice of not-sharing disadvantageous. Our models assume all individuals play a game in which the penalty probabilities depend on the spatial dispersion of the players (or their offspring) through the communal cost. Nowak and May (1992) explore how cooperation may invade and/or remain extant in a spatially structured population. Pure cooperators (rather than conditional cooperators) and pure defectors are the only types. Individuals are arrayed at the nodes of a two-dimensional lattice. Each individual interacts only with its nearest neighbors (see Caraco et al. 1997). In models without such spatially structured interactions, pure cooperation is ordinarily eliminated by defectors (see Houston 1993). But spatial structuring allows pure cooperation to persist with pure defection in an ever-changing spatial mosaic (Nowak and May 1992). Hence food-sharing among nonrelatives might originate because (1) spatial proximity imposes communal costs, or (2) spatial structure of two-player interactions can protect cooperation from elimination by noncooperative behavior.

Many instances of food-sharing can be explained as kin-directed altruism. But food-sharing between nonrelatives occurs with appreciable frequency, at least in plural-breeding communal birds and mammals (Caraco and Brown 1986).

Wilkinson (1984) discusses an interesting example of food-sharing among adult female vampire bats (*Desmodus rotundus*). The interaction suggests reciprocal altruism, since individuals alternate roles as donor and recipient of regurgitated blood (rather than benefiting simultaneously as in the PD; see Nowak and Sigmund 1994). But the phenomenon is important in that it suggests that food-sharing may be the basis of the particular species' social organization (Wilkinson 1985). Indeed, food-sharing among nonrelatives may have been an important step in the evolution of hominid social structure (Wolpoff 1980; Campeiro 1989).

3.3 Food-Calling and Cooperation

We again consider discrete games between nonrelatives. As in the preceding section, we identify conditions that might favor food-calling as an unconditional, and then as a conditional, mutualism. We restrict the analyses to direct penalties only.

We assume that two identical players search independently for food clumps. They discover clumps at combined probabilistic rate 2λ while both search. Total time available for searching during a single round of play is τ time units. No more than one clump may be discovered per play of the game. The clump remains undiscovered with probability $e^{-2\lambda\tau}$, and each player discovers the clump with probability $(1 - e^{-2\lambda\tau})/2$.

Clump size Y varies randomly; $E[Y]$ is the expected size and $V[Y]$ is the variance. Both social foragers know the probability distribution of clump size, $\Pr[Y \leq y]$. Each player has two possible pure strategies. A represents "always share" and N represents "not-share." An A-strategist that discovers food calls the other forager to join it at the clump. Each forager then consumes $Y/2$ energy units, unless a predator should attack. An A-strategist shares without knowing the realized clump size (y), an assumption drawn from observations of house sparrow behavior (see Elgar 1986a,b).

An N-strategist that locates food does not allow the other forager to join (fails to call or defends it). In the absence of predation, the N-strategist consumes the entire clump. We assume that the choice of not calling reveals that player's strategy to its opponent; this assumption permits application of the Tit-for-Tat notion to probabilistically repeated play. Food-calling, hence allowing another forager to share the food, may increase a finder's probability of an energetic shortfall. But it may also reduce the caller's chance of being taken by its own predator (Elgar and Catterall 1981; Mangel 1990). We assume that any fitness penalty due to predation (mortality, injury, or time lost from fitness-enhancing activities) occurs as a consequence of exposure while exploiting the clump and is not a direct result of allowing others to join per se.

Energy Shortfall and Predation

A forager suffers physiological impairment, hence a fitness penalty, during any round of play where it fails to consume more than R units of energy. We do not necessarily equate an energetic shortfall with starvation (see Newman and Caraco 1989). If neither player discovers a clump by time τ, each incurs an energetic failure. If an A-strategist finds food, each player's probability of failure is $\Pr[Y \leq 2R]$, since both exploit the clump. If an N-strategist finds food, its probability of a shortfall is $\Pr[Y \leq R]$, and the other player incurs an energetic failure with certainty.

We equate the penalty due to predation with mortality. Let μ_1 ($0 \leq \mu_1 < 1$) represent the probability that a solitary (hence an N-strategist) is preyed upon while exploiting a food clump. Let μ_2 ($0 \leq \mu_2 < 1$) represent the predation probability for each forager when the two exploit the same clump. When feeding together, each individual is killed or survives predation independently. We also assume that neither μ_1 nor μ_2 depends on clump size. If the foragers have no predators, $\mu_1 = \mu_2 = 0$. In general, individuals foraging together should be at least as safe as a solitary forager, especially for organisms that allow others to join (Elgar and Catterall 1981; Lendrem 1986). Therefore, we assume $0 \leq \mu_2 \leq \mu_1 < 1$.

On any given play of the game, an individual may incur a penalty in four mutually exclusive ways. An energetic failure occurs if no food is discovered, if an opponent N-strategist discovers the food clump, or if the individual's energy consumption within the clump does not meet the physiological requirement. Finally, the forager may be preyed upon while exploiting the clump. We assume the same two players continue to interact as long as both survive, and that each forager should try to minimize its penalty probability.

A versus *A*

When both individuals call to share food, each independently experiences a predation probability of μ_2 if a clump is located. Each individual also has the same probability of an energetic failure, which depends on the chance of finding a clump and clump size (given clump discovery). When both individuals are prepared to share, each has penalty probability M_{AA}:

$$M_{AA} = e^{-2\lambda\tau} + (1 - e^{-2\lambda\tau})(\mu_2 + [1 - \mu_2] \Pr[Y \leq 2R]). \quad (3.17)$$

N VERSUS *A*

An *N*-strategist discovers the clump and feeds alone with probability $(1 - e^{-2\lambda\tau})/2$. Given this event, the *N*-strategist's predation probability is μ_1. The opposing *A*-strategist is just as likely to find food but will allow joining. The penalty probability for choosing *N* against an individual choosing *A* is M_{NA}:

$$M_{NA} = e^{-2\lambda\tau} + [(1 - e^{-2\lambda\tau})/2] \bullet$$
$$(\mu_1 + [1 - \mu_1] \Pr[Y \leq R] + \mu_2 + [1 - \mu_2] \Pr[Y \leq 2R]). \quad (3.18)$$

A VERSUS *N*

An *A*-strategist must discover the clump to obtain food when its opponent is an *N*-strategist. If the *N*-strategist locates food, the *A*-strategist must incur an energetic shortfall. Food-calling implies that the only associated predation probability is μ_2. Therefore, the penalty probability for choosing to call and share against the opposite strategy is

$$M_{AN} = e^{-2\lambda\tau} + [(1 - e^{-2\lambda\tau})/2](1 + \mu_2 + (1 - \mu_2) \Pr[Y \leq 2R]). \quad (3.19)$$

N VERSUS *N*

In this interaction a player either locates and attempts to consume the entire clump, or else obtains no food; the only predation probability associated is μ_1. The penalty probability is M_{NN}:

$$M_{NN} = e^{-2\lambda\tau} + [(1 - e^{-2\lambda\tau})/2][1 + \mu_1 + (1 - \mu_1) \Pr[Y \leq R]). \quad (3.20)$$

Penalty Matrix

The four M_{rc} $(r,c = A,N)$ are the elements of the penalty matrix M for the food-calling game. Note that $(M_{AA} + M_{NN}) = (M_{NA} + M_{AN})$. Then M is decomposable, and the ESS analysis is simple (see Math Box 3.3).

Not-calling will be the pure ESS when $M_{NN} < M_{AN}$, which implies

$$\mu_1 + (1 - \mu_1) \Pr[Y \leq R] < \mu_2 + (1 - \mu_2) \Pr[Y \leq 2R]. \quad (3.21)$$

Suppose that the chance of falling to a predator is the same whether the individual feeds alone or with its opponent. Then $\mu_1 = \mu_2$, and expression (3.21) reduces to $\Pr[Y \leq R] < \Pr[Y \leq 2R]$, which must be true for reasonable clump-size distributions. Therefore, not-calling will be a pure ESS unless a solitary suffers a sufficiently greater predation hazard than a group member.

Food-calling and sharing will be the pure ESS when $M_{AA} < M_{NA}$, which becomes

$$\mu_2 + (1 - \mu_2) \Pr[Y \leq 2R] < \mu_1 + (1 - \mu_1) \Pr[Y \leq R]. \quad (3.22)$$

That is, reversing the previous inequality makes A the Nash solution and ESS. In this case, sharing is stable because the finder's reduced probability of predation overcomes the greater chance of an energetic failure (Newman and Caraco 1989; Houston et al. 1993). Food-calling and sharing is an ESS, without the complications of conditional cooperation or repeated play, when μ_1 exceeds μ_2 sufficiently. Since the individual attracted to the food otherwise incurs a penalty with certainty, food-calling qualifies, in this case, as an unconditional mutualism.

Depending on whether expression (3.21) or (3.22) holds, the associated solution will be an ESS for single and repeated play. To ask if conditional cooperation can promote sharing via food-calling, suppose that not-calling is the ESS (so that $M_{NN} < M_{AN}$) and let the matrix M conform to a PD. That is, we assume

$$M_{AA} < M_{NN} \text{ and } M_{AA} < (M_{AN} + M_{NA})/2.$$

Neither of these conditions need violate expression (3.21), and each of the PD-conditions reduces to

$$(1 - \mu_2) \Pr[Y > 2R] > (1 - \mu_1) \Pr[Y > R]/2. \quad (3.23)$$

M has the form of a Prisoner's Dilemma when a group forager's conditional probability of averting a penalty, given access to the clump, exceeds one-half a solitary forager's conditional probability of averting a penalty, given access to the food clump. The conditional probability of not being penalized is the chance of both surviving predation and meeting the physiologically required level of energy intake, given that the individual enters a food clump. Interestingly, the PD-condition does not depend on the probability that the foragers locate a clump, but does depend on predation and the distribution of clump size, once food is discovered (Newman and Caraco 1989).

If $\mu_1 = \mu_2 \geq 0$, the inequality defining the PD becomes

$$\Pr[Y > 2R] > \Pr[Y > R]/2. \quad (3.24)$$

Comparing conditions for the Prisoner's Dilemma with and without predation indicates that the likelihood of food-sharing as a conditional mutualism over repeated play increases as the relative danger of predation faced by a solitary increases. That is, as μ_1 increases relative to μ_2, the condition for the PD is more easily met for any given $\Pr[Y \leq y]$.

Summarizing briefly, not-calling and hence not sharing, is an ESS when

$$\mu_1 - \mu_2 < (1 - \mu_2) \Pr[Y \leq 2R] - (1 - \mu_1) \Pr[Y \leq R]. \quad (3.25)$$

As μ_1 increases with the other quantities kept constant, not-calling remains an ESS, but sharing by food-calling becomes the cooperative behavior in a Prisoner's Dilemma. Food-sharing might then be maintained as a conditional mutualism through TFT, or something similar, when

$$(1 - \mu_2)/(1 - \mu_1) > \Pr[Y > R]/(2 \Pr[Y > 2R]). \quad (3.26)$$

With a further increase in μ_1, inequality (3.25) is reversed. At this point, solitary foraging becomes sufficiently hazardous with respect to predation so that calling and sharing is an unconditional mutualism and the only ESS. It seems reasonable that food-sharing would be promoted by a solitary forager's relatively greater hazard of predation, compared to foraging with conspecifics (Elgar 1986b; Newman and Caraco 1989; Mangel 1990). The effects of the distribution of clump sizes, $\Pr[Y \leq y]$, are not so obvious, and we consider them next.

Clumps of Discrete Items

To develop a simple but instructive example, we assume that each clump contains one or more discrete food items. Specifically, let the number of items per clump (Y) have a geometric probability function

$$\Pr[Y = y] = p\,(1 - p)^{y-1} \quad 0 < p < 1;\ y = 1, 2, \ldots .$$

The mean and variance are, respectively,

$$E[Y] = 1/p;\ V[Y] = (1 - p)/p^2. \quad (3.27)$$

Both the expectation and variance decline as the parameter p increases.

Given that an N-strategist finds food, its probability of an energetic failure is

$$\Pr[Y \leq R] = \sum_{y=0}^{R} p\,(1 - p)^{y-1} = 1 - (1 - p)^R. \quad (3.28)$$

Given that an A-strategist finds food, replacing R by $2R$ in equation (3.28) gives the corresponding probability of an energetic failure for each of two group foragers.

Not-calling is an ESS when inequality (3.21) holds. If clump size has a geometric distribution, (3.21) reduces to

$$\frac{1 - \mu_1}{1 - \mu_2} > (1 - p)^R. \quad (3.29)$$

Stability of not-calling can be promoted by decreasing μ_1, increasing μ_2, increasing the physiological requirement R, and by decreasing the mean and variance of the number of items per clump (i.e., by increasing p). Essentially, not-calling is more likely advantageous when solitary foraging is not much more hazardous than group foraging, and when sharing the clump's food is likely to induce an energetic shortfall. Note that if $\mu_1 = \mu_2$, not-calling is an ESS if $1 > (1 - p)^R$, which is always true.

Now suppose that not-calling remains an ESS, but the penalty matrix conforms to a Prisoner's Dilemma with calling as the cooperative behavior. We apply the geometric distribution to condition for a PD, expression (3.26), and combine the result with the ESS condition for N, expression (3.29). The result is the condition where food-calling qualifies as a conditional mutualism:

$$(1 - p)^R < \frac{1 - \mu_1}{1 - \mu_2} < 2(1 - p)^R. \tag{3.30}$$

When expression (3.30) holds, two players that call and share have a lower penalty probability than foragers that do not call, but each of the former is tempted to defect selfishly and not call when it discovers food. Since we assume repeated interaction between the same individuals, given that both survive, conditional mutualism might increase the frequency of food-sharing under conditions where expression (3.30) is likely to hold. For example, a decrease in the physiological requirement R increases the range of parameter combinations where cooperative sharing could be advantageous.

Finally, suppose that food-calling and sharing is the Nash solution and an ESS. Inequality (3.29) is reversed, so that

$$\frac{1 - \mu_1}{1 - \mu_2} < (1 - p)^R, \tag{3.31}$$

and food-calling is a simple, unconditional mutualism. Note that if $\mu_1 = \mu_2$, expression (3.31) cannot hold, and food-calling cannot be an ESS. In general, group foraging's greater safety from predation, large food clumps, and a relatively small physiological energy requirement all can increase the likelihood that food-calling and sharing are advantageous.

Newman and Caraco (1989) extend the analysis by varying mean clump size and clump-size variance independently. They assume Y has a normal probability density and evaluate penalty probabilities numerically. They identify a region of mean-variance combinations where the game conforms to a PD with calling and sharing as the cooperative behavior. For a given physiological requirement, increasing either the mean or the variance of clump size increased the likelihood of food-calling as a conditional mutualism. However, increasing the requirement while holding the clump-size distribution constant decreased the extent of the region where the penalty matrix qualified as a Prisoner's Dilemma.

The assumptions and results of the food-calling game are listed in Summary Box 3.2. Further discussion of the topic can be found in Newman and Caraco (1989) or Mangel (1990).

SUMMARY BOX 3.2 A FOOD-CALLING GAME

ASSUMPTIONS

Currency: Probability of paying a fitness cost

Decision: Calling others to location of food or not

Constraints:

1. Only two players interact.
2. Players search independently.
3. Handling time is negligible.
4. Only one food clump available per round of play.
5. Clump size varies randomly; players know probability distribution of clump size.
6. Penalty related to starvation incurred if food intake fails to meet physiological requirement.
7. Food-clump size does not affect predation probabilities.
8. Not-calling after finding clump induces greater predation hazard but lower chance of energetic failure.

PREDICTIONS

1. Food-calling and sharing can become an ESS only when a solitary's predation hazard sufficiently exceeds predation hazard as a group member.
2. When food-calling is an ESS, it is a mutualism. If not-calling remains an ESS, calling may be favored as a conditional mutualism over repeated play. If a solitary's predation hazard is great enough, food-calling becomes an unconditional mutualism and the sole ESS.
3. Qualitative conclusions remain similar for geometric distribution of discrete items per clump and normally distributed, continuously divisible clump size.
4. Large mean clump size, high clump-size variance, and reduced physiological requirement tend to favor food-calling and sharing.

DURATION OF REPEATED PLAY

In the Prisoner's Dilemma, the chance of repeated play must be sufficiently large for TFT, or some strategy of cooperation conditional upon like behavior, to qualify as stable against pure defect (pure N in this chapter). That is, the expected duration of play must be large enough that TFT becomes a better response to itself than noncooperation, in a game with two pure strategies.

Axelrod and Hamilton (1981) take the probability of repeated interaction between the same two players as a constant, independent of the players' actions. They then add payoffs over the (geometric) distribution of the duration of play. Feldman and Thomas (1987) retain the assumption that payoffs accrue additively over iterations of the PD. But they let the probability of terminating play depend on the most recent actions of the players. The chance that an individual chooses to continue play depends on whether its opponent cooperated or defected during the previous round; the probability that play continues is the product of each player's probability of maintaining the interaction. The advantage of mutual cooperation will be increased if it not only implies a greater payoff (lower penalty) but also increases the expected duration of play (Feldman and Thomas 1987). When players attempt to minimize the probability of a penalty, the economics of repeated play might be characterized by a product (Newman and Caraco 1989). Penalties may be true survival probabilities, so the expected duration of play itself (a product of probabilities) may reasonably serve as a currency of fitness.

3.4 Concluding Remarks

This chapter examines two food-sharing games where concepts concerning cooperation between nonrelatives offer a basis for understanding an interesting set of foraging interactions. Given sufficient proximity of social foragers, sharing of defensible food resources, or sharing information on the availability of food, should be more likely as food abundance or communal costs increase, and as time constraints on searching for food relax. In some cases, food-sharing requires only an unconditional mutualism; in others conditional cooperation is required.

Implications for Future Work

An extensive series of theoretical studies addresses the iterated Prisoner's Dilemma and closely related problems. The literature on conditional cooperation will no doubt continue to grow, especially as computational tools for evaluating population-level consequences of spatially localized behavioral interactions become readily available (e.g., Maniatty et al. 1998). Applications of the IPD, such as the models in this chapter, are less common but will likely increase as well.

Identifying conditional cooperation and discriminating between conditional mutualism (immediate reciprocity) and reciprocal altruism (delayed reciprocity) present significant empirical challenges; see discussion in Noë (1990) or Mesterton-Gibbons and Dugatkin (1997). In this chapter's food-sharing games, we assume each player chooses its behavior before a round of play, based on a matrix of probabilities for the various outcomes. We treated the game as a PD since penalty probabilities could be reduced cooperatively before foraging began. Hence we equated conditional cooperation with conditional mutualism; the players' penalty probabilities were reduced simultaneously. An observer viewing foragers following the game's definition of conditional mutualism would see the realization of the random process and might interpret the behavior as reciprocal altruism. One forager locates a clump and calls a second individual to feed; later the second forager discovers food and calls the first. Many social foraging processes, perhaps to a greater degree than most social interactions, have a probabilistic nature. This uncertainty may influence the chance that conditional cooperation can evolve and make it difficult to recognize where it occurs.

Implications for Scaling Up

Another set of significant challenges requires predicting population-level patterns in terms of local behavioral processes (e.g., Sutherland 1996; Levin et al. 1997; Caraco et al. 1998). Therefore, we present a simple model where sharing food or information can influence population dynamics. We invite more complex models linking selfish versus cooperative behavior and population growth.

Let V_t represent the size of a consumer population at the beginning of year t. The population first enters a period of foraging during which starvation may occur. During the foraging period, individuals engage in r rounds of play of a food-sharing game; to survive the entire period, the individual must survive each play of the game. Thereafter, surviving individuals reproduce; the survivors and their offspring constitute the population at the beginning of year $(t + 1)$. We assume that survival can depend on whether or not foragers share food, and that individual reproduction is density independent. The discrete-time growth equation is

$$V_{t+1} = (1 + b)\, s_t V_t. \tag{3.32}$$

s_t $(0 < s_t < 1)$ is the proportion of the V_t consumers surviving the foraging period; we shall equate s_t with the product of survival probabilities derived for a single play of a food-sharing game. Each survivor has, on average, b offspring; reproduction does not depend on food-sharing.

If s_t were constant, the population would decline or grow geometrically depending on whether the proportion of foragers surviving was smaller or larger than $(1 + b)^{-1}$. However, we assume that food availability varies randomly among years, so that food-clump density is constant within years, but s_t exhibits random variation among years.

We assume the "constant encounter rate" version of the food-sharing game (see Math Box 3.1; Caraco and Brown 1986). For simplicity, assume no communal cost for failing to share. A population is composed entirely of individuals that forage as pairs and share each food clump discovered, or it is composed entirely of individuals that never share food. We ask which population has the greater expected long-term growth rate when food availability varies temporally. We ignore phenotypic plasticity; each individual takes its parents' phenotype. For brevity and simplicity, the model compares a cooperative population with a selfish population, without letting the latter exploit the former.

We let λ_t represent food availability during year t. λ_t varies randomly and independently between years. From Math Box 3.1, λ_t is the stochastic rate at which each forager discovers food. In a population where all food clumps are shared between paired foragers, we obtain the probability of surviving a single play of the game from equation (3.1.4) in Math Box 3.1, with τ (foraging time) equal to 1. Surviving the annual foraging period requires surviving r independent rounds of play. This yields the survival proportion for year t when food is shared between foragers:

$$s_t = \left[1 - e^{-2\lambda_t}(1 + 2\lambda_t)\right]^r, \qquad (3.33)$$

so that s_t is a function of λ_t.

In a population where no food is shared, equation (3.1.1) in Math Box 3.1 with $C = 0$ gives the probability of surviving a single play of the game. Then the survival proportion for year t yields

$$s_t = \left[1 - e^{-\lambda_t}\right]^r. \qquad (3.34)$$

The two populations have the same survival per round of play, and hence the same annual survival proportion, if $\lambda_t = 1.256$ (see Caraco and Brown 1986). Food-sharing induces greater survival within a year if $\lambda_t > 1.256$, and lower survival within a year if $\lambda_t < 1.256$.

The growth equation for the food-sharing population becomes

$$V_{t+1} = (1 + b)\, V_t \left[1 - e^{-2\lambda_t}(1 + 2\lambda_t)\right]^r. \qquad (3.35)$$

The growth equation for the population without food-sharing is

$$V_{t+1} = (1 + b)\, V_t \left[1 - e^{-\lambda_t}\right]^r. \qquad (3.36)$$

Since λ_t varies randomly, the population with the greater geometric mean survival will have the greater expected long-term growth rate and the ecological advantage. Assume λ_t has a two-point distribution with expected value 1.256, where sharing and not-sharing have the same survival. That is, assume

$$\Pr[\lambda_t = 1.256 + \delta] = \Pr[\lambda_t = 1.256 - \delta] = 1/2 \qquad (3.37)$$

for $\delta \in (0, 1.256)$, so that among-year variance in food density increases with δ. The geometric mean growth rate is

$$(1 + b)[s(\lambda + \delta)]^{1/2}\, [s(\lambda - \delta)]^{1/2}, \qquad (3.38)$$

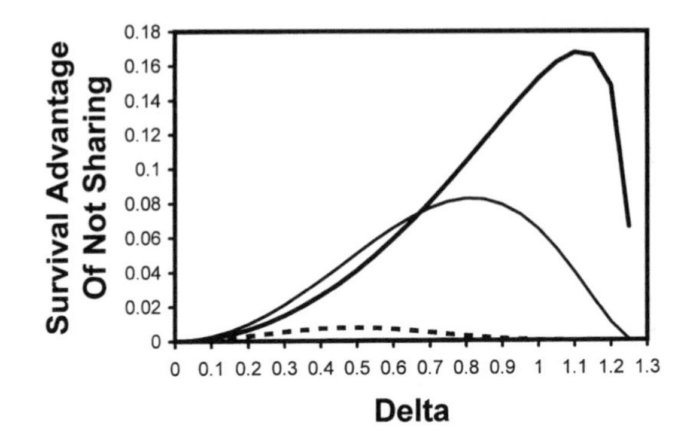

Figure 3.4 Geometric mean survival advantage for not sharing food. Ordinate is geometric mean survival for not-sharing minus geometric mean survival for sharing food; the difference is always nonnegative. Abscissa is δ; among-year variance in λ increases as δ increases. Advantage of not-sharing depends on rounds of play per year; $r = 1$ in top plot, $r = 3$ in middle plot, and $r = 10$ in lowest plot.

where only the form of the survival proportion s differs between populations. Hence, any difference between the populations' growth or decline is due to a difference in geometric mean survival.

Given $\lambda = 1.256$, the nonsharing population has the greater geometric mean survival for any δ on the interval (0, 1.256). Figure 3.4 shows the difference in geometric mean survival values, as a function of δ, for several values of r. The survival advantage of the population that does not share food can be near 0.1 for small r, but the advantage declines as r increases. The implication is that the advantage of sharing when food is relatively abundant does not compensate for the demographic disadvantage of sharing when food is scarce. Figure 3.5 shows the critical value of λ where geometric mean survival is the same for sharing and not-sharing populations; the result is plotted again as a function of δ. Since among-year variation more strongly depresses geometric mean survival in the sharing population, the critical clump-discovery rate λ increases close to linearly with δ. The food-sharing population needs a greater arithmetic-mean survival to compensate for the effect of among-year variance. In the absence of phenotypic plasticity, a temporally varying environment need not favor food-sharing.

Final Comment

Chapters 2 and 3 have used two-person games to focus on economic concepts that should help us understand why individuals join or leave foraging groups, and how they might interact with other foragers in the same group. The next two chapters apply these concepts to questions about the size of foraging groups.

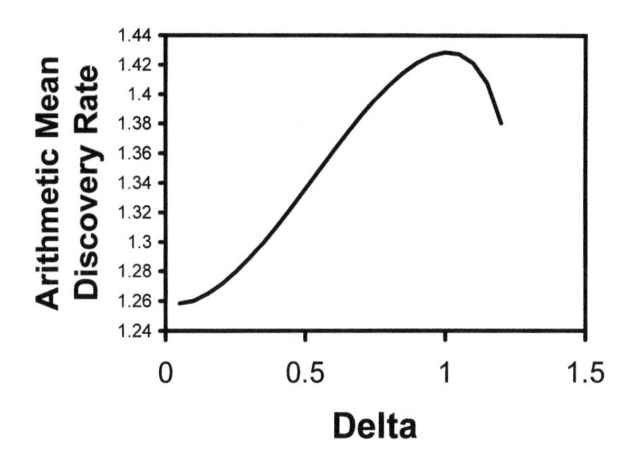

Figure 3.5 Clump discovery rate producing equal geometric mean population growth for sharing and not-sharing populations. Abscissa is δ, as in figure 3.4. Food-sharing requires an advantage in good years to compensate for its disadvantage in poor years; the necessary advantage increases as among-year variance in λ increases.

Math Box 3.1 Food Sharing: Constant Encounter Rate

The constant encounter-rate model has a simplifying property. Foragers search independently, so that one player's discovery of food does not affect the other's rate of encountering a clump.

Following the text, N represents "not-share" and A represents "always share." C is the communal cost.

N VERSUS N

Each player independently discovers food at constant probabilistic rate λ. The time required for player i ($i = 1, 2$) to locate a clump is T_i, an exponential random variable. Each player's probability of finding no food is $\Pr[T_i > \tau] = e^{-\lambda\tau}$.

If $T_1 \leq \tau < T_2$, player 1 may still incur a penalty (indirectly) with probability C. By independence, $\Pr[T_1 \leq \tau < T_2] = (1 - e^{-\lambda\tau})e^{-\lambda\tau}$. Summing over the ways to fail, each player has penalty probability

$$M_{NN} = e^{-\lambda\tau} + Ce^{-\lambda\tau}(1 - e^{-\lambda\tau}) \tag{3.1.1}$$

A VERSUS A

If each player shares, $T_1 = T_2 = T_{AA}$. Since the encounter rate is a constant, T_{AA} has a gamma density:

$$f(T_{AA}) = (2\lambda)^2\, T_{AA}\exp[-2\lambda T_{AA}] \tag{3.1.2}$$

$$E[T_{AA}] = 1/\lambda;\ V[T_{AA}] = 1/2\lambda^2. \tag{3.1.3}$$

Each player's penalty probability is

$$M_{AA} = 1 - \int_0^{\tau} f(T_{AA})dT_{AA} = e^{-2\lambda\tau}(1 + 2\lambda\tau). \tag{3.1.4}$$

N VERSUS A

Each player has probability 1/2 of discovering the first clump, given that food is found. Hence the process may develop as an A versus A interaction, or an N versus N interaction, each with probability 1/2. $T(i, j)$ is the time elapsing until the i-strategist finds the first clump.

Math Box 3.1 (*cont.*)

For $T(A, N) < T(N, A)$ we have the A versus A interaction. Each player incurs a penalty if no food is found, or if only one (shared) clump is found. The resulting unconditional penalty probability for each player is

$$\begin{aligned} &e^{-2\lambda\tau}(1 + 2\lambda\tau)\ \Pr[T(A, N) < T(N, A)] \\ &\quad = e^{-2\lambda\tau}(1 + 2\lambda\tau)/2. \end{aligned} \tag{3.1.5}$$

For $T(N, A) < T(A, N)$ we have the N versus N interaction. No food may be found. One clump may be found, so that the A-strategist is penalized with certainty and the N-strategist in penalized with probability C. The resulting unconditional penalty probability for the N-strategist is

$$\begin{aligned} &[e^{-2\lambda\tau} + 2Ce^{-\lambda\tau}(1 - e^{-\lambda\tau})]\ \Pr[T(N, A) < T(A, N)] \\ &\quad = (1/2)e^{-2\lambda\tau} + Ce^{-\lambda\tau}(1 - e^{-\lambda\tau}). \end{aligned} \tag{3.1.6}$$

The associated unconditional penalty probability for the A-strategist is obtained by setting $C = 1$ in (3.1.6).

Combining expressions (3.1.5) and (3.1.6) yields the penalty probabilities:

$$M_{NA} = e^{-2\lambda\tau}(1 + \lambda\tau) + Ce^{-\lambda\tau}(1 - e^{-\lambda\tau}) \tag{3.1.7}$$

$$M_{AN} = \lambda\tau e^{-2\lambda\tau} + e^{-\lambda\tau}. \tag{3.1.8}$$

Penalty Matrix

The penalty matrix M for a single round of play with a constant clump-encounter rate is:

$$\begin{array}{c} \\ N \\ A \end{array} \begin{array}{c} \begin{array}{cc} \qquad N \qquad\qquad\qquad & \qquad\qquad A \end{array} \\ \begin{bmatrix} e^{-\lambda\tau} + Ce^{-\lambda\tau}(1 - e^{-\lambda\tau}) & e^{-2\lambda\tau}(1 + \lambda\tau) + Ce^{-\lambda\tau}(1 - e^{-\lambda\tau}) \\ \lambda\tau e^{-2\lambda\tau} + e^{-\lambda\tau} & e^{-2\lambda\tau}(1 + 2\lambda\tau) \end{bmatrix} \end{array} \tag{3.1.9}$$

Following the text, we summarize the analysis of M by partitioning the $(\lambda\tau, C)$ space into three regions; see figure 3.2. In the first region,

$$C < (1 + 2\lambda\tau - e^{\lambda\tau})/(e^{\lambda\tau} - 1). \tag{3.1.10}$$

Math Box 3.1 *(cont.)*

Not-sharing is the pure ESS, and food-sharing cannot be favored over single or repeated play when (3.1.10) holds.

In the second region,

$$\frac{1 + 2\lambda\tau - e^{\lambda\tau}}{e^{\lambda\tau} - 1} < C < \frac{\lambda\tau}{e^{\lambda\tau} - 1}. \qquad (3.1.11)$$

M now conforms to a PD. Not-sharing remains an ESS. Food-sharing may now be favored through conditional mutualism over probabilistically repeated play, since mutual sharing incurs a lower penalty probability than does mutual not-sharing.

In the third region,

$$C > \lambda\tau/(e^{\lambda\tau} - 1). \qquad (3.1.12)$$

Sharing is now the sole ESS for single or repeated play. For more details on the development of the penalty probabilities and analysis of the game, see Caraco and Brown (1986).

Math Box 3.2 Food Sharing: Increasing Encounter Rate

X ($X = 0, 1$) is the number of clumps already discovered during a round of play. $\lambda(X)$ is the individual's rate of clump discovery,

$$\lambda(X) = \lambda(1 + X). \qquad (3.2.1)$$

See section 2.4 for a discussion of $\lambda(X)$.

N VERSUS *N*

Player i ($i = 1, 2$) locates a clump at random time T_i. The first clump is encountered at combined probabilistic rate 2λ. We have

$$\Pr[T_1, T_2 > \tau] = e^{-2\lambda\tau}. \qquad (3.2.2)$$

If only a single clump is discovered, the individual finding food still may incur a penalty indirectly if the other player fails to find food. Each individual's penalty probability is

$$M_{NN} = e^{-2\lambda\tau} + (1 + C)\lambda\tau e^{-2\lambda\tau}, \qquad (3.2.3)$$

where C is the communal cost.

A VERSUS *A*

The probability density of T_{AA}, time elapsing until two clumps are found, is (Boswell et al. 1979):

$$f(T_{AA}) = 4\lambda(\exp[-2\lambda T_{AA}] - \exp[-4\lambda T_{AA}]) \qquad (3.2.4)$$

$$E[T_{AA}] = 3/4\lambda;\ V[T_{AA}] = 5/16\lambda^2. \qquad (3.2.5)$$

The probability density for the search time assumes the first clump is encountered at combined rate 2λ. If one clump has been found, the second is encountered at combined rate 4λ. Then

$$\Pr[T_{AA} > \tau] = 1 - \int_0^{\tau}[1 - e^{-4\lambda(\tau - t)}]2\lambda e^{-2\lambda\tau}dt; \qquad (3.2.6)$$

after integrating, each player's penalty probability when both share food is

$$M_{AA} = 2e^{-2\lambda\tau} - e^{-4\lambda\tau}. \qquad (3.2.7)$$

Math Box 3.2 (*cont.*)

N versus *A*

Each player has probability 1/2 of discovering the first clump if food is found. Proceeding as in Math Box 3.1, suppose the process develops as an *A* versus *A* interaction, so that the first clump (if found) is shared. The associated contribution to each player's unconditional penalty probability is $M_{AA}/2$:

$$e^{-2\lambda\tau} - (1/2)e^{-\lambda\tau}. \qquad (3.2.8)$$

With probability 1/2, the process develops as an *N* versus *N* interaction. The first clump is encountered at rate 2λ. If a clump is found, the *N*-strategist averts a direct penalty, and the *A*-strategist continues searching at encounter rate 2λ. If the latter player fails to find food, the *N*-strategist may still incur a penalty indirectly. For this process the contribution to the unconditional penalty probability for the *N*-strategist is

$$e^{-2\lambda\tau}[(1/2) + C\lambda\tau]. \qquad (3.2.9)$$

The associated contribution to the *A*-strategist's unconditional penalty probability is obtained by setting $C = 1$ in (3.2.9). The off-diagonal elements of the penalty matrix *M* are given by summing (3.28) and (3.29), as in Math Box 3.1.

Penalty Matrix

Let $\alpha = e^{-2\lambda\tau}$. Then the penalty matrix for a single round of play with the increasing rate of clump discovery is

$$\begin{array}{c} \\ N \\ A \end{array} \begin{array}{c} \begin{array}{cc} N & A \end{array} \\ \left[\begin{array}{cc} \alpha(1 + C)\lambda\tau\alpha + \alpha(1 - \alpha)/2 & \alpha(1 + C\lambda\tau) \\ \alpha(1 + \lambda\tau) & 2\alpha - \alpha^2 + \alpha(1 - \alpha)/2 \end{array}\right] \end{array}. \qquad (3.2.10)$$

We again summarize the analysis of the penalty matrix *M* by partitioning the $(\lambda\tau, C)$ space into three regions; see figure 3.3. In the first region

$$C < (1 - e^{-2\lambda\tau} - \lambda\tau)/\lambda\tau. \qquad (3.2.11)$$

Not-sharing is the pure ESS in this region, and food-sharing cannot be favored over single or repeated play when (3.2.11) holds.

Math Box 3.2 (*cont.*)

In the second region,

$$(1 - e^{-2\lambda\tau} - \lambda\tau)/\lambda\tau < C < (1 - e^{-2\lambda\tau})/2\lambda\tau. \quad (3.2.12)$$

In this region M conforms to a PD. Not-sharing remains an ESS. Food-sharing may be favored through conditional mutualism over probabilistically repeated play, since mutual sharing incurs a lower penalty probability than does mutual not-sharing.

In the third region,

$$C > (1 - e^{-2\lambda\tau})/2\lambda\tau. \quad (3.2.13)$$

Sharing is now the sole ESS for single or repeated play.

Math Box 3.3 A Decomposable Matrix

The following 2×2 matrix defines a two-player discrete game.

	A	B
A	M(A,A)	M(A,B)
B	M(B,A)	M(B,B)

Each entry is the expected penalty for the player whose choice determines the row of the matrix. Following Mesterton-Gibbons (1992), the matrix is decomposable if

$$M(A,A) + M(B,B) = M(A,B) + M(B,A)$$

and $M(A,A) \neq M(B,B)$. That is, the trace of the matrix equals the sum of the off-diagonal elements. When our penalty/payoff matrices have this property, no mixed strategy can qualify as an ESS. But a 2×2 matrix must have at least one Nash solution qualifying as an ESS. Hence there must be a pure-strategy ESS.

Suppose the decomposable matrix produces a PD. Let *A* represent "cooperate" and *B* represent "defect." Then

$$M(A,A) < M(B,B) \text{ and } M(A,A) < [M(B,A) + M(A,B)]/2$$

from the definition of a PD. Mesterton-Gibbons and Dugatkin (1992) point out an interesting effect of decomposability. Suppose ego chooses to cooperate. Ego's penalty will be lower if its opponent also cooperates. The penalty reduction to ego when the opponent chooses cooperate, rather than defect, is

$$M(A,B) - M(A,A).$$

Now suppose ego defects. Ego's penalty reduction when the opponent chooses cooperate, rather than defect, is

$$M(B,B) - M(B,A).$$

Since the matrix is decomposable, these two quantities are identical. The opponent's choice of cooperate benefits ego to the same degree whether ego chooses to cooperate or defect.

4 Group Size in Aggregation Economies

4.1 Introduction

Killer whales (*Orcinus orca*) observed near the Pacific coast of North America come in two types: *resident* and *transient* (Baird and Dill 1996). *Transient* whales are nonterritorial animals that live in matrilineal groups called pods. Pods change size only through birth, death, or migration (Baird and Dill 1996). When these whales hunt, mostly for pinnipeds or cetacean prey, they form temporary foraging groups of one to fifteen individuals, often incorporating members of different pods (Baird and Dill 1996). An obvious question is, "Why do killer whales form multi-pod groups to attack their prey?" Baird and Dill's exceptional field study addresses this question specifically. They find that a whale's daily energy intake in a group is greater than the intake of a solitary individual. So, the whales forage under an aggregation economy as we define it. Baird and Dill conclude that killer whale hunting groups are the result of selection acting to maximize daily food intake.

If killer whales form hunting groups to maximize individual daily energetic intake, it might be possible to predict their group sizes with a simple group-membership game. Baird and Dill (1996) report that the modal whale-hunting group size is three. They estimate that a whale's daily energy intake increases initially with group size but declines past a group of three (fig. 4.1). So, groups of three maximize the individual's daily energetic return. But is three the group size expected if individuals attempt to maximize their daily energetic intake? That is, does a group of three qualify as the stable solution to a group-membership game based on foraging success? If it is, then Baird and Dill are likely correct when they state that whale-hunting groups are the result of attempts to maximize daily energetic intake. But if three is not the stable solution to the game, their conclusion requires further analysis. It turns out that predicting stable group size can be a complicated matter; the solution will likely depend on behavioral details of group formation (rules of entry) and the genetic relatedness between players. So, in this chapter we apply game theory to predict equilibrium group size in an aggregation economy.

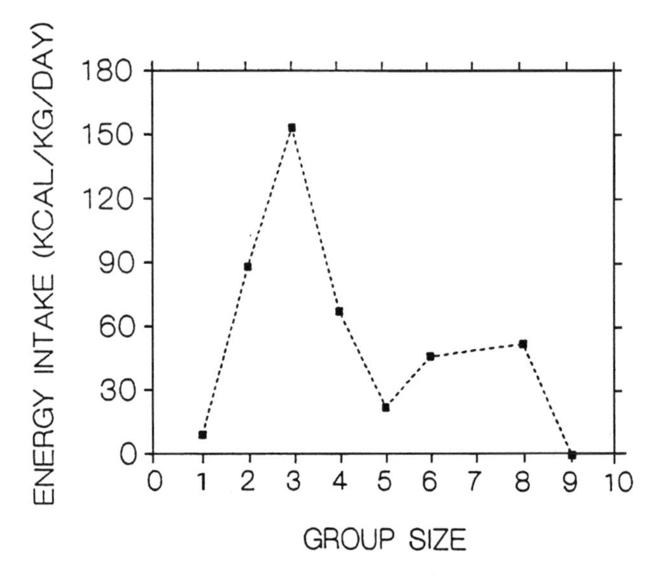

Figure 4.1 The aggregation economy for hunting groups of transient killer whales (*Orcinus orca*). The daily energetic intake increases with hunting group size up to groups of three and declines thereafter. Taken from Baird and Dill (1996). (With permission of Oxford University Press)

The Importance of the Rules of Entry

At least two studies point out that the rules of entry into groups will have a major influence on the size of those groups (Giraldeau and Caraco 1993; Higashi and Yamamura 1993). In some situations, such as the formation of sea birds' breeding colonies or the formation of nocturnal roosts in many bird species, individuals already in the group have little opportunity to prevent further individuals, intruders, from joining them. In this case, group entry has no cost to an intruder, and the size of the group is essentially under the control of intruders, particularly if group members cannot economically leave the group once they have joined (Kramer 1984). Group size then should depend on decisions made by intruding individuals.

In other circumstances, group members may be able to control group size, recruiting members when the group is too small and imposing an entry cost to prevent intruders from joining when it is too large. One example of group entry control is communal territorial defense in lionesses (*P. leo*); these lionesses collaborate to exclude strangers from joining their territorial group (Heinsohn and Packer 1995). When group control is effective, group size should depend on decisions made by group members, not intruders.

The Importance of Genetic Relatedness

Many foraging groups are composed of genetically unrelated individuals (e.g., fish schools, large flocks of seed-eating birds, etc.). However, a good number of organisms forage in kin groups. For instance, the hunting groups of *transient* killer whales are composed of individuals from just a few matrilines, so that some foragers are likely to be close kin. Many other social carnivores—lions, spotted hyenas (*Crocuta crocuta*), and wolves (*Canus lupus*), to name a few—also hunt in kin groups. The same can be said of some avian foraging groups. Snow geese (*Chen caerulescens*), for instance, form small family foraging groups within large migrating flocks. Communally breeding birds and those that defend year-round territories also forage in groups made up of genetic relatives (J. Brown 1987). Finally, many social spiders and eusocial insects forage in groups composed of kin. So, kin foraging groups are not exceptional.

There is considerable evidence that genetic relatedness of group members influences the quantity and intensity of within-group aggression. For instance, G. Brown and J. Brown (1993) show that aggression within neighborhoods of territorial Atlantic salmon (*Salmo salar*) declines as genetic relatedness of neighboring territory holders increases. Similar observations have been made for groups of brown trout (*Salmo trutta*; Olsén et al. 1996). If the level of within-group aggression depends, in part, on a potential aggresssor's relatedness to other group members (Reeve and Nonacs 1997), then genetic relatedness might affect equilibrium group size. But under more general assumptions, there is considerable disagreement concerning the effect that increased genetic relatedness between players should exert on group size. Some argue that increased genetic relatedness should enlarge group sizes (Rodman 1981); others conclude that relatedness should decrease group size (Giraldeau 1988). In this chapter, therefore, we analyze the interaction of rules of entry with genetic relatedness in an aggregation economy to predict patterns in social foragers' group size.

4.2 Which Group Size to Expect?

Defining an Aggregation Economy

An aggregation economy implies that indivduals in groups, for at least some group sizes, do better than solitaries (see section 2.1). Hence an aggregation economy promotes group formation. However, as group size increases, competition and interference will likely reduce the advantages of group membership. So, our models assume benefits increase to a maximum and decline

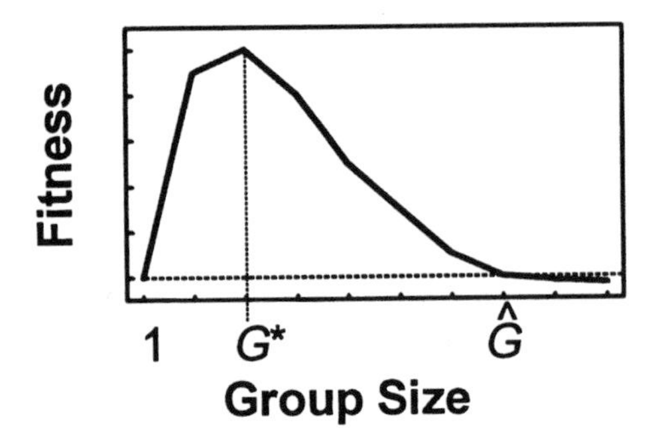

Figure 4.2 A peaked fitness function ($\Omega[G]$) of the kind that is characteristic of an aggregation economy. Fitness of group members is maximized at G^*. $\hat{G}$ gives the same fitness as solitary foraging.

thereafter. That is, we assume a peaked fitness function as described by Clark and Mangel (1986; see fig. 4.2).

To specify an aggregation economy, let G, a positive integer, represent group size. Let the fitness function $\Omega(G)$ represent the *direct* fitness of each of G group members; fitness is nonnegative for all group sizes. The first difference of the fitness function is $\Delta\Omega(G) = \Omega(G + 1) - \Omega(G)$. In an aggregation economy, fitness, at least initailly, increases with group size, so that $\Delta\Omega(G) > 0$ for small G. $\Omega(G)$ attains a unique maximum (the peak) at $G = G^*$, where $G^* \geq 2$. Therefore, we define G^* as the optimal group size (e.g., Pulliam and Caraco 1984); $\Delta\Omega(G)$ changes sign once, from positive to negative, at $G = G^*$. Note that the peak of an aggregation economy must occur at group sizes of two or more; $G^* = 1$ implies a dispersion economy (see chapter 5).

Having defined an aggregation economy, we want to predict patterns in group size expected under these conditions. A number of studies analyze a group-membership game where foragers are free to join any group that would increase their economic gain (e.g., Sibly 1983; Clark and Mangel 1984; Pulliam and Caraco 1984; Giraldeau and Gillis 1985; Giraldeau 1988). We refer to this condition as *free entry*. Other studies assume that group members regulate entry by solitaries (e.g., Janson 1988). We refer to this condition as *group-controlled entry*. We deal with each in turn.

Free Entry Games

Free entry assumes that solitaries can enter any group at no cost (Sibly 1983). Additionally, individuals leave groups only singly and hence become solitary when they do so. Free entry does not apply if, for instance, individuals within groups can collaborate to form emigration parties. Under free entry, solitaries should join a group of size G, increasing its size to $(G + 1)$

if the resulting fitness they obtain in that group size is greater than the fitness they obtain by remaining alone. That is, a solitary should join a group of G when $\Omega(G + 1) > \Omega(1)$. Since the fitness function Ω is peaked, fitness within groups first increases but declines once $G > G^*$. We further assume that

$$\lim G_{\rightarrow\infty} \Omega(G) < \Omega(1). \tag{4.1}$$

Condition (4.1) reasonably eliminates the case where solitaries can always increase their direct fitness by joining a group, no matter how large the group (see Giraldeau and Caraco 1993). Then there will be some group size $\hat{G}$ where a solitary no longer gains fitness by joining (fig. 4.2). That is, our assumptions assure there is a group size $\hat{G}$ such that $\Omega(\hat{G}) \geq \Omega(1) > \Omega(\hat{G} + 1)$. Then $\hat{G}$ qualifies as the Nash equilibrium group size for free entry; no individual is tempted to leave an equilibrium group because it cannot increase its fitness by foraging as a solitary. The Nash equilibrium group is stable if fitness in the group exceeds the fitness of being solitary ($\Omega(\hat{G}) > \Omega(1)$), or is neutrally stable if fitness in the equilibrium group equals the fitness of a solitary ($\Omega(\hat{G}) = \Omega(1)$). The shape of the fitness function implies that the equilibrium group size under free entry ordinarily exceeds the optimal group size G^* (for a special case, see Giraldeau and Gillis 1985). Consequently, $\Delta\Omega(G)$, the first difference of the fitness function, will be negative around the equilibrium group size. Hence fitness at the competitive equilibrium group size is lower than fitness at the optimal group size. Once again, the evolutionarily stable solution to the free entry group membership game does not maximize individual gains. In fact, when $\Omega(G) = \Omega(1)$, individuals in free-entry groups do not fare any better than solitary individuals, the "paradox of group foraging," which we analyze below.

Recall Baird and Dill's (1996) killer whale aggregation economy. If hunting groups form under free entry, then individuals attempting to maximize their daily energetic intake would be hunting in groups of eight, not three (see fig. 4.1). Hence, if the killer whales are maximizing daily intake, our analysis (to this point) suggests they cannot be forming groups under free entry. Alternatively, if they are forming groups under free entry, then they cannot be maximizing daily intake.

Group-Controlled Entry Games

The preceding discussion assumes that group members have no control over group formation. As mentioned in section 4.1, group members may, under some circumstances, collaborate either to recruit solitaries when group sizes are below the optimum, or to prevent solitaries from joining when the group is as large as the optimal size. The evolution of effective group control poses some serious problems, however. Heinsohn and Packer (1995) studied lionesses' group defense of communal territories. When a stranger is detected on a pride's territory, females apparently collaborate to expel the intruder. Upon closer examination, Heinsohn and Packer found that pride females did not participate equally in territorial defense. The lionesses seem to fall into at least two categories: *leaders* that always spearhead attacks, and *laggards* that stay behind, leaving *leaders* to bear the greater hazards of injury (Heinsohn and Packer 1995). It follows that group control may be costly and so might require conditions necessary for the evolution of cooperation. Notwithstanding this problem, when group control is effective, group size will reflect the interests of the group members, not the intruders. Group control, therefore, might allow groups to achieve the optimal size G^*, so that members enjoy the maximal benefit of group membership.

Returning to Baird and Dill's (1996) killer whale study, group control would imply that hunting parties of three are expected if individuals attempt to maximize daily energy returns, but only under complete group-controlled entry. We emphasize the importance of establishing a group's rules of entry before drawing conclusions about the adaptive significance of its size.

So, optimal group sizes and maximal benefits of sociality might be predicted when group members exert complete control over entry. Larger equilibrium group sizes, and little advantage over solitary foraging, should be expected under free entry. Having established that the different entry rules predict different group sizes, we analyze the way genetic relatedness may modify these predictions.

4.3 The Effect of Genetic Relatedness on Equilibrium Group Size

When group members are related, the economics involve inclusive fitness. Following Grafen (1982), we partition inclusive fitness into direct and indirect components. The direct component is simply benefits to ego. The indirect component refers to the net effect ego exerts on benefits to ego's relatives, devalued by r, the coefficient of relatedness. Note that the indirect component does not include benefits that relatives were already enjoying before interacting with ego (Grafen 1982; Giraldeau 1988; Giraldeau and Gillis 1988; Higashi and Yamamura 1993).

We use Hamilton's Rule (e.g., Grafen 1982) to show how relatedness might affect decisions governing group membership and, hence, group size. We assume that all members of a population are genetically related by coefficient r. This symmetrical genetic relatedness simplifies the analysis and is a standard assumption of analyses of the effect of genetic relatedness on group size (Rodman 1981; Giraldeau 1988).

If B is the total net benefit of ego's act on relatives, and C is the cost of that act to ego, then Hamilton's Rule implies that selection can favor an act when $rB - C > 0$. Hamilton's Rule applies to any behavior, not just the economics of kin-directed altruism. In the context of group-membership decisions, the effects of a change in group size may prove positive or negative for ego's relatives, as well as for ego. Therefore, we replace benefits and costs, respectively, by effects on relatives and effects on self. B becomes E_R, the total effect on relatives, and C becomes E_S, the effect on self. Substituting, Hamilton's Rule suggests that selection can favor an act by ego when

$$rE_R + E_S > 0, \tag{4.2}$$

where both E_R and E_S may depend on group size.

Applying Hamilton's Rule depends on the process responsible for group formation. Specifically, the application depends on the rules of entry described above. Free entry assumes that solitaries can decide whether or not to join a group. Under group-controlled entry, members of the group decide whether or not to repel a solitary attempting to join the group. We examine them in turn.

Free Entry

To calculate the effect on relatives when a solitary joins a group, we subtract the direct fitness the relatives held before the solitary joined from the fitness resulting from the solitary's act of joining. Suppose ego joins a group of $(G - 1)$ relatives and hence makes it increase to G. The effect on each of ego's $(G - 1)$ relatives is $\Omega(G) - \Omega(G - 1) = \Delta\Omega(G - 1)$. The combined effect E_R on all its $(G - 1)$ relatives is

$$E_R = (G - 1)\, \Delta\Omega(G - 1). \tag{4.3}$$

E_S, the effect of ego's act on self, is simply the fitness obtained by being in the group minus the fitness of solitary foraging:

$$E_S = \Omega(G) - \Omega(1). \tag{4.4}$$

Using Hamilton's Rule, under free entry ego should join a group when the resulting group size implies that $rE_R + E_S > 0$. Since the fitness function is peaked, the Nash equilibrium group size will now be the largest group where this inequality is not yet reversed (reversal indicates decreased inclusive fitness). Now the solitary individual's decision to join a group of $(G - 1)$ relatives depends on both the coefficient of relatedness and group size. A solitary forager should join a group of $(G - 1)$ relatives whenever (Giraldeau and Caraco 1993)

$$\Psi(r, G) = r(G - 1)\,[\Omega(G) - \Omega(G - 1)] + \Omega(G) > \Omega(1). \tag{4.5}$$

Note that Ψ varies with r and G, but that the fitness of being alone, $\Omega(1)$, is a constant. When the individuals are not related by descent ($r = 0$) the model appropriately predicts a Nash equilibrium group size of $\hat{G}$ members, since $\hat{G}$ is the largest group before $\Psi(r = 0, G) < \Omega(1)$.

We illustrate the free entry model graphically in Summary Box 4.1. Giraldeau and Caraco (1993) analyze free entry as three cases:

1. Groups currently smaller than the optimum, $2 \leq G \leq G^*$;
2. Groups currently larger than the Nash equilibrium in the absence of relatedness, $G > \hat{G}$; and
3. Group sizes on the remaining interval, $G^* < G \leq \hat{G}$.

If group size does not exceed the optimum, then clearly it pays both group members and solitary intruder to join the group. In this case there is no conflict of interest between group members and the intruder. That is, for any coefficient of relatedness between zero and one,

$$\Omega(2 \leq G \leq G^*) > \Omega(1) \text{ and } \partial\Psi(r, 2 \leq G \leq G^*)/\partial r > 0. \qquad (4.6)$$

So, as expected, a solitary forager should always join a group of $(G^* - 1)$ or fewer members, independently of the degree of relatedness.

Now suppose group size exceeds the Nash equilibrium for unrelated foragers, i.e., $G > \hat{G}$. For any group size larger than $\hat{G}$, a group member's direct fitness is, by definition, less than a solitary's, and the fitness function is declining; $\Delta\Omega(G) < 0$. Therefore, $\Psi(0 \leq r \leq 1, G > \hat{G}) < \Omega(1)$, so that joining reduces the solitary's inclusive fitness. Solitaries should never want to join groups larger than $\hat{G}$, independently of genetic relatedness. Since the free-entry model predicts that $\hat{G} = G$ at equilibrium when players are not related ($r = 0$), increasing the relatedness between players does not increase group size (see Summary Box 4.1).

The two preceding cases show that relatedness can affect group size under free entry only for groups between the optimum G^* and $\hat{G}$, the equilibrium size in the absence of relatedness. To examine the effect, suppose that fitness within a group of $\hat{G}$ members equals the fitness of a solitary; $\Omega(\hat{G}) = \Omega(1)$. Then $\hat{G}$ will not remain the equilibrium group size for any $r > 0$. Since $\hat{G}$ is larger than the optimum G^*, any increase in group size implies a decrease in direct fitness; that is, $\Delta\Omega(\hat{G} - 1) < 0$. The decline in direct fitness around $\hat{G}$ implies that the value of joining a group with $(\hat{G} - 1)$ members will, for any $r > 0$, provide lower inclusive fitness than that obtained by remaining solitary; $\Psi(0 < r \leq 1, \hat{G}) < \Omega(1)$. So, a solitary should not join a group of $(\hat{G} - 1)$ relatives; joining would not increase the solitary's direct fitness and would decrease relatives' direct fitness. Hence the equilibrium group size when players are genetically related will be no less than the optimal group size and will always be smaller than $\hat{G}$. That is, the predicted group size for $r > 0$ must satisfy $G^* \leq G < \hat{G}$. Under free entry, therefore, relatedness

SUMMARY BOX 4.1 GRAPHICAL DEPICTION OF THE EFFECT OF GENETIC RELATEDNESS ON EXPECTED GROUP SIZE

A. Free Entry Model

The second panel presents computations of Hamilton's Rule for the decision of joining the group of G individuals using the aggregation economy depicted in the top panel.

Solitary individuals should join the group so long as Hamilton's Rule is positive (the top part of middle panel). Note that the $G^* = 3$ acts as a pivot. To the left of G^* the greater the value of r the greater the value of joining. To the right of G^*, the greater the value of r the lower the value of joining. Consequently, as r increases, the value of Hamilton's Rule becomes negative at increasingly smaller group sizes. The consequence of increased genetic relatedness on expected group size is illustrated in the third panel; the expected group size moves from $\hat{G}$ toward G^* with increasing r.

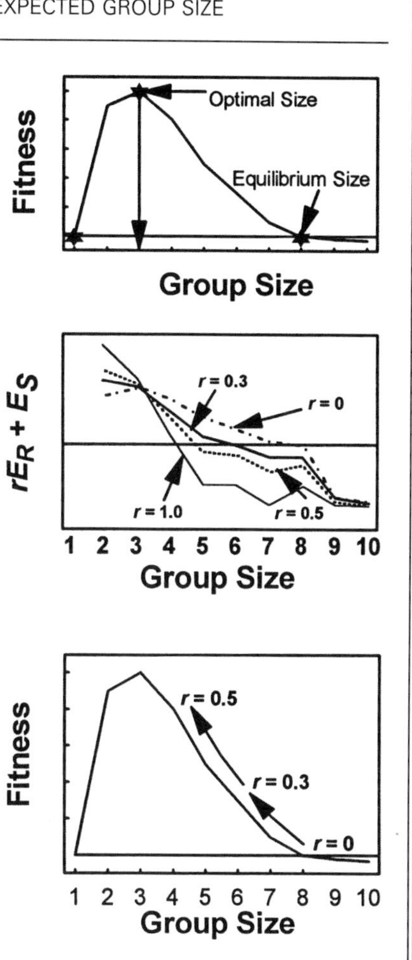

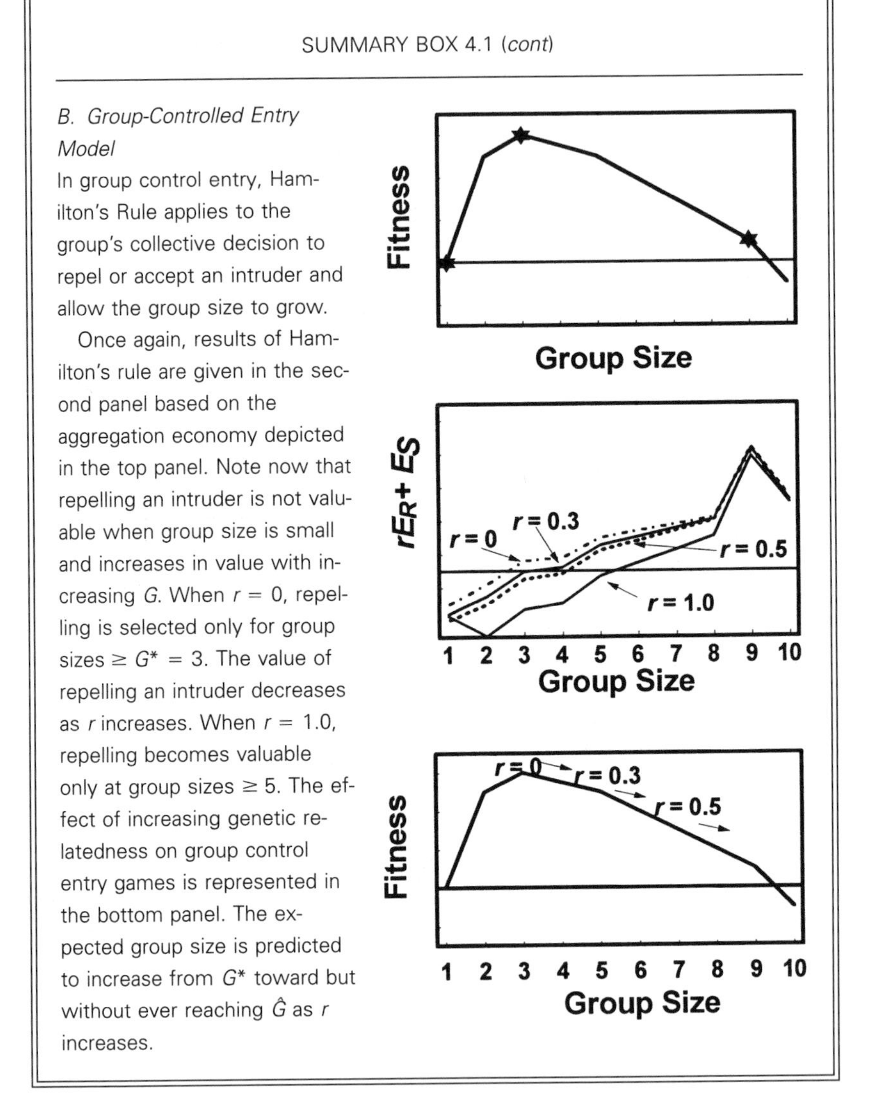

SUMMARY BOX 4.1 (*cont*)

B. Group-Controlled Entry Model

In group control entry, Hamilton's Rule applies to the group's collective decision to repel or accept an intruder and allow the group size to grow.

Once again, results of Hamilton's rule are given in the second panel based on the aggregation economy depicted in the top panel. Note now that repelling an intruder is not valuable when group size is small and increases in value with increasing G. When $r = 0$, repelling is selected only for group sizes $\geq G^* = 3$. The value of repelling an intruder decreases as r increases. When $r = 1.0$, repelling becomes valuable only at group sizes ≥ 5. The effect of increasing genetic relatedness on group control entry games is represented in the bottom panel. The expected group size is predicted to increase from G^* toward but without ever reaching $\hat{G}$ as r increases.

cannot increase, but rather can only decrease equilibrium group size. This conclusion is easily generalized to $\Omega(\hat{G}) \geq \Omega(1)$. In Math Box 4.1 (see end of chapter) we show that the equilibrium group size under free entry is the maximal G, such that

$$\frac{\Omega(G+1) - \Omega(1)}{r[\Omega(G) - \Omega(G+1)]} < G < 1 + \frac{\Omega(G) - \Omega(1)}{r[\Omega(G-1) - \Omega(G)]}, \tag{4.7}$$

where r is the coefficient or relatedness between foragers.

Returning once again to the killer whale example, our analyses suggest that under free entry with genetic relatedness, group sizes smaller than eight should be expected. As the coefficient of relatedness increases, the predicted group size approaches the optimal group size. However, the coefficient of relatedness would have to be unrealistically large to predict free-entry groups of the optimal size—i.e., groups maximizing daily energy gain per individual—that Baird and Dill (1996) observed.

Group-Controlled Entry

Group members may sometimes regulate the addition of solitaries to the group (Vehrencamp 1983; Janson 1988; Higashi and Yamamura 1993). Therefore, we now ask how relatedness will affect a group member's tendency to repel joiners. For simplicity, we assume all group members collectively expel an intruder when Hamilton's Rule favors that act over an increase in group size. Specifically, the rule implies that the group should repel a solitary attempting to increase the group size from G to $(G + 1)$ when $rE_R + E_S > 0$.

For group-controlled entry, E_R is the effect of repelling an intruder on the intruder:

$$E_R = \Omega(1) - \Omega(G + 1). \qquad (4.8)$$

The effect on the intruder per group member is E_R/G. E_S is the effect of repelling the intruder on self, where self is the entire group:

$$E_S = G[\Omega(G) - \Omega(G + 1)]. \qquad (4.9)$$

The effect per group member is $-\Delta\Omega(G)$.

Group members should repel an intruder when $rE_R + E_S > 0$, and should let a solitary join the group when the inequality is reversed. Substituting for effects on the relative and self, we find that repelling is favored when

$$[\Omega(1) - \Omega(G + 1)]\,(r/G) > \Delta\Omega\,(G). \qquad (4.10)$$

To analyze group-controlled entry, we again assume the fitness function is peaked, and $\Omega(\hat{G}) = \Omega(1)$; see figure 4.1. First consider groups smaller than the optimum; $1 \leq G \leq (G^* - 1)$. For each such group size, the fitness function is increasing, $\Delta\Omega(G) > 0$, and $[\Omega(1) - \Omega(G)] < 0$ by the definition of G^*. So, expression (4.10) indicates that repelling can never be favored for $1 \leq G \leq G^*$. A group smaller than the optimum should never repel a joiner, independently of the level of relatedness, since each addition to the group increases each group member's fitness. Again, there is no conflict of interest between a solitary and the group members.

Next suppose $G \geq \hat{G}$. Then, the fitness function is decreasing, $\Delta\Omega(G)$

< 0, and $[\Omega(1) - \Omega(G + 1)] > 0$ by the definition of $\hat{G}$. So, not surprisingly, repelling joiners will always be favored, independently of the level of relatedness, if $G > \hat{G}$.

From the two preceding paragraphs, we know that the equilibrium group size must contain at least G^* members, and less than $\hat{G}$ members. Since group-controlled entry should maintain group sizes smaller than $\hat{G}$, where fitness equals a solitary's, this entry rule predicts that a group member should obtain greater benefits, or lower fitness costs, than a solitary. For the possible equilibrium group sizes G, $G^* \leq G < \hat{G}$, the fitness function is declining ($\Delta\Omega(G) < 0$). In the absence of any relatedness between players ($r = 0$), expression (4.10) then shows that Hamilton's Rule favors repelling intruders at each of these group sizes. For unrelated players, therefore, the largest group that should accept a joiner is $(G^* - 1)$; so the equilibrium group size consequently is the optimal group size G^*.

Now, reconsider this case when foragers are related. When $G^* \leq G < \hat{G}$, and $0 < r \leq 1$, the result obtained under free entry may be reversed. That is, under group-controlled entry the equilibrium can increase as relatedness increases (see Summary Box 4.1). If neither group size nor the absolute value of the fitness function's first difference, $|\Delta\Omega(G)|$, is too large, a given value of the coefficient of relatedness is more likely to induce a group to accept a relative. That is, expression (4.10) predicts that increased relatedness can increase, and will never decrease, equilibrium group size under group-controlled entry (Giraldeau and Caraco 1993; Higashi and Yamamura 1993).

For group-controlled entry, the size of a group should increase until $rE_R + E_S > 0$. The minimal group size satisfying this inequality is the Nash equilibrium; it is the smallest group to repel, rather than accept, a joiner. Suppose $r > 0$ and the equilibrium group size G exceeds G^*. Then $\Delta\Omega(G) < 0$ for each G, where $G^* < G < \hat{G}$. The Nash equilibrium will be the largest G such that

$$\frac{r[\Omega(G + 1) - \Omega(1)]}{\Omega(G) - \Omega(G + 1)} < G < 1 + \frac{r[\Omega(G) - \Omega(1)]}{\Omega(G - 1) - \Omega(G)}. \tag{4.11}$$

The group-control equilibrium group size can never exceed the equilibrium under free entry. For the special case of $r = 1$, the definitions of the two equilibria become identical (see above), and the group-control equilibrium becomes as large as the free-entry value (Higashi and Yamamura 1993). Summary Box 4.2 collects the essential elements of this section's assumptions and predictions.

4.4 Integrating Entry Rules, Relatedness, and Aggressive Dominance

The preceding analyses show that in the absence of genetic relatedness, equilibrium group size is usually larger (and never smaller) for free entry than

SUMMARY BOX 4.2 THE GROUP MEMBERSHIP GAME IN AN AGGREGATION ECONOMY

A. Free Entry

ASSUMPTIONS

Currency: Fitness

Decision: Join foraging group of G individuals or forage alone

Constraints:

1. Perfect information about consequences of joining groups.
2. Solitaries are free to join groups.
3. Players competitively equal.
4. All players related by coefficient r.

PREDICTIONS

1. Solitaries always join when group size (G) is smaller than the optimum G^*.
2. Solitaries never join when G is at or larger than the Nash equilibrium group size $\hat{G}$.
3. For $G^* \leq G \leq \hat{G}$, solitaries join to increasingly large groups as coefficient of relatedness between players decreases.

B. Group-Controlled Entry

ASSUMPTIONS

Currency: Fitness

Decision: A group of G individuals either repels a solitary that attempts to join or allows it to enter and hence increase group size to $G + 1$

Constraints:

1. Perfect information about consequences of repelling solitary.
2. Group members can always repel solitaries and do so at no cost.
3. Players competitively equal.
4. Players related by coefficient r.

PREDICTIONS

1. Groups never repel a solitary when $G < G^*$.
2. Groups always repel a solitary when $G \geq \hat{G}$.
3. For $G^* \leq G \leq \hat{G}$, group members repel solitaries at increasingly large groups as coefficient of relatedness between players increases.

for group-controlled entry. But increasing relatedness ordinarily leads to a decrease in equilibrium group size under free entry, and an increase in equilibrium group size under group-controlled entry. This section comments on some generalizations of these predictions.

When to Prefer Relatives and Nonrelatives?

The two entry-rule models suggest group-size dependent preferences for socializing with relatives versus nonrelatives. Suppose groups form by free entry of solitaries and group sizes are currently smaller than the optimum G^*. Individuals should prefer joining a group of relatives over an equivalent group of nonrelatives. Joining relatives benefits both ego and the relatives where the fitness function is increasing, $\Delta\Omega(G) > 0$. The same conclusion results when entry is under group control. When group sizes are smaller than the optimum, group members enhance their direct fitness by accepting (not repelling) a solitary, and the total increase in inclusive fitness is greater when a more closely related individual is accepted in the group.

Now reconsider free entry for groups larger than the optimum, where the fitness function is decreasing ($\Delta\Omega(G) < 0$). Solitaries might prefer to join a group of nonrelatives. Joining nonrelatives avoids decreasing the direct fitness of a group of relatives, while still allowing the solitary to increase its own direct fitness. Interestingly, the prediction for group-controlled entry is now different. Accepting (failing to repel) a relative when group size is at least as large as the optimum G^* constitutes kin-directed altruism (see Summary Box 1.1). The group normally should repel an intruder where the fitness function is decreasing, unless the intruder is so closely related that the joiner's benefit outweighs the collective decrease in direct fitness suffered by the group members. For further discussion of preference for relatives over nonrelatives as group members, see Grafen (1984) or Giraldeau and Caraco (1993).

The Demise of the Paradox of Group Foraging under Free Entry

Most previous models of Nash equilibrium group size consider only free entry into groups of unrelated individuals (e.g., Sibly 1983; Pulliam and Caraco 1984; Giraldeau and Gillis 1985). Giraldeau (1988) identifies an apparent paradox associated with the free-entry equilibrium. The basic premise of the economics of sociality is that affiliative groups form because individuals gain a fitness advantage from group membership. But the free-entry equilibrium is defined (essentially) by fitness equality of social and solitary living. At the Nash equilibrium, an individual would gain nothing by having joined a group. How then can social foraging evolve?

The paradox is surely not real; individuals foraging in groups commonly gain greater benefits, correlates of survival and fecundity, than solitaries (e.g., Pulliam and Caraco 1984; Barnard and Thompson 1985; Giraldeau 1988). Under free entry, any genetic relatedness reduces the Nash equilibrium group size to a level where each member's fitness exceeds a solitary's. Independently of relatedness, group-controlled entry always equilibrates at a group size where each member's fitness is greater than a solitary's. In general, aggressive restriction of group entry, once the size reaches the optimum, may help ensure group members' benefits (Clark and Mangel 1986; Giraldeau 1988).

In dominance-structured groups, a high-ranking individual may regulate group size by repelling less aggressive joiners (e.g., Pulliam and Millikan 1982; Janson 1985; Vehrencamp 1983; Giraldeau 1988). A dominant's aggression may cause subordinate individuals to avoid costly encounters (e.g., Caraco 1979; Ekman et al. 1981). Then dominants could incur only a small fitness cost while still regulating subordinate individuals' access to the group; the average fitness of group members would then remain greater than a solitary's fitness (see Rohwer and Ewald 1981).

Achieving a Compromise between Free Entry and Group Control

The logic of group-controlled entry begins by assuming that a solitary tries to enter the group. Free entry predicts whether or not a solitary should attempt to join a group. Therefore, free entry has a logical precedence; a group cannot repel an intruder until a solitary tries to join it. On the other hand, group control may be the more useful model when an environmental change, such as declining food density, reduces aggregation benefits and leads to a reduction in group size through eviction.

While free entry and group control may apply separately to certain social systems, the size of some groups must depend on the outcome of a game between a solitary and group members. Suppose that $G^* \leq G < \hat{G}$, where the fitness of being in a group necessarily exceeds the fitness of being alone. A solitary's direct fitness would increase (not decrease) if it joined the group, but each current group member's direct fitness would necessarily decline. The interaction of the competing objectives, the conflict described by Higashi and Yamamura (1993), can be analyzed by modifying the inclusive fitness to include a cost of entry for the solitary and a cost of repelling an intruder for the group. There are two ways to analyze how these costs may affect group size.

First, we can fix group size and ask if a solitary should be willing to pay a bigger cost than the group can afford to offset, so that the group's size increases, or if the group should pay a cost sufficient to repel the solitary.

The cost of attempting to enter the group aggressively lowers the solitary's direct fitness and indirectly lowers the group members' inclusive fitness (for $r > 0$; Yamamura and Higashi 1992). Similarly, the cost of attempting to repel an intruder lowers group members' direct fitness and indirectly reduces the solitary's inclusive fitness. For group sizes near the optimum G^*, a solitary might gain more direct fitness by joining than the group members lose through an increase in group size. Consequently, a solitary might be willing to incur a greater cost (in terms of inclusive fitness) than the group members would. The group then should increase in size. For G near $\hat{G}$, a solitary has less to gain by joining, and the group might then be willing to pay the cost (direct and indirect) necessary to repel an intruder.

Second, one can specify group-size dependent costs for both the solitary and the group members, and ask how these costs influence group size (Higashi and Yamamura 1993). Math Box 4.2 shows how this approach can be applied to predict equilibrium group size.

Incorporating Effects of Dominance

Our analysis of equilibrium group size has assumed that each group member obtains the same direct fitness $\Omega(G)$, or that each member acts as if its direct benefit equaled the group average. For comparison, consider the dominance-structured groups described by Pulliam and Caraco (1984; also see Vehrencamp 1983). Groups form by free entry of solitaries. Each of G group members now has a different fitness $\Omega_i(G)$; $i = 1, 2, \ldots, G$. Benefits depend on social rank as suggested by a linear dominance hierarchy, so that $\Omega_i(G) > \Omega_{i+1}(G)$ for $i = 1, 2, \ldots, G - 1$; see figure 4.3. For simplicity, assume that the last individual to join a group takes the most subordinate role, so its fitness is $\Omega_G(G)$.

Even with unrelated individuals, the average fitness within the group will, at equilibrium, exceed a solitary's fitness. When the group's size has equilibrated at $\hat{G}$, we have

$$\Omega_{\hat{G}}(\hat{G}) \geq \Omega(1) > \Omega_{\hat{G}+1}(\hat{G} + 1). \tag{4.12}$$

That is, the free-entry decision rule compares a solitary's fitness with only the benefits accrued in the most subordinate role within the group. Solitaries will stop joining before the group's average fitness declines to $\Omega(1)$. Increasing relatedness might imply an even smaller free-entry equilibrium, or might favor less extreme variation in dominance-structured benefit levels (Giraldeau and Caraco 1993).

4.5 Risk-Sensitive Group Membership Games

The arguments presented above assume simple deterministic relationships between group size and direct fitness as a group member. In chapter 1 we

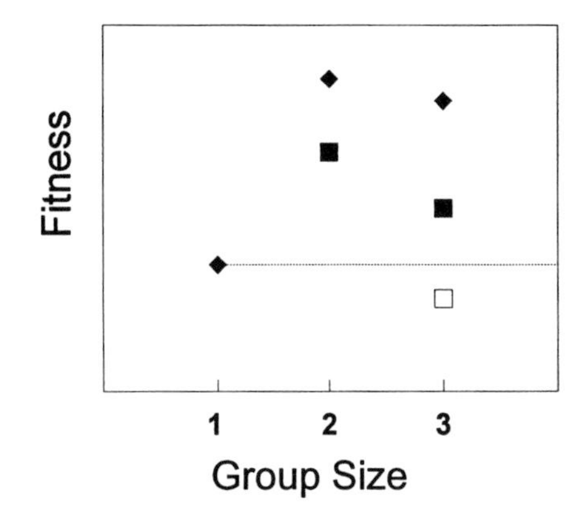

Figure 4.3 Peaked fitness functions for an asymmetric group membership game with unequal competitors (e.g., dominance). The horizontal dashed line gives the fitness obtained by any individual that forages alone. The group of two individuals has a dominance structure such that the subordinate (square) does worse than the dominant (diamond), but both still do better than remaining alone. A group of two foragers is the Nash solution because the dominance structure is such that the most subordinate individual in a group of three (open square) does worse than by remaining alone so the third individual should refuse to join.

indicated a focus on models that appreciate the random nature of many animals' foraging, and take survival probability as the currrency of fitness. Clark (1987) analyzes group membership of hunting lionesses as a risk-sensitive decision process. Specifically, he solves a dynamic optimization model for the group size minimizing each member's probability of starving (G^* in our models). Clark's (1987) prediction corresponds, in our terms, to group-controlled entry among nonrelatives. This section takes up the problem by considering risk-sensitive foragers and comparing optimal versus equilibrium group size in a stochastic environment. We develop three versions of the model. In the first analysis, we fix foraging time and we let food intake vary randomly. In the second, we fix food intake and examine random variation in the time spent searching for food clumps. In the third version, both food consumption and foraging time are random variables.

Random Food Intake and Group Membership

This subsection reviews a model presented by Ekman and Rosander (1987). Both the mean and variance of an individual's food consumption depend on group size in the model. To focus on risk sensitivity, we assume unrelated foragers.

During a single foraging day (or other, appropriately scaled period), G foragers search concurrently and independently; they locate a random number of food clumps n_G. We let c represent clump size, the total energy available in a clump. c also varies randomly; for simplicity we set E[c] equal to unity.

We assume each group member obtains the same portion (G^{-1}) of each clump discovered (cf. Caraco 1981; Mangel 1990; Caraco and Giraldeau 1991; Ranta et al. 1993). The random variable Y_G represents the total daily energy intake of a forager in a group of size G:

$$Y_G = \sum_{i=1}^{n_G} c_i/G. \tag{4.13}$$

The expectation and variance of Y_G are, respectively,

$$\mathrm{E}[Y_G] = \mu_G; \ \mathrm{V}[Y_G] = \sigma_G{}^2. \tag{4.14}$$

An individual starves, or suffers some other significant physiological penalty, if its total energy consumption at the end of the day fails to exceed its metabolic requirement. For simplicity, we ignore the possibility of starving during the foraging period (see Houston and McNamara 1986), so the forager starves if $Y_G \leq R$. We take the metabolic requirement R as fixed, hence independent of G. We assume a forager might try to minimize its probability of starvation $\Pr[Y_G \leq R]$, where the mean and variance of Y_G depend on group size.

Y_G is a "randomly stopped sum of random variables" (Ord 1972; Boswell et al. 1979). That is, both the number of clumps discovered n_G and the energy within a clump c_i vary randomly. So we say that the distribution of n_G is generalized by clump size c to give the distribution of Y_G. Realistically, foragers in patchy environments must often experience random variation in both clump size and the number of clumps found per day.

The central limit theorem for random sums allows us to approximate Y_G with a normal random variable, as long as the number of clumps n_G is sufficiently large (see Ekman and Rosander 1987). Therefore, we can express the individual group member's starvation probability via the standard normal distribution $\Phi(z)$:

$$\Pr[Y_G \leq R] \approx \Phi[z_G(R)] = \Pr\left[z \leq \frac{R - \mu_G}{\sigma_G} \right], \tag{4.15}$$

where z is a normal random variable with expectation 0 and standard deviation 1. Since $\Phi(z)$ increases strictly monotonically in its argument z, minimizing the individual's probability of starvation is equivalent to minimizing $z_G(R) = (R - \mu_G)/\sigma_G$.

When individual survival, $1 - \Phi[z_G(R)]$, is a peaked function of group size (as above), the "optimal" group size G^* is defined by

$$\min_G z_G(R) = \{[R - \mu_G)]/\sigma_G\}_{G=G^*}. \tag{4.16}$$

That is, an individual in a group larger or smaller than G^* cannot have a lower starvation probability than an individual in a group of G^* members. We define the free-entry equilibrium group size $\hat{G}$ by

$$z_G(R \mid G = 1) = z_G(R \mid G = \hat{G}), \tag{4.17}$$

where $G = 1$ again implies solitary foraging. Each member of a group of $\hat{G}$ has the same starvation probability as a solitary.

The model's predictions will depend on the way $\mathrm{E}[Y_G]$ and $\mathrm{V}[Y_G]$ vary with group size. In section 2.2 we noted that larger groups should discover food clumps faster, but the expected number of clumps discovered per individual need not increase with group size. Let the expected number of clumps found by the G group members during the entire foraging period be

$$\mathrm{E}[n_G] = \lambda G^{\alpha}. \tag{4.18}$$

The constant λ reflects clump density, and α ($\alpha > 0$) relates group size to the expected number of clumps per individual (Ekman and Rosander 1987). A solitary always expects to find λ clumps. If $\alpha > 1$, each group member expects to find more clumps as group size increases. If $\alpha = 1$, the expected number of clumps per individual does not depend on group size. If $\alpha < 1$, group members incur a flocking cost; an individual expects to find fewer clumps as group size increases. In section 2.2 we mentioned several mechanisms that could render $\alpha < 1$, so that $(\mathrm{E}[n_G]/G)$ would decline with G. As Ekman and Rosander (1987) point out, $\alpha < 1$ may be the most interesting case, since this condition implies that both the mean and variance of an individual's energy intake decrease as group size increases.

We take clump sizes c_i ($i = 1, 2, \ldots, n_G$) as mutually independent and assume the number of clumps discovered does not depend on their energy content. Then the mean and variance of an individual's energy intake in a group of size G are, respectively,

$$\mu_G = \mathrm{E}[Y_G] = \mathrm{E}\left[\sum_{i=1}^{n_G} c_i/G\right] = G^{-1}\,\mathrm{E}[n_G]\,\mathrm{E}[c] \tag{4.19}$$

$$\begin{aligned}\sigma_G^2 = \mathrm{V}[Y_G] &= \mathrm{V}\left[\sum_{i=1}^{n_G} c_i/G\right] = G^{-2}\,\mathrm{V}\left[\sum_{i=1}^{n_G} c_i\right] \\ &= G^{-2}(\mathrm{E}[n_G]\mathrm{V}[c] + \mathrm{V}[n_G](\mathrm{E}[c])^2).\end{aligned} \tag{4.20}$$

Note that the variance in total energy intake increases with both the variance in clump size and the variance in the number of clumps discovered. Letting σ^2 represent $\mathrm{V}[c]$ and recalling that $\mathrm{E}[c]$ has been scaled to unity, $z_G(R)$ is easily expressed in terms of μ_G and ${\sigma_G}^2$:

$$z_G(R) = \frac{R - G^{-1}\,\mathrm{E}[n_G]}{G^{-1}\,(\mathrm{E}[n_G]\,\sigma^2 + \mathrm{V}[n_G])^{1/2}}$$
$$= B\left(\frac{G(R/\lambda) - G^{\alpha}}{G^{\alpha/2}}\right), \tag{4.21}$$

where $B = (\lambda/[\sigma^2 + 1])^{1/2}$, a quantity independent of group size. The dimensionless ratio (R/λ) scales the individual's metabolic requirement to a solitary's expected energy intake.

If $\alpha = 1$, the mean μ_G is a constant, but the energy-intake variance σ_G^2 declines as group size increases. The value of $z_G(R)$ for $\alpha = 1$ is

$$z_G(R) = BG^{1/2}\,[(R/\lambda) - 1]. \tag{4.22}$$

Survival, $1 - \Phi[z_G(R)]$, is not a peaked function of group size. Hence, if $\alpha = 1$, the model suggests a simple energy-budget rule. When $\lambda > R$, the individual expects its energy intake to exceed its physiological requirement: a positive energy budget (Caraco et al. 1980b; Houston and McNamara 1982; Pulliam and Millikan 1982; Stephens and Charnov 1982). A positive energy budget implies that $z_G(R) < 0$, and the individual's probability of starvation decreases as group size increases. However, when $\lambda < R$, the individual anticipates a negative energy budget. Now $z_G(R) > 0$, and the forager's probability of starvation increases as group size increases (see section 2.2). So, if a forager's rate of discovering shared clumps is independent of group size, a positive energy budget predicts risk aversion and group foraging, since reduced intake variance implies a reduced chance of starvation. A negative expected energy budget predicts risk proneness and solitary foraging; greater intake variance implies reduced chance of starvation, and solitaries experience the greatest variance in energy intake (Caraco 1981, 1987; Clark and Mangel 1986; Lovegrove and Wissel 1988; Mangel 1990).

Introducing a flocking cost, so that $\alpha < 1$, makes group foraging less attractive. Individuals now require a greater food density to switch from solitary to group foraging than is the case for $\alpha = 1$ (Ekman and Rosander 1987). Specifically, the reduction in variance from group foraging becomes advantageous as soon as λ exceeds R when $\alpha = 1$. But for $0 < \alpha < 1$, paired foragers each have a lower chance of starving than a solitary only when $\lambda > R2^{1-\alpha}$.

Following Ekman and Rosander (1987), we can differentiate $z_G(R)$ with respect to group size:

$$\partial z_G(R)/\partial G = BG^{-\alpha/2}\,[(R/\lambda)(1 - \alpha/2) - (\alpha/2)G^{\alpha-1}]. \tag{4.23}$$

If $\alpha \geq 2$, $\partial z_G(R)/\partial G$ is always negative, so that the chance of starvation always declines with an increase in G. Similarly, if $\lambda > R$ and $\alpha \geq 1$, individuals in larger groups always have lower starvation probabilities. The last, and most biologically interesting case, occurs when $\lambda > R$ and $\alpha < 1$. Increased food-clump density (greater λ) tends to favor larger groups, but an increased flocking cost (reduced α) favors smaller groups. The result of this interaction is that survival, $1 - \Phi[z_G(R)]$, is a peaked function of G, and the minimal starvation probability occurs at the equilibrium group size for group-controlled entry:

$$G^* = [(2/\alpha)(R/\lambda)(1 - \alpha/2)]^{1/(\alpha - 1)}. \tag{4.24}$$

Ekman and Rosander (1987) examine how G^* might predict patterns in group size. The resulting predictions qualitatively match similar models for group foraging in stochastic environments (Caraco 1981, 1987; Clark and Mangel 1984, 1986; Ekman and Hake 1988; Ranta et al. 1993) and have been listed in Summary Box 4.3. Groups that exploit shared food clumps will be larger when:

1. An individual's required energy intake (R) is relatively low.
2. Food density (λ) is relatively high.
3. The cost of flocking is relatively low ($\alpha \rightarrow 1$).

More generally, both the mean and the variance of an individual's food intake should influence social foraging in stochastic environments.

We can extend the model by Ekman and Rosander (1987) and consider the Nash equilibrium group size under free entry ($\hat{G}$). To compare the optimal and Nash equilibrium group sizes, we assume clump density exceeds the individual requirement ($\lambda > R$) and $\alpha < 1$, so that survival is a peaked function of group size. Equating survival probabilities of a solitary and a member of a group of $\hat{G}$ foragers, we find that $\hat{G}$ is the value of G ($G > 1$) where

$$G^{\alpha/2}\,[G^{1-\alpha}\,(R/\lambda) - 1] = (R/\lambda) - 1. \tag{4.25}$$

Using this expression we can find $\hat{G}$ numerically. Figure 4.4 shows both G^* and $\hat{G}$ as a function of α, for given (R/λ). The free-entry equilibrium $\hat{G}$ can exceed G^* considerably. But, like G^*, $\hat{G}$ increases as either food density or α increases, and as the physiological requirement decreases.

Randomizing Foraging Time

We next consider the dual to the previous model. The random variable of interest is now the total time spent searching until the individual consumes a required amount of food. A group of G foragers ($G = 1, 2, \ldots$) discovers food clumps at the combined probabilistic rate λG^{α}($\alpha > 0$), and each group member obtains the same portion ($1/G$) of every clump discovered. To emphasize searching efficiency, we again neglect handling times.

To simplify the problem, let each clump contain the same amount of food;

SUMMARY BOX 4.3 THE RISK-SENSITIVE GROUP MEMBERSHIP GAME IN AN AGGREGATION ECONOMY

ASSUMPTIONS

Currency: Probability of starvation

Decision: Forage alone or join group of G foragers increasing its size to $G + 1$

Constraints:

1. Food requirement is fixed (R).
2. Food encountered as clumps that require no handling time.
3. Clump richness varies randomly.
4. The number of clumps encountered varies randomly and does not depend on clump richness.
5. Survival determined only at the end of a foraging period (not during).
6. Individuals have perfect information concerning the value of alternatives.
7. Individuals are free to join groups.

PREDICTIONS

When the group imposes a search cost ($\alpha < 1$)

When the number of clumps encountered is expected to exceed the requirement ($\lambda > R$),

1. We expect group foraging.
2. The fitness function is peaked, and therefore under free entry we expect groups of size $\hat{G} > G^*$ and groups of size G^* under group-controlled entry.

When the number of clumps is not expected to reach requirement ($\lambda < R$) we expect solitary foraging

When the group imposes no search cost ($\alpha = 1$)

The fitness function is never peaked, so simple energy budget rule applies.

1. When $\lambda > R$, we expect group foraging.
2. When $\lambda < R$, we expect solitary foraging.

When the group creates a search advantage ($\alpha > 1$)

Group foraging more likely and group foraging always favored when $\alpha > 2/[1 + (\lambda G^{\alpha - 1}/R)]$.

in terms of the previous model $c_i = 1$ for all i. A total of τ time units are available to search for food; τ does not depend on group size. To avoid starvation or a severe physiological penalty, an individual must consume the equivalent of R food clumps by time τ. So, a group of G foragers must locate GR clumps in the time available.

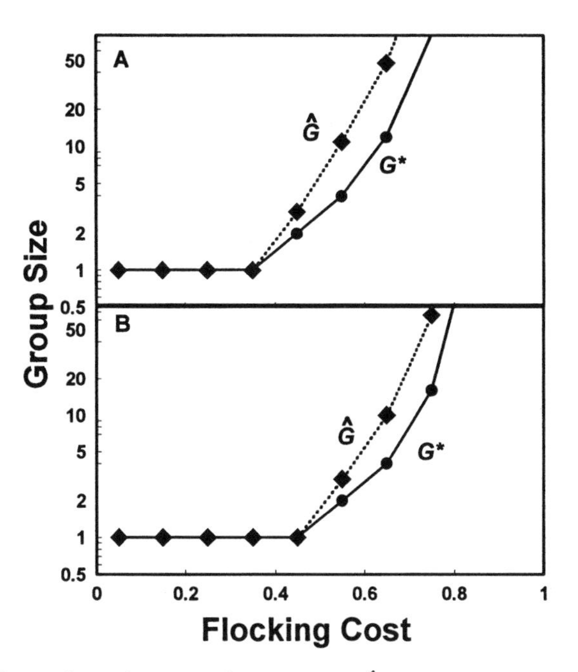

Figure 4.4 Comparison of optimal (G^*) and stable ($\hat{G}$) group sizes as a function of the flocking effect on searching efficiency. The metabolic requirement relative to a solitary's expected intake (R/λ) is 0.2 and 0.3 in (A) and (B), respectively.

Let t_G represent the total time a group of G takes to find GR clumps. t_G is a continuous random variable with probability density $f(t_G)$. Math Box 4.3 describes $f(t_G)$; the mean and variance are, respectively,

$$\mathrm{E}[t_G] = GR/\lambda G^{\alpha} = (R/\lambda)G^{1-\alpha} \tag{4.26}$$

$$\mathrm{V}[t_G] = GR/(\lambda G^{\alpha})^2 = (R/\lambda^2)G^{1-2\alpha}. \tag{4.27}$$

The probability of starvation is $\Pr[t_G > \tau]$. We analyze this probability in Math Box 4.3 and show how to find G^* and $\hat{G}$. The results lead to the same predictions resulting from the version of the model with fixed foraging time and random food consumption. Increases in either (λ/R) or α promote group foraging in a stochastic environment.

Randomizing Both Intake and Foraging Time

To complete the model, the third version examines how group size affects the probability of survival when both food intake and time spent searching vary randomly. A simple, but plausible, way to randomize both intake and foraging time is to constrain the maximal length of time that a group will search without finding a clump. That is, the foraging process continues as long as clumps are discovered quickly enough. But as soon as one inter-clump waiting time (or the first waiting time) extends to an upper limit, foragers "give up" searching for food. We call this limit T. Our analysis of the problem follows Bhat (1972).

To keep the formulas simple, we set the flocking cost $\alpha = 1$, so that a group of G foragers encounters food clumps at constant probabilistic rate λG. Each clump is divided equally among the group members.

Suppose the group discovers n clumps during a given foraging bout ($n \geq 0$). The time spent searching for the ith clump is designated $t(i)$; $1 \leq i \leq n$. $t(i)$ is conditioned by the requirement that it not exceed T. Our assumptions imply that each $t(i)$ is an upper-truncated exponential variate. Then, after the last (i.e., nth) clump has been found, the group searches for T additional time units, following which the foraging bout ends.

The total searching time leading to clump discovery is $t_g = t(1) + t(2) + \ldots + t(n)$. We let t_f represent total foraging time so that $t_f = t_g + T$. If the first clump is not found in the interval $(0, T)$, then $n = 0$ and $t_f = T$.

The discrete random variable n_G is again the number of clumps the group discovers, and $Y_G = n_G/G$ represents each group member's energy intake. In Math Box 4.4 we analyze the mean and variance of both the individual's energy consumption and the duration of the foraging period. The individual's expected energy intake is

$$\mathrm{E}[Y_G] = G^{-1}(\mathrm{e}^{\lambda TG} - 1). \tag{4.28}$$

The variance of the individual's energy intake is

$$\mathrm{V}[Y_G] = G^{-2}\, \mathrm{e}^{\lambda TG}\, (\mathrm{e}^{\lambda TG} - 1). \tag{4.29}$$

From (4.28), the expected number of clumps encountered by the group must be $E[n_G] = e^{\lambda TG} - 1$. Using this result, the total expected search time is

$$\mathrm{E}[t_g] = \{(\mathrm{e}^{\lambda TG} - 1)/\lambda G\} - T. \tag{4.30}$$

So, the expected total foraging time, $\mathrm{E}[t_g] + T$, is $(\mathrm{e}^{\lambda TG} - 1)/\lambda G$. The variance of the duration of the foraging period is

$$\mathrm{V}[t_g] = (\mathrm{e}^{2\lambda TG} - 2\lambda TG\, (\mathrm{e}^{\lambda TG}) - 1)/(\lambda G)^2. \tag{4.31}$$

The model's third version yields a mean and variance for both energy intake and foraging time. Each quantity depends on group size, so that group foraging and risk sensitivity can be linked economically. For example, if we introduce a target energy intake level R, an energy-budget rule results. That is, the probability that energetic intake of a group member falls at or below its energetic requirement declines with increasing group size if the expected number of clumps encountered by the group ($\mathrm{E}[n_G]$) is greater than the total requirement for the group (GR). However, the failure probability increases with increasing group size if the expected number of clumps is less than the group's total requirement.

4.6 Concluding Remarks

Summary of Conclusions

This chapter's models identify an important distinction between free entry and group-controlled entry, and consider how these rules for group formation interact with genetic relatedness. Two predictions of the models strike us as amenable to empirical investigation:

1. The opposing effects of increasing genetic relatedness when the group versus solitaries control group entry.
2. The pivotal importance of the effect of group size on individual patch encounter rates when predicting the group size of risk-sensitive foragers (Ekman and Hake 1988).

The first model in this chapter predicts that under group-controlled entry, optimally sized groups are predicted for unrelated players. Increasing genetic relatedness predicts an increase in group size. In contrast, under free entry by solitaries, unrelated players should form larger equilibrium groups, and predicted group size decreases with increasing genetic relatedness (see Summary Box 4.2). Our models suggest that under some circumstances kin recognition may imply assorting with kin, while other circumstances, predict avoidance of kin (see also Grafen 1984; Giraldeau and Caraco 1993).

Some situations, such as the formation of fish schools or avian roosts, provide group members little control over group membership. In those cases, an aggregation economy predicts that a large population should assort into equilibrium groups. However, if the population is composed of genetically related individuals, the predicted group size should be smaller or, alternatively, individuals should assort so as to minimize genetic relatedness within groups.

The risk-sensitive model for group membership emphasizes the significance of patch encounter rates for predicting group size. Peaked fitness functions in this model correspond to the case where group membership imposes a cost on individual searching efficiency (i.e., $\alpha < 1$). When this happens, groups should be no smaller than the optimum size under conditions promoting risk aversion. Risk-prone individuals should forage alone. If individual searching efficiency is either unaffected by or enhanced by group membership ($\alpha \geq 1$), the simple energy-budget rule applies. Large groups are predicted when conditions promote risk aversion, and solitary foraging is predicted under risk-prone conditions.

Implications for Scaling Up

Here we make a simple point about effects of consumer group size on resources. For social predators, changes in the type of prey available can induce change in hunting group size (Packer and Ruttan 1988; Gese et al. 1988; Giraldeau 1988). For example, seasonal variation in the prey available to pride lions modulates the average size of foraging groups (Schaller 1972). Extending this observation, suppose two different prey species occur simultaneously in a social predator's habitat. The prey differ in size, and so the "fitness" accruing to each of G consumers will depend on which prey they exploit.

Let $\Omega_S(G)$ represents the fitness of each of G predators exploiting small prey, and let $\Omega_L(G)$ do the same for consumers of large prey. Each $\Omega(G)$ is peaked. Large prey are invulnerable to solitary predators; $\Omega_S(1) > \Omega_L(1) = 0$. Reasonably, the group size maximizing the fitness returns for small prey will be less than the "optimal" group size for exploiting large prey; $G^*_S < G^*_L$.

Assume free entry among nonrelatives, so that the (neutrally stable) Nash equilibrium group size $\hat{G}$ is defined by equating its members' fitness to the fitness of a solitary taking small prey, $\Omega_S(1)$. Which prey is exploited when group size equilibrates? Figure 4.5 shows that the answers depends on the

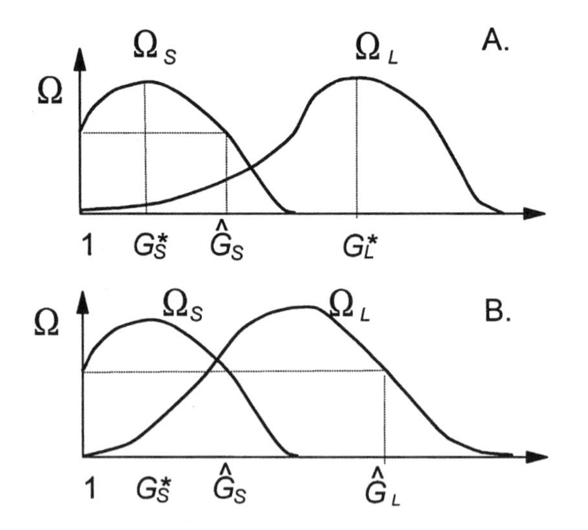

Figure 4.5 Aggregation economies where groups specialize on either small or large prey. In (A) the equilibrium group size is $\hat{G}_S$, and only small prey are exploited. In (B) the equilibrium group size is $\hat{G}_L$, and only large prey are exploited. The fitness of each group member is the same in (A) and (B), but group size differs.

difference ($G^*_L - G^*_S$). Solitaries cannot survive if only large prey are available. But solitaries can coalesce and exploit small prey advantageously until group size equilibrates. If ($G^*_L - G^*_S$) is sufficiently large, the equilibrium is $\hat{G}_S$, and only small prey are exploited. But if ($G^*_L - G^*_S$) is not too large, exploiting small prey provides a bridge to the aggregation economy summarized by $\Omega_L(G)$. At the Nash equilibrium group size $\hat{G}_L$, only large prey are exploited. In each case, group members achieve essentially the same fitness. But in the former case, a population of social predators exploits only small prey, while in the latter case social predators exploit only large prey. Any relatedness may reduce the equilibrium group sizes a bit, but the impact of the predator population on the two-resource community depends on the details of the consumers' aggregation economy.

If we assume group-controlled entry, a different prediction emerges. Each member of an optimal group for exploiting small prey incurs a fitness reduction if an unrelated solitary is admitted to the group. Hence the optimal group for exploiting small prey G^*_S will be stable under this entry rule, and consumer groups will specialize on small prey. With sufficient relatedness, however, each group member's inclusive fitness could be increased by admitting an intruder. If ($G^*_L - G^*_S$) is not too large, kinship could increase group size to the point where switching to large prey would be favored. The ultimate group size would be no less than G^*_L (Giraldeau and Caraco 1993). The ecological consequence of increased kinship, under group-controlled entry, would be a switch from small to large prey.

Final Comment

We hope this chapter clarifies particular issues that have remained cloudy, and occasionally a bit contentious, for some time. First, the "paradox of group foraging" is not real. That is, members of groups in aggregation economies are likely to gain a direct advantage over solitary foraging. Group control of entry and genetic relatedness among members of groups formed under free entry combine to predict that groups will not continue to grow in size until each group member does no better than a solitary forager.

Second, we have asked whether increased genetic relatedness predicts an increase or decrease in group size. We claim that the answer depends on the rules of entry governing group formation. Under group control, increased relatedness predicts larger groups, as Rodman (1981) anticipated. However, under free entry, increased genetic relatedness predicts a decrease in equilibrium group size, as Giraldeau (1988) argued.

This chapter focuses strictly on general properties of aggregation economies. However, in many circumstances groups commonly form even though solitaries experience the greatest foraging success. For instance, parasitic jaegers (*Stercorarius parasiticus*), during migratory stopovers on the north shore of the St. Lawrence River, Québec, often form small groups when they aggressively kleptoparasitize common terns (*Sterna hirundo*). Bélisle (1996) asked whether the jaegers derived hunting advantages from forming groups, effectively whether they foraged under an aggregation economy. He found instead that the most likely explanation for jaeger group attacks was the paucity of target terns. When terns were rare, jaegers more often attacked in groups; when terns were common, they were more likely to be attacking solitarily. This pattern suggests a dispersion economy, the topic of the next chapter.

Math Box 4.1 Genetic Relatedness, Free Entry, and Equilibrium Group Size

To find the equilibrium group size under free entry, recall that $\Omega(G) > \Omega(1)$ for $G^* < G < \hat{G}$. However, $\Delta\Omega(G - 1) < 0$ for any such G, so that

$$\partial\Psi(r, G^* < G < \hat{G})/\partial r < 0. \qquad (4.1.1)$$

$\Psi(r, G)$ initially exceeds $\Omega(1)$, but declines as relatedness increases. For $G = (\hat{G} - 1)$, and possibly for all G on$[G^* + 1, \hat{G} - 1]$, $\Psi(1, G) < \Omega(1)$. Then there must be a Nash equilibrium group size. For any G on this interval, there is a critical level of relatedness r_G:

$$r_G = \frac{\Omega(G) - \Omega(1)}{(G - 1)[\Omega(G - 1) - \Omega(G)]}. \qquad (4.1.2)$$

If $r < r_G$, $\Psi(\text{r}, G^* < G < \hat{G}) > \Omega(1)$. Therefore, a solitary increments its inclusive fitness by joining and so increasing the group size to G. Equivalently, the equilibrium group size must have at least G members. If $r > r_G$, $\Psi(r, G^* < G < \hat{G}) < \Omega(1)$. In this case, a solitary should not join; G exceeds the equilibrium group size. Clearly, greater relatedness may imply a reduced equilibrium group size under free entry when $G^* < G < \hat{G}$. Indeed, if $r > r_G$ at $G = (G^* + 1)$, the equilibrium will be as small as the optimal group size G^* (Giraldeau and Caraco 1993; Higashi and Yamamura 1993).

Suppose $r > 0$, and the equilibrium group size G exceeds G^*. Since $G^* < G < \hat{G}$, $\Delta\Omega(G) < 0$ for each group size. Since G is the Nash equilibrium,

$$\Psi(0 < r \leq 1, G) > \Omega(1) > \Psi(0 < r \leq 1, G + 1). \qquad (4.1.3)$$

Substituting from above, these inequalities indicate that the equilibrium group size is the maximal G, where

$$\frac{\Omega(G + 1) - \Omega(1)}{r[\Omega(G) - \Omega(G + 1)]} < G < 1 + \frac{\Omega(G) - \Omega(1)}{r[\Omega(G - 1) - \Omega(G)]}. \qquad (4.1.4)$$

The general point of this analysis is that increasing relatedness will ordinarily decrease, and will never increase, equilibrium group size under free entry.

Math Box 4.2 Interaction of Free Entry and Group Control

Suppose a group's size has reached the equilibrium size under group control; the group would repel any further joiners in the absence of conflict costs. Further, suppose a solitary would increase its inclusive fitness if the group allowed it to join. The outcome of the conflict should depend on costs of an aggressive encounter. Should aggressive conflict occur, the solitary inflicts a cost of $\theta\beta G$ inclusive-fitness units on the group, and the group inflicts a cost of βG units on the solitary. θ ($\theta > 0$) scales the relative aggressiveness of the solitary versus the group.

The group should avoid the conflict and admit the solitary if the cost of conflict (direct and indirect) is great enough to offset the inclusive fitness saved by keeping the group's size at G. That is, the group of G should be prepared to admit a solitary if

$$\theta\beta G + r\beta G \geq G[\Omega(G) - \Omega(G + 1)] + r[\Omega(1) - \Omega(G + 1)]. \tag{4.2.1}$$

Given this relationship, the group will increase to $(G + 1)$ members if the solitary would gain sufficient inclusive fitness to more than offset the cost of conflict:

$$\beta G + r\theta\beta G < [\Omega(G + 1) - \Omega(1)] + rG[\Omega(G + 1) - \Omega(G)]. \tag{4.2.2}$$

Beginning at the group-controlled Nash equilibrium, the size of the group now increases until a new equilibrium, determined by the costs of conflict, is attained. Simplifying a bit, group size increases from G to $(G + 1)$ as long as these two conditions hold:

$$\beta(\theta + r) + \Delta\Omega(G) \geq (r/G)[\Omega(1) - \Omega(G + 1)]. \tag{4.2.3}$$

and

$$\beta(1 + r\theta) - r\,\Delta(G) < (1/G)[\Omega(G + 1) - \Omega(1)], \tag{4.2.4}$$

where $\Delta\Omega(G) < 0$, and $[\Omega(1) - \Omega(G + 1)] \leq 0$. Otherwise, G is an equilibrium.

Math Box 4.3 Random Duration of Time Spent Searching

From the text, a group of G discovers food clumps as a Poisson process at combined probabilistic rate λG^{α} $(\alpha > 0)$; food clumps are identical. τ time units are available to search for food; τ does not depend on G. An individual must consume the equivalent of R food clumps by time τ, so a group of G must locate GR clumps in the time available.

t_G is the total time a group of G takes to find GR clumps. t_G is a continuous random variable with probability density $f(t_G)$; $t_G > 0$. The waiting time until the first clump is found has an exponential density with expectation $1/(\lambda G^{\alpha})$. Since clumps are discovered as independent events, the total waiting time t_G has a gamma density (e.g., Boswell et al. 1979):

$$f(t_G) = (\lambda G^{\alpha})^{GR}\, (t_G)^{GR-1} \exp[-\lambda t_G G^{\alpha}]/(GR - 1)! \quad (4.3.1)$$

$E[t_G] = GR/\lambda G^{\alpha} = (R/\lambda)G^{1-\alpha}$, and $V[t_G] = GR/(\lambda G^{\alpha})^2 = (R/\lambda^2)$ $G^{1-\alpha}$. The probability of starvation is the chance that the total waiting time t_G exceeds τ; $\Pr[t_G > \tau]$ is

$$\Pr[t_G > \tau] = \int_{\tau}^{\infty} f(t_G)\, dt_G. \quad (4.3.2)$$

Let $L = \lambda \tau G^{\alpha}$, the expected number of clumps the group will find; then

$$\Pr[t_G > \tau] = \Gamma(GR, L)/(GR - 1)!, \quad (4.3.3)$$

where

$$\Gamma(GR, L) = \int_{L}^{\infty} e^{-t}\, t^{GR-1}\, dt. \quad (4.3.4)$$

The effect of group size on the expected waiting time depends on α:

$$\partial E[t_G]/\partial G = (1 - \alpha)(R/\lambda)G^{-\alpha}. \quad (4.3.5)$$

Then $E[t_G]$ declines as G increases when $\alpha > 1$. If $\alpha < 1$, expected search time increases as G increases (the cost of flocking). α also governs the effect of group size on the waiting-time variance:

Math Box 4.3 (*cont.*)

$$\partial V[t_G]/\partial G = (1 - 2\alpha)(R/\lambda^2)G^{-2\alpha}. \quad (4.3.6)$$

If α 1/2, the waiting-time variance declines as G increases, and if $\alpha < 1/2$, $V[t_G]$ increases as G increases.

In the simplest case $\alpha = 1$. Then $E[t_G]$ does not depend on group size, and the associated variance strictly decreases as a function of group size. Risk-aversion always promotes group foraging, and risk-proneness always promotes solitary foraging, for $\alpha = 1$. Whenever $1/2 < \alpha < 1$, the mean searching time increases with group size, but the waiting-time variance decreases. Whether the effects combine to favor or disfavor group foraging depends on the change in $\Pr[t_G > \tau]$.

Starvation probabilities $\Pr[t_G > \tau]$ can be calculated numerically. Under either free entry or group entry, individuals begin forming groups when each member of a pair has a lower chance of starving than a solitary has. For simplicity, suppose $R = 1$. Then a solitary's probability of starving reduces to

$$\Pr[t_1 > \tau] = \Gamma(1, \lambda\tau) = e^{-\lambda\tau}. \quad (4.3.7)$$

For each member of a pair we have

$$\Pr[t_2 > \tau] = \Gamma(2, \lambda\tau 2^\alpha) = (1 + \lambda\tau 2^\alpha) \exp[-\lambda\tau 2^\alpha]. \quad (4.3.8)$$

A member of a pair should do better than a solitary as food density (λ), available foraging time (τ), or individual searching efficiency (α) increases.

For $1/2 < \alpha < 1$, $E[t_G]$ increases with G, but $V[t_G]$ decreases with group size. If $(\lambda\tau/R)$ is sufficiently large, then survival $(1 - \Pr[t_G > \tau])$ is a peaked function of group size. $\Pr[t_G > \tau]$ reaches a minimum at $G^* > 1$, and the equilibrium under group-controlled entry is defined by

$$\min_G \Gamma(GR, \lambda\tau G^\alpha)/(GR - 1)!. \quad (4.3.9)$$

The equilibrium group size under free entry, $\hat{G}$, is defined by

$$\Gamma(R, \lambda\tau)/(R - 1)! = \Gamma(\hat{G}R, \lambda\tau(\hat{G})^\alpha)/(\hat{G}R - 1)!, \quad (4.3.10)$$

where the left-hand side is a solitary's probability of starving.

Math Box 4.4 Random Variation in Energy Intake and Foraging Time

We assume the foraging process continues as long as clumps are discovered quickly enough. While searching, a group of G foragers encounters food clumps at constant probabilistic rate λG. Each clump is divided equally among the group members.

The group discovers n clumps during a foraging bout ($n \geq 0$). $t(i)$ is the time spent searching for the ith clump; $1 \leq i \leq n$. $0 < t(i) \leq T$. Since the probabilistic encounter rate is constant, each $t(i)$ is an upper-truncated exponential variate (Bhat 1972). After the nth (the last) clump has been found, the group searches for T additional time units and the bout ends.

The total searching time leading to clump discovery is t_g; $t_g = t(1) + t(2) + \ldots + t(n)$. t_f represents total foraging time; $t_f = t_g + T$. If the first clump is not found in $(0, T)$, then $n = 0$ and $t_f = T$.

The discrete random variable n_G is the number of clumps the group discovers, and $Y_G = n_G/G$ is the individual group member's energy intake. If no clumps are located, $n_G = 0$ and

$$\Pr[n_G = 0] = \Pr[t(1) > T] = e^{-\lambda TG} \tag{4.4.1}$$

Therefore, $\Pr[t(1) \leq T] = 1 - e^{-\lambda TG}$. Since each clump is discovered independently,

$$\Pr[n_G = n] = (1 - e^{-\lambda TG})^n \, e^{-\lambda TG};\; n = 0, 1, \ldots . \tag{4.4.2}$$

n_G has a geometric distribution with mean $E[n_G] = e^{\lambda TG} - 1$, and variance $V[n_G] = (e^{\lambda TG})\, E[n_G]$. Then the mean and variance of an individual's energy intake Y_G are, respectively,

$$E[Y_G] = G^{-1}(e^{\lambda TG} - 1);\; V[Y_G] = G^{-2}\, e^{\lambda TG}\, (e^{\lambda TG} - 1). \tag{4.4.3}$$

Recall that $t_f = t_g + T$. For t_g 0,

$$t_g = \sum_{i=1}^{n_G} t(i);\; t(i) < T \tag{4.4.4}$$

the $t(i)$ vary independently; each has an exponential probability density truncated at T:

$$f[t(i) \mid t(i) \leq T] = \lambda G e^{-\lambda G t(i)} \mid (1 - e^{-\lambda TG}). \tag{4.4.5}$$

Integration by parts gives the expected waiting time for $t(i)$:

$$E[t(i) \mid t(i) < T] = \frac{1}{G\lambda} - \frac{Te^{-\lambda TG}}{1 - e^{-\lambda TG}}. \tag{4.4.6}$$

Math Box 4.4 (*cont.*)

Since $\mathrm{E}[n_G] = \mathrm{e}^{\lambda TG} - 1$, total expected search time is

$$\mathrm{E}[t_g] = \{(\mathrm{e}^{\lambda TG} - 1)/\lambda G\} - T. \tag{4.4.7}$$

The expected foraging time is then $(\mathrm{e}^{\lambda TG} - 1)/\lambda G$. Since T is a constant, $\mathrm{V}[t_f] = \mathrm{V}[t_g]$ which, after a lengthier analysis, is

$$\mathrm{V}[t_g] = (\mathrm{e}^{2\lambda TG} - 2\lambda TG(\mathrm{e}^{\lambda TG}) - 1)/(\lambda G)^2. \tag{4.4.8}$$

The mean and variance for both energy intake and foraging time depend on group size, so that group foraging and risk sensitivity can be linked. If we assume a target energy intake level R, $\Pr[Y_G \leq R]$ declines with group size if $\mathrm{E}[n_G] > GR$, and increases with group size if $\mathrm{E}[n_G] < GR$.

5 Predicting Group Size in Dispersion Economies

5.1 Introduction

Many foraging groups apparently lack the advantages we associate with an aggregation economy. Consider, for instance, group foraging in parasitic jaegers (*Stercorarius parasiticus*). At their migratory stopover on the north shore of the St. Lawrence River, Québec, they prey on a number of invertebrates, ducks, and shorebirds (Bélisle and Giroux 1995). They also kleptoparasitize a number of seabird species, notably common terns (*Sterna hirundo*), transporting fish they captured to their nesting colony (Bélisle and Giroux 1995; Bélisle 1998). A jaeger kleptoparasite patrols an area through which terns fly. At some point a tern that is carrying a fish either in its bill or crop is singled out and a chase involving rapid, often aerobatic flight starts. The chase ordinarily ends when the tern drops the fish and the jaeger pursues it. The individual that initiates the chase, however, is frequently joined by other kleptoparasitic jaegers that were patrolling in the vicinity. The kleptoparasites then line up behind the tern, switching positions as the chase repeatedly changes direction. But only one of the jaegers in the line will catch the fish released by the tern.

Bélisle (1998) asked whether the jaegers derived any benefits by chasing terns in groups. He tested a number of potential alternative sources of benefits for group attacks including per capita probability of obtaining a fish, gross and net rates of energy intake, and an energetic efficiency estimate. But none of these four likely currencies of fitness yielded peaked fitness functions as we defined them in chapter 4, and so we conclude that the jaegers were not foraging under an aggregation economy (fig. 5.1). In all cases, the maximum benefit was obtained by solitary foraging. We term this a *dispersion economy* because fitness, or a foraging currency, strictly declines as group size increases. In the example, if enough feeding patches (i.e., fish-carrying terns) are available, consumers (i.e, jaegers) would maximize their foraging gains by dispersing one per patch. Yet jaegers nonetheless regularly attack terns in groups of up to nine (Bélisle 1998). Why? The answer is because there are likely not enough fish-carrying terns to allow the maximum dispersion. Bélisle (1998) notes, to support this claim, that jaegers indeed disperse into smaller groups, tending toward one, as the relative number of terns per jaeger increases (fig. 5.1). So, in a dispersion economy the

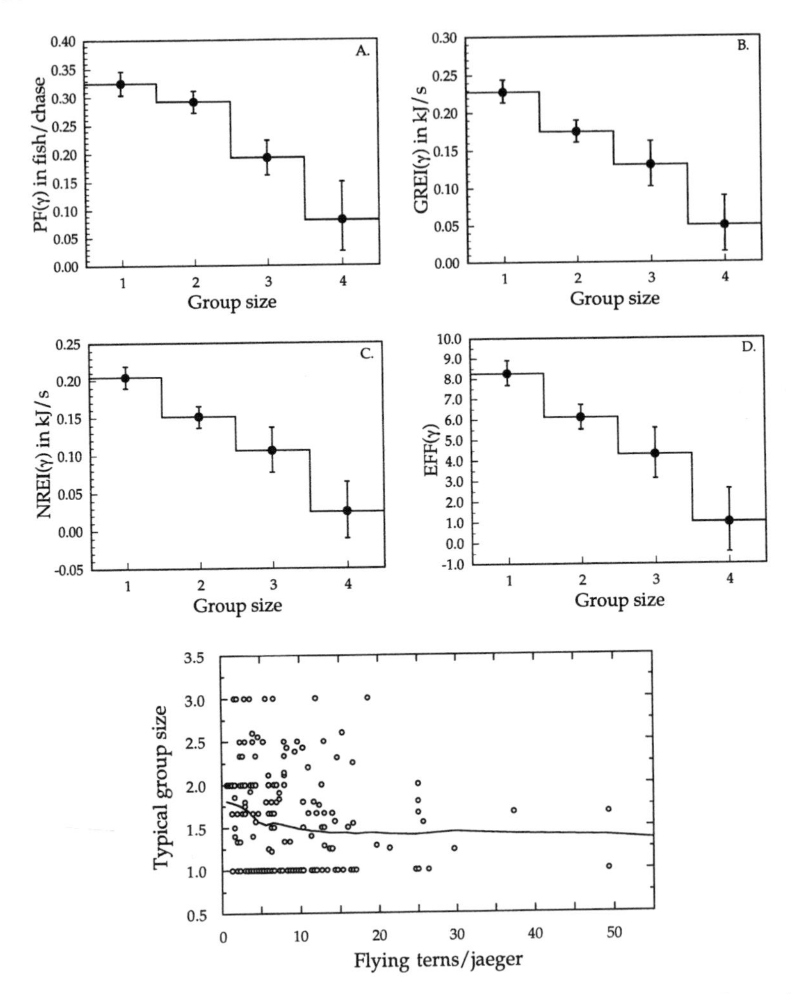

Figure 5.1 *Top panel:* Four different measures of kleptoparasitic success of parasitic jaegers (*S. parasiticus*) attacking common terns (*S. hirundo*) at a migratory stopover as a function of the number of jaegers involved in the chase (Group Size). Each point is a mean, and its 95% confidence interval is based on two-thousand bootstrap replicates. (A) probability of obtaining a fish; (B) gross rate of energy intake; (C) net rate of energy intake; and (D) energetic efficiency of a chase. None of the functions are *peaked.* Maximum benefits always occur at a group size of one: a dispersion economy. *Bottom panel:* Variation in parasitic jaeger typical group size as a function of the ratio of flying terns per jaeger in the area. Each point gives the number of jaegers in separate 15 min observation periods. The line is fitted using a locally weighted regression procedure. Note a decrease in jaeger group size as relatively more terns become available, an indication of a dispersion economy. Both panels taken from Bélisle (1998). (With permission of Ecological Society of America)

distribution of the resource is partly responsible for the aggregation. However, as was true of aggregation economies, the influence that each individual has on the other foragers is also important. The major difference is that in dispersion economies, joining a group is always a selfish behavior (see section 1.2). The individual joining both increases its benefits and decreases current group members' benefits through either exploitative or interference competition.

A body of theoretical and empirical research, originating from Fretwell and Lucas's (1970) theory of *ideal free distributions* (the IFD) has been devoted to analyzing how different forms of competition affect expected group sizes (see Tregenza 1995; Tregenza et al. 1996a for recent reviews). In this chapter we review the models of ideal free distribution theory. We take a historical approach, starting with the first published models based on continuous input of resources into habitats, and then move on to so-called interference models of animal distributions (Sutherland 1983).

5.2 Introduction to Continuous Input Models

The first habitat distribution models assume that resources arrive continuously at habitat-specific rates. This assumption would apply to animals feeding on drift food in streams, or to male dung flies waiting for the arrival of females at a fresh cow pat (Parker 1970). This type of continuous input, however, is not likely to be common in nature. Generally all resources are present in a habitat at the arrival of consumers, as, for instance, when starlings land in a lawn to forage for earthworms. It is also likely that resources in an area deplete. Yet, possibly because the first distribution models were based on the continuous input assumption, these models have generated the most interest.

To describe continuous-input assumptions more formally, assume that an environment contains Z habitats. The basic suitability of each habitat can be characterized by its total rate of resource arrival k_i ($i = 1, 2, \ldots, Z$), where each resource-arrival rate is a constant. In its simplest formulation, the consumer's expected intake rate in habitat i (W_i) increases with the resource input rate in that habitat (k_i) and varies inversely with the density of competitors G_i:

$$W_i = k_i/G_i. \tag{5.1}$$

This expression describes a much simpler relationship between total intake and consumer density than was used originally in Fretwell and Lucas's (1970) distribution model. Equation (5.1) assumes equal resource division among the G_i consumers. Pursuit and handling times are so short that a consumer's resource intake rate increases linearly with the habitat's input rate (assuming competitor number stays constant). Recall that resources requiring nonnegligible handling and/or pursuit times can render the intake rate a nonlinear function of resource input rate (e.g., McNamara and Houston 1990).

The continuous input model assumes that all resource units are the same quality, and all are exploited as they arrive, so that resources never accumulate within a habitat. The model further assumes that consumers are "ideal" in the sense that they instantaneously choose, without error, habitats that offer the greatest experienced intake. The model's consumers are "free" to enter the habitat they prefer without incurring costs due either to travel or aggression.

The IFD's central assumption is that all consumers experience the same resource-intake rate at equilibrium. That is, when the distribution of individuals across habitats equilibrates, we have

$$W_i(G_i) = C$$

for all individuals where C is a constant. Since an individual's intake varies inversely with the number of competitors, any individual changing habitat at the equilibrium will decrease its rate of resource acquisition. Then the ideal free distribution of consumers qualifies as a stable Nash equilibrium and an ESS (Parker and Sutherland 1986; Fagen 1987a; Recer et al. 1987).

From equation (5.1), equilibration of intake rates implies that

$$k_i/G_i = C \tag{5.2}$$

for all individuals. Then the Nash equilibrium property predicts that the ratio of competitor numbers in habitats i and j must be:

$$G_i/G_j = k_i/k_j, \tag{5.3}$$

This form of the IFD has been called the *habitat matching* rule (e.g., Pulliam and Caraco 1984). Its application is more general than equation (5.1); see Fagan (1987b).

Pairwise application of equation (5.3) to different habitats predicts the equilibrium proportion of consumers in each habitat:

$$G_i \,/ \sum_{i=1}^{Z} G_i = k_i \,/ \sum_{i=1}^{Z} k_i. \tag{5.4}$$

Then we can rewrite equation (5.2) as

$$G_i = k_i \left(\sum_{i=1}^{Z} G_i \,/ \sum_{i=1}^{Z} k_i \right), \tag{5.5}$$

which is the *input matching* rule (Parker and Stuart 1976; Parker and Sutherland 1986).

The input-matching and habitat-matching equivalents predict that relative consumer densities will exhibit the same among-habitat variance as do the relative suitability values (cf. Houston and McNamara 1987). Proportional habitat matching is a simple prediction based on assumptions that need not hold in nature. The general empirical success of the model is impressive, but a number of tests of the predicted consumer distribution indicate consistent deviations from proportional habitat matching; see Tregenza (1995). When consumers choose between a rich and a poor continuous-input habitat, we often find fewer competitors in the rich habitat, and more in the poor habitat, than predicted (see Houston et al. 1995). This discrepancy has led to refining the original model's assumptions. The next section describes some useful revisions to the continuous-input model of habitat choice. The revisions and their consequences are listed in Summary Box 5.1.

5.3 Changing the Assumptions of Continuous Input Models

Relaxing the Assumption of Instantaneous Uptake

The continuous input model assumes that prey are consumed immediately upon their arrival in the habitat. A number of factors can lead to violations of this assumption. For instance, any need to pursue or handle items for any substantial amount of time implies that items are not consumed immediately. Yamamura and Tsuji (1987) have, in a different context, proposed equations defining a consumer population where continuously input resources require significant handling time and migrate out of the habitat if not exploited after some time. Essentially, they argue that an equilibrium will be reached where the density of resources will reflect the balance between input and both consumption and loss from the system. Given the equilibrium resource state, the equilibrium number of consumers searching for resources and the number exploiting resources follow. McNamara and Houston (1990) consider similar assumptions and include the possibility of mortality due to predation while foraging. They calculate an IFD where each consumer's patch choice depends on its current level of energy reserve: we review these predictions in a section on complex currencies.

To date, only Lessells (1995) has considered explicitly the consequences of violations of the assumption of immediate food consumption for continuous-input ideal free distribution models. Her model is straightforward and assumes that a predator's consumption rate is a function of the standing crop of prey and the number of consumers. Competition can occur through one or two processes. *Exploitation competition* is an "indirect" interaction among

SUMMARY BOX 5.1 MAIN ASSUMPTIONS AND PREDICTIONS OF CONTINUOUS INPUT HABITAT MODELS

THE BASIC MODEL

Main assumptions: Resources consumed immediately, no loss, no accumulation in habitat. Consumers are ideal, free, and equal.

Main predictions: Habitat matching, equal payoffs across habitats and consumers.

Observation: Some overuse of poor habitat, payoffs lower in poor habitat (see Kennedy and Gray 1993 and Tregenza 1995).

PROPOSED MODIFICATIONS	CONSEQUENCE
Resources not exploited immediately	Depends on source of competition and whether prey dies without having been consumed (i.e., alternative source of mortality). In absence of alternative mortality, input matching expected whenever exploitation competition occurs, either alone or with interference competition. Adding alternative mortality disrupts input matching; exact effect depends on whether interference competition occurs, and when it does, if alternative mortality or consumption rate is more strongly density dependent (Lessells 1995).
Consumers not ideal	
Perceptual constraints:	Deviations from habitat matching increase with intensity of constraint. Payoffs greater in rich habitat but equal across consumers within each habitat (Abrahams 1986).
Not omniscient:	Habitat matching and equal payoffs (Regelmann 1984).
Consumers not free	
Travel costs:	Habitat matching and equal payoffs (Regelmann 1984).
Consumers not equal	
Constant comp. weights:	Competitive weight habitat matching. Same as habitat matching for small *G*, payoffs equal within phenotype, unequal across habitats. Deviations from habitat matching increase with

SUMMARY BOX 5.1 (*cont.*)

	group size, good competitors overrepresented in rich habitat, payoffs higher in rich habitat (Parker and Sutherland 1986).
Variable comp. weights:	Payoffs unequal across habitats and phenotypes. Truncated phenotype distributions, heaviest phenotype weights found in habitats where competitive weight most strongly affected by phenotype (Parker and Sutherland 1986).
Input rates stochastic	Habitat matching to average resource level if consumers cannot track state of stochastic habitat. If consumers can track state, then fewer consumers found in stochastic habitat than predicted by its mean input rate (Recer et al. 1987).
Currency is risk of starvation	According to Recer and Caraco (1989a) matching when resources occur in indivisible items. When each arriving resource can be divided among consumers in habitat, distribution will match standard deviations of habitats' input rates. According to McNamara and Houston (1990), risk-sensitive choice even for resources arriving as items.
Including predation hazard	When consumers are doing poorly, ignore predation and use habitats according to resources. When consumers doing well, avoid any increased level of predation hazard (McNamara and Houston 1990).

individuals caused by a reduction of the standing crop resulting from resource exploitation. *Interference competition* is a direct effect caused by the presence of competitors in the same patch (Lessells 1995). The model assumes a stable equilibrium standing crop s^*. When this standing crop increases beyond the equilibrium value, consumers feed more efficiently and consequently return the standing crop back down to equilibrium. When the standing crop is less than the equilibrium value, consumer efficiency declines, allowing the standing crop to increase back to equilibrium. The effect of standing crop on individuals' foraging efficiency establishes the equilibrium between resource arrival and consumption by consumers. Lessells's (1995) conclusions are as follows.

In the absence of any *interference* competition or any other source of mortality for the prey items, the resource input matching rule still holds. That is to say, the ratio of consumers at the patches should match the ratio of the patches' input rates. When consumers suffer from both *exploitation* and *interference* competition but, once again, there is no alternative source of prey mortality (death other than through being consumed), the input matching rule continues to hold. The main difference here is that the patches with higher input rates, and hence more consumers, have higher-equilibrium standing-crop values than patches with lower input rates, despite the fact that the per capita consumption rate remains constant across patches. When predators show only interference competition and prey suffer some other source of mortality, input matching no longer holds. Similarly, when both sources of competition operate and prey are subject to an additional source of mortality, the input matching rule does not hold.

Relaxing the Assumption of Ideal Consumers

The ideal assumption of the basic continuous input model requires that consumers move to the most profitable habitat. But the effectiveness of habitat choice may often be constrained by the quality of the information on which it is based. The ideal assumption therefore implies that all foragers have perfect information about all habitats, and that the information is updated when conditions change. The ideal assumption can be violated in two ways:

1. Consumers are limited in their ability to distinguish between habitats varying in resource input rates and competitor densities (perceptual constraint); and
2. Information about the quality of habitats must be acquired during the foraging process (consumers not omniscient).

Perceptual Constraints

Abrahams (1986) investigated the consequences of perceptual constraints on the distribution of otherwise free individuals competing in a continuous input system. He notes that as perceptually unconstrained (i.e., ideal) foragers choose habitats with the highest expected intake rate, differences among habitats in those rates decline until they are equal at the IFD. At this point, choice of habitat becomes random, since the intake experienced in each habitat is the same. However, if the consumers' ability to distinguish among intake rates in different habitats is constrained, then habitats with different intake rates will be perceived as equal. Therefore, at some point during the population's occupation of available habitats, this perceptual constraint will result in premature random choice.

Abrahams's (1986) perceptual-constraints model has important consequences for the predicted consumer distribution. Perceptually constrained individuals will reach an equilibrium distribution where more consumers occupy a poor habitat, and fewer occupy a rich habitat, than expected by the IFD. As a result, the equilibrium distribution of perceptually constrained consumers allows intake rates to remain unequal across habitats. The perceptual constraint hypothesis predicts that a decline in consumers' ability to discriminate increases the fraction of the population that will erroneously choose to assort randomly. Hence decreased perception generates a greater deviation from the proportional habitat-matching rule (Abrahams 1986; fig. 5.2). In the worst possible situation, consumers will choose uniformly among habitats that differ in total input. Moreover, for a given perceptual constraint, it is easy to show that the hypothesis predicts deviations from proportional habitat-matching that will ordinarily increase as the total density of consumers increases. This density-dependent effect follows because as the total population size increases, the similarity between the habitats' input rates will reach the critical discrimination levels when a greater fraction of the population still will not have chosen a habitat. Hence, all other factors kept constant, an

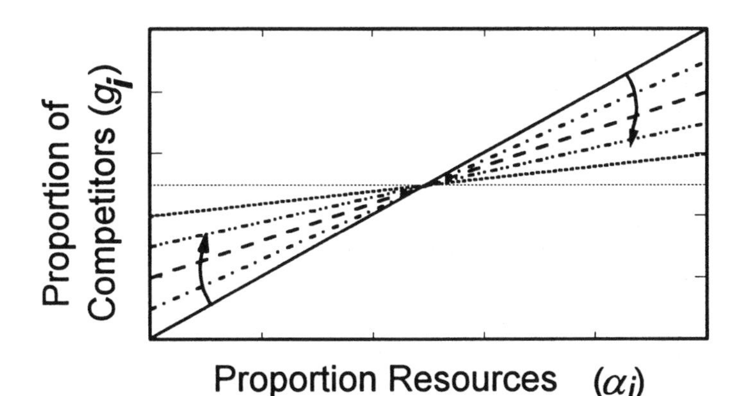

Figure 5.2 The relationship between the predicted proportion of consumers in a habitat and the proportion of resources in the same habitat for different intensities of perceptual constraints. The solid line shows habitat matching for consumers with perfect perceptual abilities. The family of dashed lines gives the predicted competitor distribution when consumers in a two-habitat environment have different levels of perceptual constraints. The arrows indicate the direction of increasing perceptual constraint. Greater constraints increase the deviation from habitat matching. When consumers are entirely incapable of distinguishing between two habitats, we expect 50% of the foragers in each habitat, independent of the proportional resource distribution.

increasing fraction of the population will assort itself randomly as population size increases.

CONSUMERS NOT OMNISCIENT

Foragers can violate the ideal assumption, despite having perfect discrimination abilities, if the information concerning average habitat quality can be obtained only after sampling. Regelmann (1984) presented a simulation model in which foragers gain accurate information on habitat value only after visiting the habitats. He assumed that consumers update their estimate of the habitats in the area following the relative pay-off sum learning rule (Harley 1981; see Houston 1987; Ollason 1987). Regelmann's (1984) simulation indicated that despite the need to sample habitats in order to gain an accurate estimate of their relative values, consumers eventually were distributed in a way that matched the habitats' profitabilities. Hence consumers reached the ideal free distribution. Violation of the omniscient assumption, therefore, appears to have little effect on the predicted consumer distribution provided animals can learn the relative profitabilities of each habitat.

Relaxing the Assumption of Free Consumers

Consumer freedom can be constrained by the presence of significant travel costs between habitats. Similarly, habitat entry may require a cost as occurs, for instance, when some consumers attempt to prevent others from occupying a habitat (the "ideal despotic distribution"; Fretwell 1972).

TRAVEL COSTS

Regelmann's (1984) simulation mentioned above also investigated whether travel costs between habitats affect the final consumer distribution. His simulations predict proportional habitat matching, which could be achieved in two ways. In one, after a period of sampling and switching between habitats, foragers settle across habitats, and each individual remains in the habitat it chooses. In the other, foragers constantly switch between habitats, but the ratio of waiting time spent by consumers in each habitat matches the habitats' resource input ratio. Regelmann's (1984) simulation suggests that even with nonnegligible travel costs, consumers will assort according to proportional habitat matching.

HABITAT ENTRY COSTS

Few studies of IFD theory have dealt specifically with habitat entry costs. Most likely, such a cost would involve aggression by residents directed toward potential recruits (e.g., group-controlled entry, chapter 3). Residents, however, may also use aggression to reduce the success of more subordinate individuals already recruited to the habitat (see Grand and Grant 1994). Distinguishing between these two forms of aggression may be difficult in practice; they likely fall on a continuum of resource defense tactics. The effect of preventing some individuals from entering habitats will result in overrepresentation of consumers in poor habitats and underrepresentation in rich habitats, assuming that the most aggressive consumers monopolize the best habitats (Fretwell 1972). Aggression that does not exclude recruits from habitats likely has different consequences on consumer distributions (see below). The economic determinants of despotic behavior are usually studied under the heading of resource defense theory (Grant and Kramer 1992; Grand and Grant 1994).

Relaxing the Assumption of Equal Competitors

Most models for unequal competitive abilities assume that consumers are free to enter a habitat. But once in it, consumers obtain resource shares that vary according to their differing competitive abilities (Milinski 1988; Milinski and Parker 1991). Factors that promote unequal allocation of food can sometimes be related to resource defense, and hence to the spatial and temporal resource predictability promoting defense (Grand and Grant 1994). However, in the models discussed here unequal allocation is not necessarily due to aggression and is often modeled as a result of any phenotypic difference that affects an individual's foraging rate.

Introducing unequal competitors makes finding a stable distribution more complicated. According to Sutherland and Parker (1985; and Parker and Sutherland 1986), the stable distribution depends on whether phenotypes vary discretely or continuously, whether the differences in the phenotypes' competitive abilities remain constant across habitat types, and whether there are two or several habitat types. The importance of this competitive asymmetry has been discussed repeatedly. Sutherland and Parker (1985), Parker and Sutherland (1986), Milinski (1988), Milinski and Parker (1991), Huntingford (1993), and Tregenza (1995) offer excellent analyses. We briefly present the main conclusions.

Constant Competitive Weights, Two Phenotypes, and Two Habitats

Assume there are only two phenotypes, *A* and *B*. Phenotype *A* has a competitive advantage over *B*; say its trophic apparatus allows it to feed faster. Under identical resource availabilities an *A* individual gains q ($q > 1$) times the resources of a *B* individual. Then the constant q quantifies the asymmetry between the phenotypes.

Next assume the area has two habitats, and that k_1 and k_2 are their respective resource input rates. An equilibrium distribution is reached when the sum of the competitive abilities of the consumers in each habitat matches the associated habitat's profitability ratio: the *competitive-weight matching* rule. Different linear combinations of competitive abilities can produce the same sum, therefore a number of different distributions of the two phenotypes can satisfy the equilibrium condition. Prediction can be complicated. Moreover, most of the possible equilibrium distributions result in the resource intake being higher in one of the two habitats. Houston and McNamara (1988) demonstrate that not all equilibrium combinations satisfying the competitive-weight matching rule are equally likely to occur, in a probabilistic sense. When total population size is moderately small (< 60), then unequal competitor ability will induce little difference between the expected distribution and proportional habitat matching, although payoffs will be unequal across habitats. The most likely equilibrium distribution exhibits proportional habitat matching within phenotypes; each phenotype assorts independently following the proportional habitat matching rule. The intake rates of individuals of one phenotype are equal to those of the same phenotype in all other habitats. Unlike the usual expectation of proportional habitat matching, however, the average intake experienced in each habitat, pooling phenotypes, differs between habitats at equilibrium. As the total population size increases (> 60) proportional habitat matching may disappear. In this case the most likely distributions exhibit overrepresentation of consumers in the poorer habitat and an increased ratio of good to poor competitors in the rich habitat (Houston and McNamara 1988).

VARIABLE COMPETITIVE ABILITIES, CONTINUOUS PHENOTYPES, AND SEVERAL HABITAT TYPES

Assume now that the difference in competitive ability of phenotypes A and B depends on the habitat in which they are found, but not the density of competitors in those habitats. For instance, phenotype *A* is twice as good as *B* in habitat 1 but three times as good in habitat 2, no matter what the number of competitors is in either habitat. Parker and Sutherland (1986) demonstrate that for a distribution of consumers over habitats to qualify as an ESS it must obey two rules: the boundary phenotype and extrapolation rules.

To apply the boundary phenotype rule, we first order the habitats according to the strength with which competitive weights are affected by the consumers' phenotypic trait (say age; see fig. 5.3A). Then order consumers by the phenotypic trait (age) which is related to their competitive weights, and assume that individuals initially join habitats randomly relative to their phenotype. Call the individuals with the largest and smallest phenotype (not competitive weight) within each habitat *boundary phenotypes*. The boundary phenotype rule specifies that the distribution is an ESS if the payoff of the high boundary in each habitat equals the payoff of the low boundary in the habitat ranked immediately above it in terms of phenotypic effect on relative

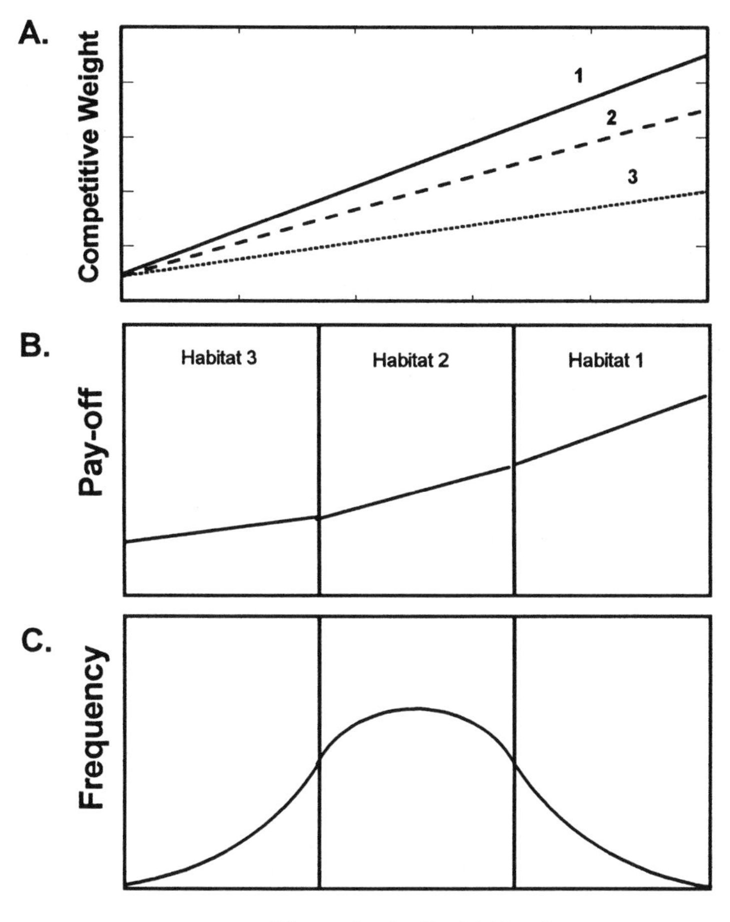

Figure 5.3 The relationship between a phenotypic trait, age in this case, and competitive weight, measured in three different habitats. In (A) we show that in habitat 1 (solid line), age has a stronger effect on competitive weight than in habitat 2 (long dash), where, in turn, age has a stronger effect than in habitat 3 (short dash). The predicted distribution of consumers across these three habitats will place older individuals in habitat 1, middle-aged individuals in habitat 2, and the youngest in habitat 3. Panel (B) illustrates the payoffs obtained by each phenotype within each of the three habitats at equilibrium. The boundary phenotype rule requires that the oldest consumer of habitat 3 experience the same payoff as the youngest consumer of habitat 2. Similarly, the oldest consumer of habitat 2 should experience the same payoff as the youngest consumer of habitat 1. The extrapolation rule requires that none of the phenotypes in habitat 3 could do better by moving to another habitat. Panel (C) illustrates a hypothetical frequency distribution of phenotypes within the three habitats. Note that within each habitat the distribution is truncated.

competitive abilities (fig. 5.3B). This rule can be met only if the individuals with the highest phenotypes, and hence competitive weights, end up in the habitat where the effect of phenotype on competitive weight is strongest. Conversely, individuals with the lowest phenotypes will end up in the habitats where phenotype has the weakest effect on competitive weight.

The extrapolation rule assumes that the relationship between an individual's phenotypic attribute (age) and its competitive weight is continuous and monotonic. Moreover, it assumes that there are large numbers of consumers in each habitat so that when one individual switches from one to the other it exerts negligible effects on the total competitive weights in each habitat. The extrapolation rule states that a distribution is an ESS when no phenotype can do better by switching to another habitat (fig. 5.3B).

This version of the habitat model is potentially more realistic. However, quantitatively predicting the expected distribution can be difficult without knowledge of both phenotypic frequencies and the explicit function relating phenotypes to their competitive weights. Once both relationships are known, quantitative predictions can be obtained numerically (see Parker and Sutherland 1986). Qualitatively, however, the habitat model predicts that equilibria can be reached with unequal payoffs across both habitats and phenotypes, while the phenotypic distribution within each habitat will be truncated (fig. 5.3C).

Relaxing the Assumption of Deterministic Resource Input Rates

The continuous input model assumes constant resource arrival rates, and therefore admits no temporal variance. In natural circumstances, a habitat's input rate likely varies about a mean over time; that is, input rates are likely to be stochastic rather than deterministic. It is reasonable to ask if such stochasticity would have any effect on the predicted consumer distribution over a set of habitats.

Recer et al. (1987) model a two-habitat environment where the constant habitat, c, has a deterministic input rate k_c , while a variable habitat v has a stochastic input rate k_v. Let the variable habitat have only two states, a low (k_1) and a high (k_2) input rate. Assume that the state of the variable habitat varies temporally according to a Markov chain governed by one-step transition probabilities. For the time interval Δt there is a fixed probability θ that the state will change. Assume that the probability of transition from state k_1 to k_2 is the same as changing from k_2 to k_1. Then, over a large number of transitions, each state should occur the same number of times. The expected duration of each state is $\mu = \Delta t/\theta$ time units.

If the input rates to both habitats c and v were constants, we would simply apply the proportional habitat-matching rule to predict g_v, the proportion of consumers in habitat v. Then:

$$g_v = \frac{G_v^*}{G_c^* + G_v^*} = \frac{k_v}{k_c + k_v}, \tag{5.6}$$

where G_c^* and G_v^* are the number of consumers in the respective habitats at equilibrium. This form of the habitat-matching rule shows that the predicted proportion of consumers in a habitat (g_v) increases as a decelerating function of the resource density in that habitat (k_v). Specifically, $\partial g_v / \partial k_v \geq 0$ and $\partial^2 g_v / \partial k_v^2 \leq 0$; see figure 5.4.

Compare that result with what happens when the resource density in a habitat varies randomly. The first consequence is that the number of consumers in the variable habitat is also a random function. The expected fraction of consumers in the variable habitat is denoted $E[g(k_v)]$. Since the relationship between the resource availability and the proportion of consumers

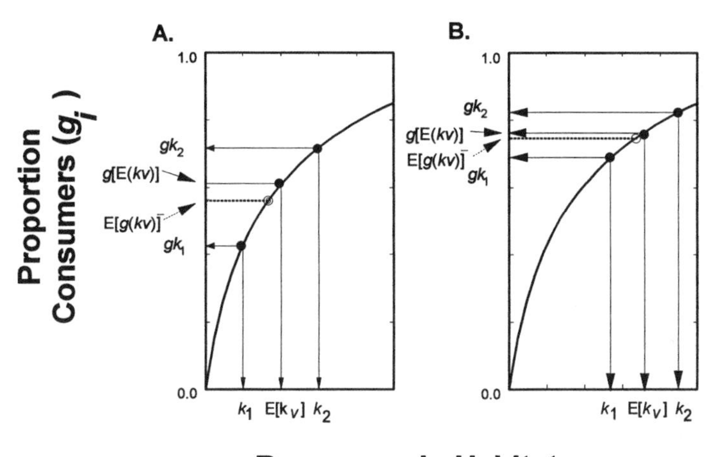

Figure 5.4 The relationship between resource availability in the variable habitat and the proportion of consumers expected in that habitat. Since each state of the variable patch is equally likely, mean resource level of the variable patch is given by $E[k_v] = (k_1 + k_2)/2$. Panels (A) and (B) are identical except that $E[k_v]$ in (A) is smaller than $E[k_v]$ in (B). $g(k_1)$ and $g(k_2)$ are the expected proportions of tracking consumers in the variable habitat when that habitat is in state 1 and 2, respectively. Since each state is equally likely, the average number of tracking consumers in the variable habitat is $E[g(k_v)] = [g(k_1) + g(k_2)]/2$. Note that this value is smaller than $g[E(k_v)]$, the proportion predicted if consumers, instead of tracking, simply matched $E[k_v]$. The difference between the tracking and nontracking proportions of individuals declines as the average quantity of resources in the variable patch increases [compare (A) and (B)].

in a habitat is nonlinear, the predicted proportion of consumers in the variable habitat ($E[g(k_v)]$) will be different from ($g(E[k_v])$) the fraction of consumers evaluated at the mean of the random resource density. That is, $E[g(k_v)] \neq g(E[k_v])$ (Recer et al. 1987; see fig. 5.4). The predicted consumer density should depend on whether consumers can track the fluctuating resource levels in the variable habitat.

Suppose that consumers are ideal, so that they track the changing resource levels in the variable habitat. That is, they immediately readjust their distribution between habitats whenever the variable habitat's input rate changes state. The expected proportion of tracking consumers occupying the variable habitat over some extended period of time is:

$$E[g(k_v)] = [g(k_1) + g(k_2)]/2. \tag{5.7}$$

Substituting equation (5.6) for the $g(k_v)$ yields:

$$E[g(k_v)] = \frac{1}{2}\left[\frac{k_1}{k_c + k_1} + \frac{k_2}{k_c + k_2}\right]. \tag{5.8}$$

Now suppose that the ideal assumption is violated so that consumers require some response delay τ in order to reach a new equilibrium once the state of the variable habitat has changed. The effect of stochastic input to the variable habitat now depends on the length of the response delay compared to the duration of the variable habitat's state. Assume the response delay is shorter than the time during which the variable habitat stays in one state (i.e., $\tau < \mu$), so that the alternate equilibrium can be reached before the variable habitat changes state again. Then consumers can track the changing local input rate of the stochastic habitat, and the long-term proportion of individuals in the variable habitat is given by equation (5.8).

But this is no longer true when the response delay is so long that it takes more time to reequilibrate than the average duration of the variable habitat's state (i.e., $\tau < \mu$). In this case, consumers cannot track the variable habitat's input rate. The best they can do is match the variable habitat's expected state $E[k_v]$. Given the symmetry in the transition probabilities mentioned above, the expected state of the variable habitat is $E[k_v] = (k_1 + k_2)/2$. The resulting density of consumers in the variable patch g_v is:

$$g(E[k_v]) = \frac{k_1 + k_2}{k_1 + k_2 + 2k_c}. \tag{5.9}$$

This corresponds to the prediction based on individuals using the average value of the habitat, which in essence is equivalent to the expectation from a deterministic model. Given the nonlinear relationship between the proportion of consumers and resource density we noted above, it follows that the predicted proportion of tracking consumers in the variable habitat ($E[g(k_v)]$) is always smaller than the predicted proportion of nontracking (i.e., averaging) foragers in the habitat ($g(\mathrm{E}[k_v]$; fig. 5.4). So, when foragers can track the state of the variable habitat, the average number observed in that habitat will always be smaller than the number predicted on the basis of proportional matching to the average. In fig. 5.4 we show that as the mean quantity of resources in the variable habitat increases, the difference between tracking and averaging decreases. In addition, for any mean input rate the extent of the discrepancy between tracking and averaging increases as the variance of the variable habitat's input rate increases. So, stochastic input rates can lead to overestimates of consumer density in the variable habitat when consumers can track the changes in its input rates.

Changing Currencies: Risk Sensitivity

The previous section considers the effect of stochastic input rates on the distribution of consumers that attempt to maximize their mean reward. So, it ignored any effect that stochastic input can have on reward variance and any consequent effect that reward variance can have on fitness. Stochastic reward levels may imply that both the mean and the variance of a given strategy's payoff can influence fitness. So we summarize Recer and Caraco's (1989a) development of a risk-sensitive habitat model. Below we review McNamara and Houston's (1990) state-dependent ideal free model which makes qualitatively similar predictions.

The currency of fitness now is the risk of starving, so that at equilibrium consumers will have equal probabilities of starvation within and among habitats. Retaining all other assumptions of the continuous input model, Recer and Caraco (1989a) predict that a population of risk-sensitive foragers will match relative resource availabilities (as in the deterministic model). Therefore, mean maximizing and minimizing the probability of starvation may predict the same consumer distributions.

Recer and Caraco (1989a) next explored the consequences of habitats delivering a resource that allows sharing between consumers. When a resource occurs as divisible clumps, the intake of each consumer varies continuously. This continuity has an effect on the predicted stable consumer distributions. The risk-sensitive habitat model now predicts that the number of individuals in habitat 1 will be given by

$$G_1 = \left(\frac{\sqrt{k_1}}{\sqrt{k_1} + \sqrt{k_2}}\right)\left(G + \frac{T}{R}\right)[\sqrt{k_1 k_2} - k_2], \tag{5.10}$$

where G is total population size ($G = G_1 + G_2$), and T is the duration of the foraging period. If the food intake needed to avert starvation (R) and total competitor number (G) are not too large, then consumers are likely to meet their daily energy requirement. In this case G_1 will approach G when $k_1 > k_2$. In other words, richer habitats will attract a disproportionate number of consumers while the poorer habitats will have close to none. The extent to which this "overmatching" is expected increases with foraging time T. However, if G and/or R are large, so that consumers are likely to fall short of their requirement, the model predicts matching to the square root of the input rates:

$$G_1/G_2 = \sqrt{k_1}/\sqrt{k_2}. \tag{5.11}$$

Expressing the problem in terms of total food eaten over a fixed time period (rather than foraging rate) allowed Recer and Caraco to separate effects of mean and variance on the predicted distribution. Suppose now that the total number of clumps eaten in habitat 1 is normally distributed with mean e_1 and variance $\sigma_1{}^2$. Then each forager's food consumption will be normal with expectation e_1/G_1 and variance $= \sigma_1{}^2/G_1{}^2$. At equilibrium:

$$G_1 = \left(\frac{\sigma_1}{\sigma_1 + \sigma_2}\right)\left[G + R^{-1}\left(\frac{\sigma_2 e_1}{\sigma_1} - e_2\right)\right]. \tag{5.12}$$

When the probability of starvation is low (small G and/or small R), the predicted proportion of foragers in habitat 1 increases as either e_1 increases or σ_2 decreases. So, consumers prefer high mean intake and avoid variance (they are risk averse). However, if foragers are likely to starve (G and/or R are large), consumer densities will tend to match the standard deviations of the habitats' input rates. In the limit

$$G_1/G_2 = \sigma_1/\sigma_2. \tag{5.13}$$

The model predicts that if otherwise ideal and free consumers use resources that can be shared as they arrive in the habitat, they will occupy space differently than when they exploit indivisible items. Moreover, sharing clumps allows for risk-sensitive assortment of consumers.

Complex Currencies: Trade-Offs with Predation Hazard

The habitat models we have reviewed to this point all assume that habitats differ only in terms of their resource input rates. In nature, habitats will likely vary in several fitness-related attributes. In particular, habitats may present consumers with different levels of exposure to predation. Since selection acts on fitness, consumers are expected to respond to trade-offs between danger from predation and benefits from foraging.

Two modeling approaches developed for dealing with trade-offs between predation hazard and foraging benefits are

1. Use of simple optimality following conversion of predation hazard into foraging units (Gilliam and Fraser 1987; Abrahams and Dill 1989), and
2. Use of stochastic dynamic programming after converting both food gain and predation hazard into survival (Houston et al. 1988; Mangel and

Clark 1988; McNamara 1990; Newman 1991; McNamara and Houston 1990, 1993).

The first type of habitat model does not predict consumer distributions but rather the amount of resource that must be added to a hazardous habitat in order to get consumers to use it. McNamara and Houston's (1990) dynamic habitat model, on the other hand, does predict consumer distributions. Its complexity, however, makes it more useful as a means to identify qualitative trends rather than predict quantitative distributions.

Abrahams and Dill (1989) argue that it is possible to use the basic habitat model to measure (F) the quantity of food resources that consumers are willing to give up in order to avoid the danger of predation. Since the habitat model predicts that, at equilibrium, all consumers achieve a constant intake rate C, the food-equivalent predation hazard of a habitat can be measured as a deviation from this expected value of C based on proportional habitat matching. Hence, adding F resources to a habitat with a higher level of danger should return consumers to proportional habitat matching.

McNamara and Houston (1990) present a stochastic dynamic programming model for the distribution of foragers over two habitats that differ either in reward variance or in predation hazard. Ignoring predation hazard at first, the model predicts that consumers will behave relative to input variance much as predicted by Recer and Caraco's (1989a) risk-sensitive model. However, unlike Recer and Caraco, the dynamic risk-sensitive habitat model is not based on equalizing risks across habitats and predicts risk-sensitive distributions even when resource items are not shared. Basically, when overall consumer abundance is low, all foragers are doing well and should be risk averse and prefer the habitat with lower variance. As the consumer density increases, foraging rates decline and eventually lead to negative energy budgets and hence to risk-proneness. Preference for habitats with variable input rates follows.

When McNamara and Houston (1990) consider the effect of danger they keep all other foraging parameters constant between habitats. When the density of competitors is high, so that all are doing poorly in terms of foraging gains, their dynamic model predicts that consumers will be distributed independently of danger. However, at low consumer densities, even very small levels of danger are sufficient to have all consumers forage in the safer habitat.

5.4 Introduction to Interference Models

This version of the habitat model may apply more generally than the previous version. The interference model expresses each habitat's basic suitability not on the basis of resource abundance per se, but as an estimate of a consumer's potential foraging rate when alone in that habitat. The model, like its predecessor, assumes that consumers are ideal and free. Despite the ideal free assumptions, this version of the habitat model has been called the *interference* model (Sutherland 1983). Sutherland's use of the word "interference" refers to a reversible depression in foraging rate resulting directly from the presence of competitors. Interference is distinct from decreased foraging rates resulting from exploitation competition. The latter is the indirect result of competitors reducing prey density, an effect that was already included in the continuous input models (Lessells 1995).

The interference model's most interesting assumption is that it permits the effect of competitors to vary under different sets of conditions. For instance, in cases of strong interference, it is possible that when two individuals occupy the same habitat, each obtains less than half the intake rate it would have achieved on its own. Similarly, in situations of weaker interference, the effect could be such that each obtains more than half of the rate achieved in the same habitat when foraging alone. Neither case was possible in the continuous input model because all resources appearing in a habitat were exploited (but see Lessells 1995).

In the interference model, a habitat's basic suitability is expressed as the total number of prey (K) found during some time interval T. The model uses Holling's (1959) disk equation to relate basic suitability to the habitat's prey density (κ):

$$K = \frac{a'\kappa T}{1 + a'\kappa T_h}, \tag{5.14}$$

where a' is the consumer's instantaneous search rate, and T_h is the handling time per item. A consumer's searching efficiency (a) will often change as a function of the number of competitors present such that

$$a = Q\,G^{-m}, \tag{5.15}$$

where Q is a constant related to searching efficiency when alone, G is the number of consumers in the habitat, and m a parameter scaling the effect that consumers have on each other's search rate. Hence m scales interference. Defining $a = a'T_s$, where T_s is time spent searching, expression (5.15) can be substituted into (5.14) to obtain the individual's expected resource intake:

$$K = \frac{(QG^{-m}/T_s)\kappa T}{1 + (QG^{-m}/T_s)\kappa T_h}. \tag{5.16}$$

An ESS condition requires that prey be eaten at the same rate by all consumers in all habitats so that $K/T = c = \kappa_i\,(G_i)^{-m}$, where κ_i is the prey density in habitat i. Therefore, independently of the prey density κ in a habitat, the same number of prey per forager will be consumed there by time T as will be consumed in any habitat. The predicted ideal free consumer number in habitat i is then

$$G_i = (\kappa_i / c)^{1/m}. \tag{5.17}$$

If one knows the total population of consumers and the total resources over all habitats of interest, then one can express the above relationship as the interference habitat-matching rule

$$g_i = \gamma \alpha_i^{1/m}; \quad 0 < g_i < 1, \tag{5.18}$$

where α_i is the proportion of resources in habitat i and γ is a "normalizing constant" given by

$$\gamma = \sum_{i=1}^{Z} g_i / \sum_{i=1}^{Z} \alpha_i^{1/m} \tag{5.19}$$

for the Z habitats in the area. γ assures that the g_i values sum to unity. Note that γ depends on both m and the relative distribution of resources among the Z habitats. Consequently, there is a single value of γ for each array of resources among habitats and the potential value of m. This makes generalizing about the effect of m on the relationship between α_i and g_i difficult.

Examples of potential resource distributions for an area with three habitats are given in fig. 5.5. In general, the interference model shows that proportional matching is expected only in the special case where the interference constant (m) is unity, which implies that $\gamma = 1$. If interference is strong (i.e., $m > 1$), the interference model predicts more consumers in poor habitats and fewer in rich habitats than does proportional habitat matching (fig. 5.5). When interference is weak (i.e., $m < 1$), the opposite trend is expected. Consumers will tend to aggregate more than expected in good habitats and therefore be underrepresented in poor habitats (fig. 5.5). In all cases, intake is constant across habitats and consumers. Naturally this prediction requires knowing the values of m and hence γ in advance. Sutherland and Parker (1985) suggest that one could estimate m by measuring the number of prey items consumed per unit search time (excluding handling time) for a given prey density and for a range of consumer densities. The slope of the log intake against log consumer density is m and the intercept is Q.

The interference model has not been followed by the diversity of variations and extensions associated with the continuous input model. According to Sutherland and Parker's (1985) review, data relevant to the interference model come mostly from field studies. Few such studies appear to support the basic interference model (but see Tregenza 1995; Tregenza et al. 1996a). We list the variations of the interference model and their consequences in Summary Box 5.2 and compare them to those proposed for continuous input models. We now review some of the interference model's variations.

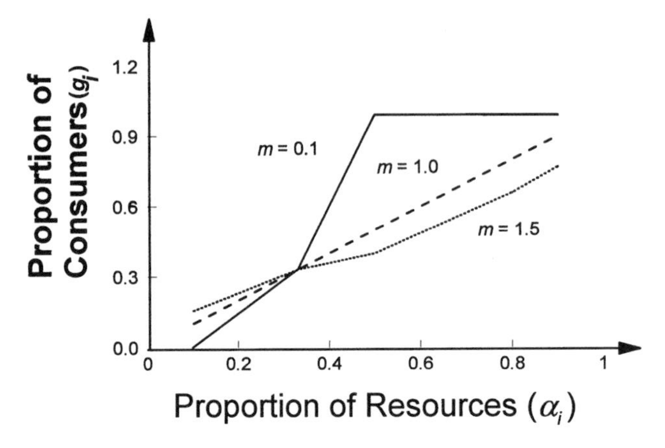

Figure 5.5 Interference matching between the proportion of consumers (g_i) and proportion of resources (α_i) in an environment with three habitats. Interference (m) may be weak (solid line), intermediate (dashed), strong (dotted). When interference is weak, consumers are predicted to aggregate in the rich habitats and avoid poor ones. Strong interference leads to the opposite pattern. Note that at intermediate levels of interference, where in this case $m = 1.0$, interference matching is identical to habitat matching.

5.5 Changing the Assumptions of Interference Models

Relaxing the Assumption of Ideal Consumers

The ideal assumption is violated when decisions are based on imperfect information. In the continuous input models, information quality could be reduced by perceptual constraints preventing consumers from distinguishing profitability levels. Perceptual constraints have not been considered within the context of interference models. However, some studies have investigated the consequences of assuming consumers must learn the average habitat quality through sampling; Bernstein et al. (1988) developed an interference habitat model that deals with this violation of the assumption of ideal consumers.

Bernstein et al. (1988) model a population of consumers that forages on different quality patches within a habitat and ask whether a stable distribution of consumers over a habitat's patches exists. They assume that consumers have some initial estimate of the average-habitat quality in an area, but that they periodically update this estimate based on current success within the habitat's patches. This learning process differs from that modeled by Regelmann (1984) for continuous input systems. In Regelmann's model, consumers had to learn the input rate of the habitat they currently exploited. Bernstein et al. (1988) allow consumers to recognize instantaneously the

SUMMARY BOX 5.2 MAIN ASSUMPTIONS AND PREDICTIONS OF INTERFERENCE HABITAT MODELS

THE BASIC MODEL

Main assumption: Resources all present concurrently, habitats do not deplete, consumers are ideal, free, and equal.

Main prediction: Payoffs constant across consumers and habitats. Interference matching appears linear only when $m = 1$. If interference weak, $m < 1$, then interference matching leads to overrepresentation in rich habitats. If interference strong, $m > 1$, then interference matching leads to underrepresentation in rich habitats.

Usual observation: Payoffs rarely constant across habitats (see Parker and Sutherland 1986).

PROPOSED MODIFICATIONS WHEN HABITATS DO NOT DEPLETE

Consumers not ideal	CONSEQUENCES
Perceptual constraints:	Not applied to this version of model. Will likely affect distribution, especially in cases when interference is weak.
No omniscience:	Expect interference matching, hence payoffs equal across habitats and consumers (Bernstein et al. 1988).
Consumers not free	
Travel costs:	Not applied to this version of model.
Consumers unequal	
Constant competitive weight:	Not applied to this version of model.
Variable competitive weight:	Assumes phenotypes vary in the way interference affects them. Only one stable distribution, which is difficult to predict quantitatively. Payoffs unequal across phenotypes and habitats (Parker and Sutherland 1986).

PROPOSED MODIFICATIONS WHEN CONSUMERS DEPLETE RESOURCES IN HABITAT

Consumers not ideal	
Perceptual constraint:	Not applied to current version of model.
No omniscience:	Depletion has little effect on distribution (i.e., expect interference matching) until habitats depleted to 50% of initial resources. When depletion increasingly severe, expect increasing deviations from interference matching (Bernstein et al. 1988).

SUMMARY BOX 5.2 (*cont.*)

Consumers not free	
Travel costs:	As travel cost increases, expect increasing deviations from interference matching (Bernstein et al. 1991).
Consumers unequal	Not considered in current version of model.

quality of the patch they exploit. However, the quality of the exploited patch is then used to update the estimate of the habitat (i.e., among patch) quality.

Bernstein et al. (1988) also assume that foragers stay within a habitat over successive foraging periods. As long as their current success in a patch exceeds the expected success based on their updated estimate for the habitat, they stay (a rule borrowed from patch models). If the current foraging rate is less than the updated estimate for the average patch in the habitat, the consumer chooses to switch patches. It then samples another, possibly with a different foraging rate which, after some specified time interval, is used to update its estimate for the habitat once again. Bernstein et al. (1988) ask whether this sort of learning biases the distribution of consumers away from the ideal interference model's equilibrium.

The model assumes that a consumer's intake during a foraging period of duration T is given by a slightly different version of equation (5.16):

$$K = \frac{QG\kappa^{1-m}}{1 + QT_hG\kappa^{-m}}. \tag{5.20}$$

Habitats in this case are characterized by a constant (nondepleting) prey density (κ), and prey items require handling time (T_h). Let β represent the *profitability* of a patch that is being exploited. The profitability is given by the total number of prey captured per consumer within the patch (K/G) divided by the maximal attack rate possible (T/T_h): $G > 0$ and $0 < \beta \leq 1$. All foragers have equal competitive abilities and so experience the same profitability within a given patch. ϵ_t is the estimate of average profitability of the patches available to consumers at time t. The estimate is updated at the end of each simulation cycle. Updating is achieved by a linear combination:

$$\epsilon_{t+1} = \beta\psi + (1 - \psi)\epsilon_t, \tag{5.21}$$

where ψ is a memory factor ($0 \leq \psi \leq 1$). At the end of each simulation cycle, consumers for which the profitability is less than the updated estimate (i.e., $\beta < \epsilon_{t+1}$) leave the patch and join another randomly, starting a new simulation cycle. Note that the choice of the next patch need not be done randomly. In fact, one could expect foragers to look around to see if others have found a resource and join those that have, in effect a producer-scrounger game (see chapter 6; Beauchamp et al. 1997). Bernstein et al. (1988) simulated this process for forty cycles to predict the distribution of consumers over a large number of patches, and to compare predictions with the basic interference model. Consumer distributions generated by the model approached those expected on the basis of *interference matching* by the fifteenth simulation cycle.

A related property of the model is that increasing the range of profitability values experienced across patches predicts consumer distributions closer to interference matching predictions. This result is qualitatively similar to the prediction of Abraham's (1986) perceptual constraint model (see above).

An interesting prediction in Bernstein et al. (1988) is that consumers will conform to interference habitat matching only if the initial profitabilities, estimated before they start foraging, are not too low compared to the environment's actual profitability. If consumers start with a pessimistic estimate of profitability, they will find themselves in situations where the first patch encountered exceeds expectations ($\beta > \epsilon_t$). So, they will not switch patches and therefore they will be unable to obtain realistic updated estimates of the habitat's profitability. Pessimistic consumers become trapped in patches they erroneously assume to be better than the average in the habitat. This suggests that selection might either favor optimism in prior estimates of average patch quality or at the very least skepticism concerning prior estimates (so that sampling is never eliminated). If individuals in one habitat can learn the success of consumers in other habitats, erroneous prior estimates may be unlikely (Clark and Mangel 1984; Valone 1989, 1993; Templeton and Giraldeau 1995a,b).

We have not found an explicit interference model that includes travel costs (hence consumers that are not free) when habitats do not deplete. However, Bernstein et al. (1991) investigate the effect of travel costs when habitats deplete (see below). As was true for continuous input models, little work has focused specifically on the consequences of habitat-entry costs. Instead, most theoretical developments assume free entry and consider the effect of asymmetric resource division.

Relaxing the Assumption of Equal Consumers

Continuous input models include versions assuming that consumers' competitive weights differ in one of two ways. First, the relative competitive weights for two phenotypes remain constant across habitats. Second, the consumers' relative competitive weights change as a function of the habitat exploited. Asymmetric interference models deal solely with a third type of competitive limitation which assumes that consumers differ in the degree to which local competitors depress their foraging rate (Sutherland and Parker

1985; Parker and Sutherland 1986). A poor competitor's foraging rate declines more quickly with increasing competitor density than does a better competitor's rate. An implicit assumption is that the phenotypic composition of the group of competitors will influence the way consumer density affects an individual's foraging rate.

The stable distribution of consumers can be found through simulation if, once again, both the continuous phenotypic frequency distribution and the function relating phenotype to competitive weights are known. The model's predictions are qualitatively similar to those of the continuous input version; the stable phenotypic distribution over each habitat is always truncated and obeys both the phenotype boundary and extrapolation rules. The quantitative predictions, however, are different; the interference model allows only one possible stable distribution for a given resource distribution, and the payoffs at equilibrium will be unequal across both habitats and consumers.

Allowing Habitats to Deplete

The models presented to this point all assume that the basic suitability of a habitat, whether expressed as resource input rate (continuous input models) or resource density (interference models) is unaffected by consumers' exploitation. That is, the previous models let resources renew as they are exploited. However, the assumption of no resource depletion limits the applicability of habitat models to consumer distributions across arrays of drift food (e.g., Milinski 1979), or distributions among resource units so large that exploitation exerts negligible effects on resource density. Habitat models allowing for depletion extend the applicability of the general approach to distributions across resources where consumers may reduce food availability significantly.

Bernstein et al. (1988) also developed an interference model predicting a stable distribution of consumers over a number of depleting patches within a habitat. The model is as described in the previous section, but now includes resource depletion. At the start of each simulation cycle the suitability of all patches, K in equation (5.20), is lowered based on the intensity of resource consumption during the previous cycle. Because it is an interference model, it allows for both handling and search times. The model also assumes that consumers are not ideal and must learn about the profitability of the habitat. Consumers attempt to maximize their expected resource intake rate; they are equal competitively and free to enter any patch within the habitat (for consequences of travel costs, see below).

The results indicate that patch depletion has little effect on consumer distribution when depletion is weak (specifically, when only 4.15% of the initial level of resources was consumed through forty simulation cycles). Even when as much as 50% of the resources are consumed during the simulation, the model generates distributions that do not differ appreciably from interference matching. However, when depletion involves more than 50% of the initial resource levels, the simulated consumer distribution deviates increasingly from interference matching.

Bernstein et al. (1988) argue that depletion reduces the fit to the interference matching predictions for two reasons. First, depletion reduces the among-patch variance in prey density, since richer patches are more heavily exploited than poorer ones. They showed in their nondepleting version of the model that a reduced variance in experienced profitability results in poorer fit between consumer distributions and expectations from the interference model. The second reason involves an interaction between depletion and learning rates. When the rate of depletion exceeds the rate at which the information is updated, it is impossible for consumers to have an accurate estimate of the average habitat quality. Hence they are expected to choose habitats randomly. On the other hand, the interference model can apply, at least approximately, when information is updated faster than resources deplete. Again, monitoring other consumers' success across patches can increase the speed with which consumers update their information (Clark and Mangel 1984; Valone 1989, 1993; Templeton and Giraldeau 1995b, 1996).

Relaxing the Assumption of Free Consumers

The only impediment to freedom considered in interference models for depleting habitats has been travel costs (Bernstein et al. 1991). Consumers do not forage during travel. So when a consumer has just spent T_T time steps traveling and enters a patch, it will have updated its estimate ϵ of the average patch in the habitat (as defined above) according to

$$\epsilon_{t+1} = \epsilon_t(1 - \psi)^{T_T}. \tag{5.22}$$

Bernstein et al. (1991) analyze the effect of travel time on consumer distribution through computer simulations. The results indicate that consumer distributions deviate increasingly from the associated interference habitat matching as travel time increases. Since the cost of travel is subtracted from the gains that could be achieved by moving to another habitat, this prediction follows logically. Simulated consumers are increasingly reluctant to leave a habitat, even though profitability is lower than the estimated profitability when travel time is long. Only those competitors in a very poor habitat tend to move. Hence travel costs in a depleting habitat system generate strong deviations from interference habitat matching.

5.6 Concluding Remarks

Implications for Future Work

Of all social foraging theory, research on consumer distributions across patches of resource is probably the most active and productive. The number of published models and tests of these models continues to increase regularly (Sutherland 1996; Tregenza 1995; Tregenza et al. 1996a,b). Most recent theoretical developments have moved away from the less realistic continuous input formulations to the more general interference type models. However, it is somewhat ironic that empirical research appears to have remained attached to the simpler but less realistic continuous input models (Tregenza 1995). We feel that experimental work should focus more on interference models. Developments by Lessells (1995) and more recently Tregenza et al. (1996a) provide more readily testable versions of interference-type models. Researchers might consider estimating the interference parameter m in a number of different systems. Only then could interference models be tested effectively.

When comparing Summary Boxes 5.1 and 5.2, one cannot help but notice the extent to which developments and modifications to continuous input models have not been echoed in interference models. For instance, risk-sensitive continuous input models exist, but there is no equivalent interference model. Another example is the examination of the consequences of perceptual constraints in continuous input models but not in interference models. A final example is sufficient to illustrate the point; while the effect of unequal competitors with constant competitive weights across habitats has been examined for continuous input models, it has not been in interference models. Surely modifications called for by increased realism in one modeling tradition should be relevant to the other.

Implications for Scaling Up

In this chapter we reviewed a series of models for social foragers' choice of habitats. Habitat-selection models readily suggest links between individual behavior and ecological pattern or process (Huey 1991; Rosenzweig 1991; Houston 1996; Sutherland 1996). Behavioral interactions can directly influence consumers' use of habitats; ecological effects may include spatial patterns in prey mortality (Kacelnik et al. 1992) and spatiotemporal pattern in consumer population densities (e.g., Pulliam 1988).

All models we reviewed are based on habitat selection theory introduced by Fretwell and Lucas (1970). The original formulation of both the ideal free and ideal despotic distributions make density-dependent reproductive success a direct consequence of habitat choice (Fretwell 1972). Some recent extensions of these models also link habitat choice and population dynamics directly (e.g., Pulliam and Danielson 1991). Applications of ideal free theory to social foragers' patch use hypothesize mean feeding rate or short-term survival as currency of fitness, without quantifying the consequences for population growth. Social foragers may adjust their numbers to match food availability in different patches within minutes (e.g., Recer et al. 1987). The impact of responses at this scale on population dynamics remains uncertain, but it is clear that consistent differences among individuals' foraging success can induce predictable variation in their life histories (e.g., Morse and Fritz 1987; Blanckenhorn 1992). Therefore, both short-term (e.g., choice of a food patch) and long-term (e.g., choice of breeding site) behavioral responses to spatial heterogeneity in resources can have demographic effects. Furthermore, responses to resource abundance over different scales of spatial averaging may interact with the temporal scale of habitat choice. Short-term behavior can track fine-grained spatial variation (Recer and Caraco 1989a). Long-term behavioral responses presumably average over extended spatial scales. The former processes are logically nested within the latter (Orians and Wittenberger 1991), suggesting complex interdependence of competitive habitat-choice decisions and population growth.

Conclusions

We began this chapter by asking why parasitic jaegers attacked terns in groups. Evidence suggests that jaegers forage in dispersion economies. That means that the group size in which we should expect them to attack terns depends on two factors:

1. The availability of terns carrying fish back to their colony, and
2. The type and intensity of competitive interactions within the jaeger groups.

All the models we presented, both continuous input and interference, develop relationships among consumer density, resource distribution, and the type and intensity of competitive interactions within groups.

No doubt readers are now in a better position to appreciate the subtle differences between the research traditions involved in predicting group size for aggregation and dispersion economies. Hopefully, the value of dealing with both as part of the same question becomes obvious. For instance, a major advance in the study of groups under aggregation economy has been to replace the optimal group size with the equilibrium group size, a direct application of the Nash equilibrium expected in dispersion economies (Sibly 1983). The idea of testing alternative currencies of fitness in a dispersion economy (e.g., Bélisle 1998), on the other hand, comes directly from studies of grouping under aggregation economies.

Dealing with aggregation and dispersion economies as two facets of the same problem can contribute even more to understanding why animals are in groups. It suggests that it may be important that researchers interested in aggregation economies also consider the subtle yet important effects of different forms and intensity of competition on the shape of their peaked fitness functions. Similarly, researcher interested in animal distributions, hence dispersion economies, should not focus too narrowly on competition within resource parcels. For instance, when groups of animals forage over a habitat composed of many separate patches, it may be unrealistic to suppose that they do not pay any attention to one another's foraging successes (Templeton and Giraldeau 1995a; Beauchamp et al. 1997). An improved understanding of social foraging calls for a closer integration of studies designed for aggregation and dispersion economies.

PART TWO Producer-Scrounger Decisions

6 An Introduction to Producer-Scrounger Games

6.1 Introduction

Social foragers can obtain food as solitary foragers do: by spending time and effort in search, pursuit, and capture of prey. Alternatively, they can avoid some costs of the full foraging cycle by parasitically exploiting food that another forager's efforts have made available: kleptoparasitism. Parasitic interactions over food resources are, according to Barnard (1984b), one of the most widespread forms of exploitation both within and among species. Not surprisingly, therefore, a vast literature describes the many forms that parasitic foraging can take in a number of taxa, including insects and spiders (Vollrath 1984; Field 1992), fish (Pitcher 1986), birds (Brockmann and Barnard 1979), mammalian carnivores (Packer and Ruttan 1988), as well as primates, both nonhuman and human (Winterhalder 1996, 1997).

Despite the large number of descriptive accounts of the phenomenon, no structured body of theory and tests, comparable, say, to prey or patch models of conventional foraging theory, has emerged to analyze kleptoparasitic behavior as a foraging decision. Initial attempts to study parasitic feeding behavior as a conventional prey problem (Kushlan 1978; Dunbrack 1979; Thompson 1986; Ens et al. 1990) generally had difficulty accounting for observed foraging behavior (but see Thompson 1986). The proposal that the economics of parasitic foraging be modeled as a frequency-dependent game (Barnard and Sibly 1981; Vickery et al. 1991; Caraco and Giraldeau 1991; Ranta et al. 1996) has received some recent empirical testing (Barnard and Sibly 1981; Hansen 1986; Ens et al. 1990; Giraldeau et al. 1994b; Koops and Giraldeau 1996; Giraldeau and Livoreil 1998), but there is currently no consensus concerning its relevance to kleptoparasitism in general (Ruxton et al. 1995; Beauchamp and Giraldeau 1996; Ranta et al. 1996).

In this chapter we first briefly review the diversity of kleptoparasitic foraging, suggest a nomenclature, and then examine the assumptions and predictions of a game-theoretic model of kleptoparasitic behavior. To help synthesize current views, we carefully compare our producer-scrounger game with its ecologically similar, but economically distinct, counterpart, often called the "information-sharing" model.

6.2 The Diversity of Kleptoparasitism

The forms of kleptoparasitic foraging are diverse, but they have given rise to an even greater variety of names. Many of these terms, such as "piracy," "robbing," and "stealing," carry strong anthropomorphic connotations. We feel it important to use terminology that presents fewer human referents and describes more precisely the behavior we wish to depict. We therefore refer to all forms of exploitation of others' food discoveries or captures as *kleptoparasitism*. There are three distinct ways in which a kleptoparasite can gain all or some of a host's food. It can use overt aggression, competitively scramble for a share of the food, or use stealth. We consider examples of each.

Aggressive Kleptoparasitism

An aggressive kleptoparasite, not too surprisingly, uses force or threat of force to gain exclusive access to food. Typically, only one individual exploits the resource at a time. For example, bald eagles (*Haliaeetus leucocephalus*) supplant conspecifics from the carcasses of chum salmon (Hansen 1986). Similarly, dominant Harris' sparrows (*Zonotrichia querula*) displace subordinates from seed clumps and so obtain exclusive use of the clump (Rohwer and Ewald 1981). Subtle variants of this type of kleptoparasitism have been called "robbing" (Kushlan 1978; Ens et al. 1990), "piracy" (Hatch 1973; Vollrath 1984), "usurpation" (Vollrath 1984), "stealing," "snatching" (Barnard and Sibly 1981), "cleptoparasitism," "scavenging" (Curio 1976), "sponging" and "using others as truffle-pigs" (Czikeli 1983). Aggression and exclusive use of the resource characterize all of these variants.

Scramble Kleptoparasitism

Some resources are exploited simultaneously by two or more competitors. For example, when a water cricket (*Velia caprai*) captures a small (< 7.9 mg) soft-bodied insect, it can exploit it exclusively. However, whenever it captures a larger item, up to six other competitors can exploit it simultaneously, scrambling for a share (Erlandsson 1988). Observing a small flock of spice finches (*Lonchura punctulata*), a southeast Asian species of estrildid waxbills, searching for clumps of seeds hidden on an aviary floor, it soon becomes obvious that when one individual discovers a clump, its feeding alerts others that food has just been uncovered. Inevitably a number of individuals converge at the clump to partake of the discovery (Giraldeau et al. 1990, 1994b; Giraldeau and Livoreil 1998; Beauchamp and Giraldeau 1997; Beauchamp et al. 1997b). This form of kleptoparasitism, characterized by little or no aggression and simultaneous exploitation of the resource by several individuals, has also been called "local enhancement" (Thorpe 1956), "area copying" (Krebs et al. 1972; Barnard and Sibly 1981), "social facilitation" (Curio 1976), "contest kleptobiosis" (Vollrath 1984), "joining" (Lefebvre 1986; Giraldeau and Lefebvre 1986, 1987; Giraldeau et al. 1990), and "tolerated theft" (Blurton Jones 1984; Winterhalder, 1986, 1987).

Stealth Kleptoparasitism

Some kleptoparasites take food while avoiding interactions with the host, a form of parasitism Vollrath (1984) termed "pilfering." For example, some kelp gulls (*Larus domenicanus*) appropriate mussels (*Mesodesma donacium*) that other gulls have dropped to break their shell (Hockey et al. 1989). Eastern chipmunks (*Tamias striatus*), a North-American terrestrial squirrel, feed on their rich seed and nut larder during the winter months when they periodically come out of torpor (Elliott 1978). They build this larder every fall by hoarding large quantities of food in their underground burrow (Elliott 1978; Giraldeau et al. 1994a). While an individual is away from its burrow, to collect seeds or engage in any other activity, one of its neighbors may enter the vacant burrow, fill its cheek pouches with food, and leave, adding the food to its own reserves: stealthful kleptoparasitism (Elliott 1978). Similar forms of stealth kleptoparasitism occur in Myriam's kangaroo rat (*Dipodomys merriami*), where individuals commonly kleptoparasitize scatter hoards made by other kangaroo rats (Daly et al. 1992).

6.3 Kleptoparasitism: A Game-Theoretic Approach

The Difference between Information Sharing and Producer-Scrounger Models

A common way of thinking about food kleptoparasitism is to imagine animals that search independently for their own food while, at the same time, keeping a look out for any other group member's discoveries. When a group member discovers food, all remaining individuals join in. This observation has been made countless times in group-feeding animals of almost any taxa. This way of portraying group foraging is particularly common in models of group membership. Several models presented in Part One adopt this *information-sharing* view of social foraging (see Clark and Mangel 1984, 1986; Ranta et al. 1993; Ruxton et al. 1995). We did this because group-membership models predict decisions about joining or leaving groups; further prediction of whether some or all individuals should use kleptoparasitism would complicate the analysis needlessly. Kleptoparasitism is, in a sense, a constraint in these group-membership models; any individual uses it when the opportunity arises. In this chapter, however, we analyze the decision to use kleptoparasitism or not. In this case, complicating the model by allowing individuals to join or leave a group would obscure the focal question. So the model we discuss assumes that group membership is constrained: no individual can leave a group.

The more recent versions of information-sharing models allow for varying detectability of kleptoparasitic opportunities. Consequently, not all group members use the option, only those that detect the food discovery of a group member (Ruxton et al. 1995). Other information-sharing models allow competitive and foraging-efficiency asymmetries among individuals (Ranta et al. 1996). None of these models, however, analyze whether individuals ought to use kleptoparasitism or not. Information-sharing models fix conditions under which individuals can use kleptoparasitism, and then assume they always do.

A game-theoretic view, first proposed by Barnard and Sibly (1981), provides a useful alternative to information-sharing. Individuals play a two-strategy game where at *any one time* they can *choose* to play either *producer* (i.e., search for food) or *scrounger* (i.e., wait for another to find food and then kleptoparasitize). A clearly specified constraint of information-sharing models becomes the question the game-theoretic approach asks: How much kleptoparasitic behavior is evolutionarily stable?

Consequences of the Game-Theoretic Approach

An important element of Barnard and Sibly's (1981) game-theoretic model of kleptoparasitism was its novel view of the way a kleptoparasite's payoff should depend on the frequency of kleptoparasitism in the group (or population). Under information-sharing assumptions, foragers can simultaneously search for food *and* look for others' food discoveries (Vickery et al. 1991). An important consequence is that varying the frequency at which individuals use kleptoparasitism has no effect on the total rate at which food is discovered, hence no effect on the rate at which opportunities for kleptoparasitism arise. A kleptoparasite's payoff is still affected by the number of individuals that compete for shares of each food discovery, but the number of food discoveries is independent of the frequency of kleptoparasitism (see section 1.3; Math Box 6.1 at the end of the chapter).

The frequency-dependence of a kleptoparasite's payoff in a producer-scrounger game is quite different. Producer and scrounger are distinct alternatives in the sense that a forager cannot *simultaneously* play producer (i.e., search) and scrounger (i.e., keep a lookout for others' discoveries) (Barnard and Sibly 1981; Vickery et al. 1991). So, as the frequency of scrounger changes, it not only affects the number of individuals competing at each food discovery, but also the rate at which kleptoparasitic opportunities arise. That is, increasing the frequency of kleptoparasitism now implies a reduction in the total rate at which food is discovered; increased scrounging must imply reduced production. Hence the payoff to scrounger is now affected by both an increase in the number of competitors and a decrease in the number of producers (see Math Box 6.1).

Assumptions of The Producer-Scrounger Game

TYPE OF GAME

The producer-scrounger game is an alternative-option, *N*-person interaction. Alternative option means that on any round of play an individual chooses one of two actions (producer or scrounger, with uninformed play). *N*-person means that the success of the chosen option depends on the actions taken by all individuals that economically interact with the focal player. The number of interactants may extend to an entire population or may be restricted to a single group's membership.

FREQUENCY-DEPENDENCE

The game assumes that producer and scrounger are true alternatives, i.e., they cannot co-occur in the same individual at the same time (West-Eberhard 1989). The significance of this property has been emphasized by several authors including Field (1989), Ens et al. (1990), and Vickery et al. (1991), who termed it the assumption of "complete incompatibility." This incompatibility is crucial because it generates the distinction between producer-scrounger and information-sharing models. The assumption does *not* mean that an individual in an incompatible system must use producer or scrounger invariantly (i.e., among plays of the game). Individuals observed to alternate between the producer and scrounger alternatives can be modeled by a producer-scrounger game if the incompatibility is such that when a forager is playing producer (e.g., searching in the substrate with its head down), it cannot at the same time be playing scrounger (e.g., holding its head up to monitor others' success).

Parker (1984a) pointed out that the general producer-scrounger game allows a number of different types of frequency-dependent relationships (fig. 6.1). Note that scrounger fitness in all cases declines as the frequency of scrounger increases, but the effect of scrounger frequency on producer varies. Producer fitness might increase with scrounger frequency (fig. 6.1A), but this situation does not correspond to food parasitism, so we ignore it. Some games where scrounger imposes no or very little cost to producer (fig. 6.1B) can be considered parasitic, but scrounger is undoubtedly a parasitic alternative when it imposes some costs on the producer strategy (fig. 6.1C).

6.4 A Symmetric Rate-Maximizing Producer-Scrounger Model

Preliminary Considerations

Here we review an *N*-person, deterministic, mean-maximizing model. It is the simplest producer-scrounger model we analyze, and hence may be the

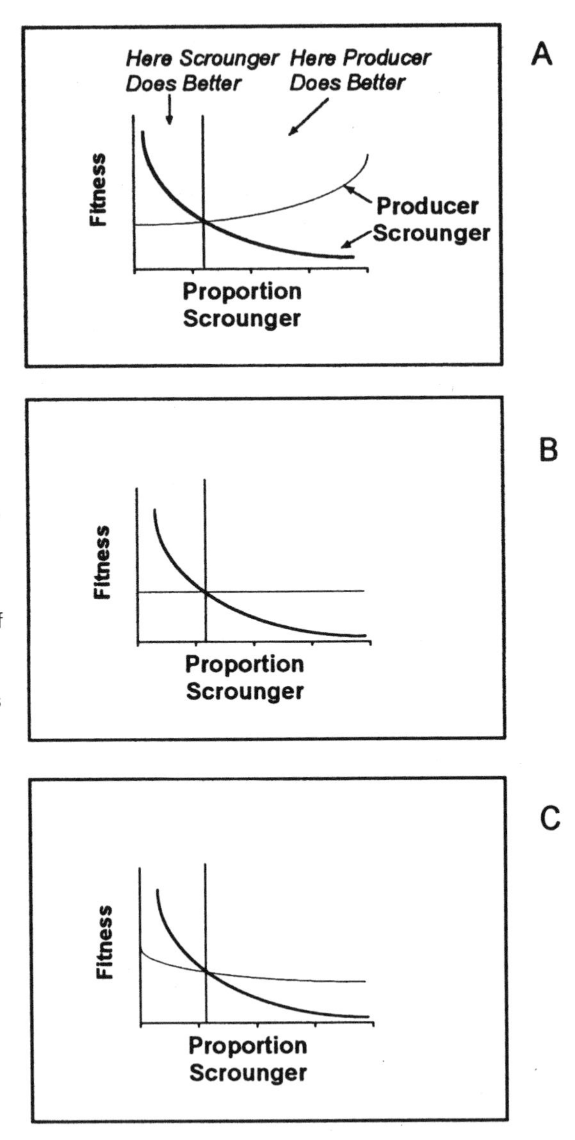

Figure 6.1 Fitness of producer and scrounger alternatives as a function of the proportion of scrounger within the population. In all three panels the fitness of the scrounger alternative decreases as the frequency of the scrounger increases (i.e., negative frequency dependence). The panels differ in the potential effect of the scrounger on the producer. In the top panel, the fitness of producer increases with the frequency of the scrounger alternative. In the middle, scrounger frequency appears to have no effect on producer fitness. The bottom panel presents the more likely outcome of food parasitism: the fitness of the producer is negatively affected by increasing scrounger frequency, but less so than the scrounger. In all three cases, the functions' intersection marks the stable equilibrium.

least realistic. However, it is a direct derivative of information-sharing models that several authors have used to analyze social foraging (e.g., Clark and Mangel 1984; Ruxton et al. 1995; Ranta et al. 1996). The deterministic model takes a first step toward understanding how various ecological factors interact to promote kleptoparasitism.

We model a symmetric game. Symmetry of players implies that every individual playing producer obtains the same payoff and that every individual playing scrounger gets the same payoff at equilibrium. The symmetry assumption suits empirical results for a variety of social foragers (Giraldeau and Lefebvre 1986, 1987; Giraldeau et al. 1990).

When players are symmetric, stable coexistence of the behavioral alternatives occurs when rarity implies economic advantage. We let π represent the frequency of producers in a population such that $W(\pi|P)$ and $W(\pi|S)$ are the frequency-dependent payoffs of producers and scroungers, respectively. $\hat{\pi}$ is the stable frequency of producing when the following conditions hold (Caraco and Giraldeau 1991):

$$\lim_{\pi \to 1} [W(\pi|P) - W(\pi|S)] < 0$$

$$\lim_{\pi \to 0} [W(\pi|P) - W(\pi|S)] > 0$$

$$[W(\hat{\pi}|P) = W(\hat{\pi}|S)]$$

$$\left.\frac{dW(\pi|P)}{d\pi}\right|_{\pi=\hat{\pi}} < \left.\frac{dW(\pi|S)}{d\pi}\right|_{\pi=\hat{\pi}}.$$

Note, once again, that a given individual might strictly produce or strictly scrounge without temporal variation; or, an individual might produce during one round of play and scrounge during another.

The Model

The model we present here is derived in Vickery et al. (1991). It applies to the following scramble-kleptoparasitic scenario. A total of G individuals forage in sufficient proximity that group members playing scrounger all detect and exploit the food uncovered by any of the group's producers. A food clump contains F items. The producer obtains a *finder's advantage*, a portion a of the patch $(0 \leq a < F)$ that it can use exclusively before the scroungers arrive. The magnitude of the finder's advantage is fixed; it is a constraint, not a decision of the forager. The finder's advantage is not an index of social dominance; it is an attribute of the order of arrival at a patch, not of specific individuals.

The proportion of individuals playing producer is p, and the proportion playing scrounger is $(1 - p)$. Once a producer finds a clump, $(1 - p)G$ scroungers arrive in unison and divide the remaining $(F - a)$ food items equally among the $[1 + (1 - p)G]$ individuals now present (all the scroungers plus the producer). I, the currency of fitness, is energy intake. Patch discovery is rare, occurs sequentially, and patch exploitation time is negligible. Each producer's expected intake I_P after T time units of foraging is

$$I_p = \lambda T\left(a + \frac{F - a}{1 + (1 - p)G}\right),$$

where λ is the producer's encounter rate with food patches. The rate of encounter with scrounging opportunities is a function only of the number of producers. The scrounger's expected intake I_S is

$$I_S = \lambda GTp\frac{F - a}{1 + (1 - p)G}.$$

The stable frequency of producer $\hat{p}$ can be found by setting I_P equal to I_S and solving for p:

$$I_p = \lambda T\left(a + \frac{F - a}{1 + (1 - p)G}\right) = I_S = \lambda GTp\frac{F - a}{1 + (1 - p)G}. \quad (6.1)$$

Both mean intake levels vary inversely with $[1 + (1 - p)G]$ and hence increase with producer frequency p. At equilibrium,

$$a/F + 1/G = \hat{p} \quad (6.2)$$

for $a < F$ and $G \geq 2$. This means that if $(F/G) \geq (F - a)$ then $\hat{p} = 1$ and, additionally, that $\hat{p} > (a/G)$, $(0 < \hat{p} < 1)$.

To demonstrate the stability of $\hat{p}$ consider the difference in the expected feeding rates $D(p) = I_P - I_S$. By definition $D(\hat{p}) = 0$; stability requires $\partial D(p = \hat{p})/\partial p < 0$. Differentiating and substituting, we obtain:

$$\left(\frac{\partial D(p)}{\partial p}\right)_{\hat{p}} = \frac{GF(F - a)}{F + a}\left(\frac{a}{F + a} - 1\right) < 0. \quad (6.3)$$

Hence $\hat{p}$ is stable since $a/(F + a) < 1$.

Note that the corporate rate of patch encounter by the producers $(pG\lambda)$ drops out of the solution for the stable frequency of producers. This occurs because the model assumes that scroungers have no effect on an individual producer's rate of encounter with patches. Under these conditions, the stable frequency of producer is independent of the rate at which producers encounter clumps. Thus, in the deterministic mean-maximizing model, the stable frequency of producers—and therefore the stable frequency of scroungers—depends on only group size (G) and the *finder's share* (a/F), the proportion of each patch consumed as the finder's advantage (fig. 6.2).

To examine the interaction of these two factors, note that $\partial\hat{p}/\partial(a/F) > 0$, but $\partial\hat{p}/\partial G < 0$. Hence, the finder's share of the food in clumps it discovers must increase as the size of the group increases, or else the frequency of scrounging is likely to increase. Figure 6.2 gives an example of how the stable frequency of producer is affected by the finder's share and group size.

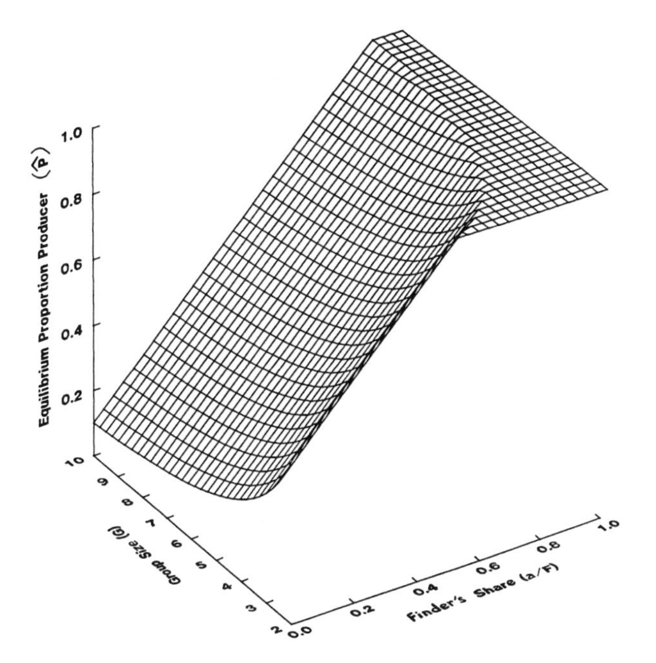

Figure 6.2 The stable frequency of producer strategists ($\hat{p}$) as a function of group size (G) and the finder's share (a/F), the fraction of the patch to which producers have exclusive access. As group size increases, a larger finder's share is required to prevent the scrounger alternative from being profitable. Taken from Giraldeau and Livoreil (1998). (With permission of Oxford University Press)

No scrounging occurs when $\hat{p} = 1$, a situation likely for small group sizes. For instance, for a group of two, all producers (i.e., no scrounging) is stable whenever the finder's share represents 50% or more of each discovered patch. However, as group size increases, the finder's share must rise disproportionately to exclude scrounger from the stable solution. So, as group size increases, all-producer may become unstable. For instance, in a group of ten foragers, all-producer is the stable solution only if the finder's share represents 90% or more of each patch discovered. Factors that regulate the finder's share, therefore, are important influences on the economics of scramble kleptoparasitism.

Determinants of the Finder's Share

Some food clumps require very little processing time before a forager can exploit its items, e.g., unfolding a leaf to gain access to an insect's egg mass. For this kind of food, the producer can commence exploitation almost immediately upon encounter and quickly reduce the number of items available to scramble kleptoparasites (high a/F). Other types of food, such as seeds within conifer cones or live prey that require significant pursuit and capture times, reduce the amount of resource a forager can monopolize before the scramble kleptoparasites arrive. As a result, scroungers may, in some cases, be attracted to a discovery of food even before the producer has actually begun to consume the resource ($a/F = 0$). Food items that can only be consumed slowly will also promote the incidence of scramble kleptoparasitism. Hence, clumps that require extensive or slower processing may advance the frequency of scrounging.

Since the finder's share is a ratio, large values may be more likely when clumps contain relatively few items. The expected number of items obtained from a clump, however, may not always depend strictly on depletion effects. In some cases, clumps can be ephemeral, as occurs when prey disperse once attacked. When prey flight is rapid, the number of items consumed from a clump depends not on competitor density but on the time spent exploiting the clump (Millikan and Pulliam 1982; Clark and Mangel 1986). The finder's share for highly ephemeral patches must be based on the duration of patch exploitation, which influences the proportion of that time the producer spends in the patch. Extreme ephemerality, therefore, may lead to a large finder's share and low scrounger frequency even when clumps initially contain a large number of prey items.

Other candidate factors influencing the finder's share may involve the position of the clump's producer relative to the food when certain positions are advantageous (Ward and Enders 1985). The finder's advantage may also depend more complexly on the geometry of the flock (Barta et al. 1997) and the distance between individuals, both of which influence the delay between a producer's discovery and the arrival of the scroungers (Giraldeau et al. 1990).

Relaxing the Incompatibility Assumption

Summary Box 6.1 lists the deterministic producer-scrounger model's assumptions concerning the foragers and the foraging process. Among these assumptions, complete incompatibility has frequently been challenged (see above, Caraco and Giraldeau 1991; Ruxton et al. 1995; Ranta et al. 1996; Beauchamp and Giraldeau 1996). This assumption implies, for instance, that an animal currently playing producer cannot (or will not) respond to opportunities to scrounge food. Conversely, an individual playing scrounger cannot (or will not) attempt to produce food. Complete incompatibility can occur when producing and scrounging involve distinctly different behaviors (e.g., Field 1989) or different locations within the group or environment. But in many foraging groups, producing and scrounging opportunities can be simultaneously available to all members.

Vickery et al. (1991) relax the assumption of complete incompatibility by allowing a third strategy, the *opportunist* (see section 1.3; Math Box 6.1).

SUMMARY BOX 6.1 THE MEAN-MAXIMIZING, DETERMINISTIC P-S FORAGING MODEL

ASSUMPTIONS

Decision: The frequency of playing *producer* and *scrounger*

Currency: Long-term average energy intake

Constraints:

1. Complete incompatibility. Producer and scrounger are true alternatives (i.e., cannot occur concurrently in the same player).
2. All scroungers detect and scrounge at each scrounging opportunity.
3. Players are symmetric and scroungers use scramble kleptoparasitism.
4. The rate of appearance of scrounging opportunities is set entirely by the number of producers in the group (i.e., scroungers do not interfere with producers' search efficiency).
5. Food clump exploitation time is negligible, patch discovery is sequential.
6. Producers encounter food at a constant rate.
7. The finder's advantage a and the finder's share a/F are set by foraging conditions and are not under control of the foragers (i.e., they are *constraints*, not *decisions*).
8. Group $G \geq 2$.

PREDICTIONS

1. The equilibrium frequency of producer is $\hat{p} = a/F + 1/G$.
2. $\hat{p} = 1$ when $(F/G) \geq (F - a)$.
3. $\hat{p} > (a/G)$ when $0 < \hat{p} < 1$.

IMPORTANT GENERAL POINTS

1. The model predicts the average population frequency of alternatives, not each individual's decision at a given time.
2. The model assumes that the prevailing conditions favor scramble kleptoparasitism over other forms; it cannot be used to predict when scramble kleptoparasitism will be favored over other forms of kleptoparasitism.

The opportunist uses both producer and scrounger alternatives *simultaneously*. The opportunist closely corresponds to the forager assumed in information-sharing models. But the opportunist is more general; because opportunists attempt to find food and attend to kleptoparasitic opportunities simultaneously, some interference between the two resource-acquisition roles is possible. In particular, the detectability of a kleptoparasitic opportunity may be reduced by searching for one's own food (e.g., head up versus head down).

To depict possible interference between an opportunist's clump-searching effort and its kleptoparasitism, we assume that an individual employing one alternative exclusively attains the maximal possible efficiency, unity, for that alternative. An opportunist's overall efficiency combines its success in each role. The opportunist has efficiency $0 \leq \kappa \leq 1$ when it produces and efficiency $0 \leq \eta \leq 1$ when it scrounges. If there is no interference between resource-acquisition roles within the opportunist, then it has combined efficiencies of a pure producer and a pure scrounger: $\kappa + \eta = 2$. For any $\kappa + \eta < 2$, there is some interference between the two activities. When $\kappa + \eta < 1$, opportunists do worse than either producer or scrounger. In this case, opportunists should be eliminated, so the game described in the previous section applies reasonably.

Vickery et al.'s (1991; see fig. 6.3) analysis indicates that the condition $\kappa + \eta \geq 1$ is necessary for the occurrence of opportunists, but not sufficient. The range of finder's share predicting a stable frequency of opportunists depends on the extent of interference between roles within the opportunist. When $\kappa + \eta = 2$ (no interference), we expect opportunists at any finder's share. However, increased interference decreases the range of finder's share values for which opportunists can succeed (fig. 6.3). Generally, at large finder's share values, opportunists can coexist with pure producers. For small finder's share, opportunists can coexist with pure scroungers (fig. 6.3).

A group made up of only opportunists would differ from an equilibrium mixture of producer and scrounger in a number of ways. A group of only opportunists is equivalent to the group composition assumed by information-sharing models: each individual has the identical phenotype. In such a group, each forager would appear to act as a producer since all actively search for food. In addition, once a clump of food was discovered, each of $(G - 1)$ group members would kleptoparasitize (assuming all detected the opportunity, i.e., $\eta = 1$). The predicted frequency of scramble kleptoparasitism is then $[(G - 1)/G]$, independently of clump size (Beauchamp and Giraldeau 1996). Finally, the number of clumps an individual finds and the number of clumps it parasitizes in an opportunist group should vary as a function of the efficiencies of the alternatives, κ and η.

Ranta et al. (1996) provide an information-sharing foraging model based on entirely compatible alternatives. It goes farther than any previous deterministic information-sharing model in that it allows for

1. Variation among individuals in tendencies to search and to act as a kleptoparasite.
2. Competitive asymmetries in obtaining food from a discovered patch.
3. The possibility of individuals choosing to leave foraging groups.

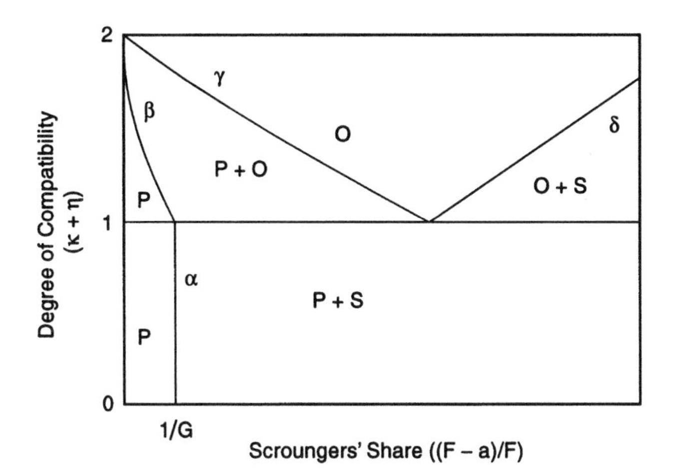

Figure 6.3 The transition thresholds between various ESS combinations of the producer (P), scrounger (S), and opportunist (O) strategies. Thresholds depend on the compatibility between the opportunist's producing and scrounging abilities and the proportion of the patch that scroungers compete for (the complement of the finder's share). Below compatibility values of 1, no opportunists are expected. When the scrounger's share is smaller than $1/G$ (i.e., finder's share is larger than $(G - 1)/G$), only producers are expected. Beyond this value of scrounger's share, a mix of producer and scrounger is stable (region α). Above compatibility values of 1, opportunists are part of almost all stable solutions. At low scroungers' share, a mixture of producer and opportunist is stable (region β). At intermediate scrounger's share, only opportunists are expected, especially as the compatibility values increase (region γ). At high scrounger's share, a mixture of opportunist and scrounger is stable (region δ). Producer, scrounger, and opportunist mixtures are technically possible but unlikely because they require compatibility values of exactly 1 (the horizontal line). Modified from Vickery et al. (1991). (With permission of University of Chicago Press)

In their model, each individual uses both producer and scrounger alternatives; they are opportunists. But two discrete categories of individuals exist. They call "producers" individuals that obtain most of their food from that alternative, while the opposite is true for "scroungers." For clarity we prefer Barnard and Sibly's (1981) terminology and refer to these biased opportunists as *searchers* and *copiers*, respectively, because each type uses producer and scrounger alternatives simultaneously (which violates an assumption of our producer-scrounger game). Both types of individuals were assigned a relative competitive efficiency in relation to searching and competing for food discoveries. Ranta et al. (1996) note that very few combinations of search and competitive efficiencies across individuals lead to intersecting searcher and copier payoff functions. They use this result to question the generality of the producer-scrounger game for systems with competitive asymmetries. However, it is important to point out that the lack of intersection they note refers to payoffs for *searcher* and *copier* individuals, not *pro-*

ducer and *scrounger* tactics. Both searchers and copiers obtain food by using producer and scrounger alternatives simultaneously, so the calculated payoffs do not correspond directly to those of producer and scrounger alternatives.

6.5 Empirical Tests of the Rate-Maximizing Producer-Scrounger Model

Evidence for the Right Kind of Frequency Dependence

A convincing demonstration of frequency-dependent payoffs to each tactic in a producer-scrounger system requires manipulation of the producer:scrounger (P:S) ratio and estimation of the payoff obtained by each alternative under the changing P:S ratio (Mottley and Giraldeau, in press). Both of these tasks may pose practical problems. Measuring payoffs to each alternative will be simple in dimorphic systems where individuals can use only one or the other alternative. However, when individuals switch between tactics over time, estimating the payoffs to each requires discriminating which alternative tactic is being used by a forager at the time of observation. This may be straightforward in some circumstances but difficult in others. When the solitary wasp, *Ammophila sabulosa* searches for a place to dig her nest, she occasionally encounters another female's burrow. When this happens, she may ignore it. But more commonly she parasitizes the nest (Field 1989). When attempting to estimate payoffs to each alternative, Field (1989) found it impossible to identify when a wasp was looking for her own nest site or for another female's nest and wondered whether these were indeed true alternatives (i.e., were incompatible). In another situation, however, he had no such difficulty. When the same solitary wasp searches for prey to provision her nest, she hunts away from sites where nests have been dug. Thus, it was simple to determine when a wasp was hunting for her own prey (producer) versus attempting to sequester other females' already captured prey (stealth kleptoparasitism). Each foraging system will present its own difficulties. It is important to recognize that estimating payoffs will require identifying each alternative unambiguously.

Manipulating a group's P:S ratio will also present considerable practical problems. The exercise will be easiest when animals cannot alter their investment in each alternative opportunistically. Although most individuals within house sparrow (*Passer domesticus*) flocks use both producer and scrounger alternatives, among-individual distributions of producer and scrounger were bimodal. Some individuals, *searchers*, obtained more (0.60–0.89) of their food by finding it, while others, *copiers*, obtained more (0.51–0.80) food by interacting with others (Barnard and Sibly 1981). Much like Ranta et al. (1996), Barnard and Sibly (1981) report the payoffs obtained by searcher and copier individuals. For each of these types of individuals, the payoffs represent the summed rewards from both producer and scrounger alternatives. Hence, the payoffs to searcher and copier individuals cannot be used directly to illustrate the frequency dependence of the scrounger payoff.

The task of establishing frequency dependence will be even more difficult when, as is likely to be true in many cases, foragers can adaptively alter their allocation to each alternative through assessment. Under those conditions, the P:S ratio in a group remains under the foragers' control, and it will be difficult to maintain a reward asymmetry. Adaptive plasticity does not vitiate the concept of a producer-scrounger game but makes the payoff pattern governing behavioral responses difficult to discern.

Limiting some individuals to the scrounger role is possible, for instance, by preventing them from acquiring a special skill required to discover food. This approach was used in experiments with spice finches, which exhibit scramble kleptoparasitism (Giraldeau et al. 1994b). Food was hidden under lids, and lid-lifting was taught only to certain individuals (producers). Unfortunately, although individuals who were not trained to lift lids were effectively limited to the scrounger alternative, skilled lid-lifters used both alternatives and obtained 35% of their food by using scrounger. Despite this problem, the results are consistent with the frequency-dependence assumption; when scrounger was common, it did worse than producer. The payoff to producer declined as the frequency of scrounger increased (fig. 6.4). When scrounger was rare (i.e., one scrounger in the flock), it fared equally well, not better, than the so-called producer individuals who now could alternate temporally between producer and scrounger. Finding a way to constrain individuals to each alternative would no doubt generate clearer results (Mottley and Giraldeau, in press).

Testing the Effect of the Finder's Share

One powerful test of the deterministic model requires keeping group size constant and manipulating the finder's share of patches. Producer frequencies are then compared with those predicted by equation (6.2). Giraldeau and Livoreil (1998) used Giraldeau et al.'s (1990) result that the finder's share, at least in small flocks of spice finches, can be altered by changing the spatial aggregation of the food. Small finder's shares are expected when food occurs in a few large clumps; larger finder's shares are expected when the clumps are small but numerous.

They presented three flocks of five spice finches with a different sequence of three seed distributions. In flock A (fig. 6.5), for instance, the birds first experienced six days in an environment with few clumps, each containing many seeds (small finder's share). Then they experienced six days with environments of intermediate clumping, and finally six days of environments

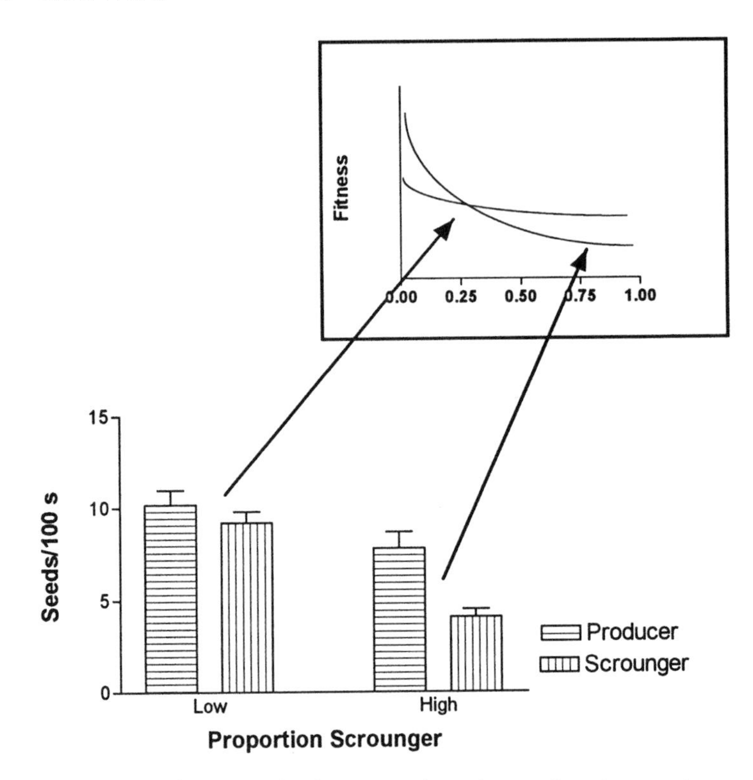

Figure 6.4 A test of the negative frequency dependence of producer and scrounger payoffs in flocks of spice finches, *Lonchura punctulata*. Bars give the mean; vertical lines are one SE above the mean. As expected, scrounger does worse than producer when the scrounger proportion is high. Both producer and scrounger foraging rates decrease with increasing frequency of scrounger, but scrounger does not do better than producer when scrounger is rare (taken from Giraldeau et al. 1994b). The lack of difference when scrounger is rare is attributed to producer using the scrounger alternative when it pays more to do so. The insert shows how the results can be interpreted as consistent with the assumption of negative frequency dependence.

with many clumps, each containing just a few seeds (large finder's share). The other two flocks experienced the same conditions but in a different order (fig. 6.5). Counting the average number of seeds eaten per patch both when a bird discovers and when it scrounges estimates, by subtraction, the finder's advantage (a) and hence the finder's share (a/F). Equation (6.2) can then be used to predict the stable frequency of producer for each of the three food distributions.

The model succeeded qualitatively in predicting the change in the frequency of producer over the three food distributions in all three flocks (fig. 6.5; Giraldeau and Livoreil 1998). It is important to note that information-

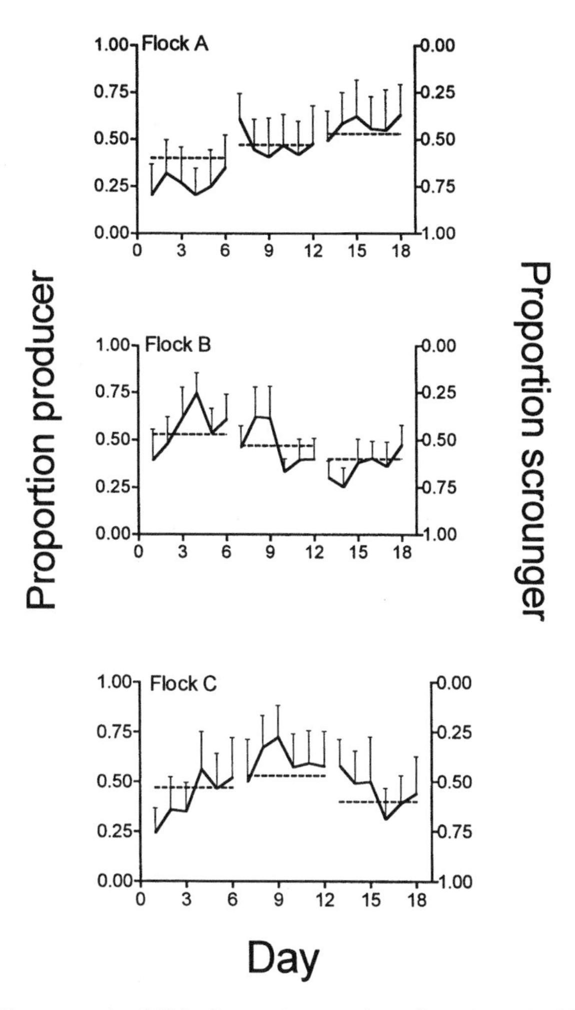

Figure 6.5 The mean (+ SEM) observed proportion of producer (solid line) within three flocks of five spice finches *Lonchura punctulata* foraging on three levels of food patchiness compared to predictions (dashed line) of the mean rate maximizing producer-scrounger game. The flocks were tested five times a day for six consecutive days on each food-clumping condition. Clumping levels were low (200 seeds in 10 patches), medium (200 seeds in 20 patches) and high (200 seeds in 40 patches). The effect on proportion using producer was significant using repeated measures ANOVA. Taken from Giraldeau and Livoreil (1998). (With permission of Oxford University Press)

sharing models do not predict such a change. But the simple producer-scrounger model's prediction that the proportion of producers should increase with the finder's share was upheld.

6.6 Concluding Remarks

Implications for Future Work

Experimental work on producer-scrounger games is just starting; much still needs to be done. Results obtained to date suggest that rate-maximizing games may be effective approximations of the economics of scramble kleptoparasitism in one species of seed-eating bird. Testing the model in other species, and in a more quantitative way, is necessary in order to assess the model's generality. The finder's share prediction may be particularly amenable to testing. The group size prediction remains untested; its empirical analysis likely will prove challenging. Nonetheless, if this could be done, we would have an entirely different way of testing the rate-maximizing game.

Future work might concentrate on examining factors that govern the finder's share. Distance between individuals, size of the foraging area, group size, and detectability of scrounging opportunities are all likely to be important. The issue of incompatibility between the producer and scrounger alternatives will likely remain an important topic. It would be extremely useful, for instance, to identify behavior that reveals whether individuals are currently playing producer or scrounger or are foraging as opportunists. Moreover, it would be useful to document instances where the compatibility of information-sharing models is applicable. In the end, some foraging systems may more appropriately be analyzed using information-sharing models, while others are more clearly the domain of producer-scrounger games.

Finally, it would be important to understand the relationship between flock geometry and kleptoparasitism. Using a genetic algorithm approach, Barta et al. (1997) have shown that groups of foragers without scroungers adopt different flock geometries than equivalent groups with scroungers. Producers spread out more and tend to occupy peripheral positions forming a ring of searching individuals. When scroungers occur, the flock is more densely packed as scroungers fill the inner ring, a position that minimizes their distance from any one producer (Barta et al. 1997). If flock position influences the effectiveness of producer and scrounger, then it may be that an individual's relative position within a flock reveals the alternative it currently favors. No study has explored whether information-sharing systems should exhibit equivalent geometric effects.

Conclusions

In this chapter we compare the game-theoretic approach to analyzing kleptoparasitic foraging behavior with the information-sharing approach. The rate-maximizing producer-scrounger model we reviewed makes a number of quantitative predictions about the use of scramble kleptoparasitism. It predicts that the finder's share and group size will be principal influences on the stable frequency of scrounger and producer alternatives within groups. Evidence obtained so far provides qualitative support for the model. It suggests that the frequency dependence of the payoffs to producer and scrounger are consistent with the assumptions of producer-scrounger games. Foragers appear flexible in their use of alternatives and seem to alter their relative investment in kleptoparasitism as predicted by the model.

In the next chapter we return to our emphasis on environmental unpredictability and model producing and scrounging as a risk-sensitive game.

Math Box 6.1 Producers, Scroungers, and Opportunists: Comparing Frequency-Dependent Feeding Rates

Definitions

Producer searches and encounters food clumps at rate λ; a producer feeds only at clumps it discovers.
Scrounger does not search directly but feeds at every clump any producer or opportunist discovers.
Opportunist searches and encounters food clumps at rate $\beta\lambda$ $(0 < \beta \leq 1)$; each opportunist also feeds at every clump any producer or other opportunist discovers.
Feeding Rate is the expected long-term rate of food intake; units are clumps/time; a deterministic currency suffices to demonstrate frequency dependence across group compositions.
G is group size.

Assumptions

1. Search time dominates handling time, so latter is ignored.
2. Individual finding a clump consumes a fraction α $(0 < \alpha < 1)$; remainder $(1 - \alpha)$ is divided equally among individual finding clump and those playing scrounger or opportunist.

Group Members Play Producer Only

1. Each producer consumes entirely each clump it finds.
2. Feeding rate $= \lambda$, basis of comparison for other rates.

Group Members Play Producer and Scrounger

1. Frequency of scroungers in the group is s $(0 \leq s < 1)$; group composition is $(1 - s)G$ playing producer and sG playing scrounger.
2. Each producer finds clumps at rate λ; fraction $(1 - \alpha)$ of each clump divided equally among the producer and sG scroungers.
3. The feeding rate of producer is

$$\alpha\lambda + [(1 - \alpha)\lambda/(1 + sG)] \leq \lambda. \qquad (6.1.1)$$

4. Each individual playing scrounger takes fraction $[(1 - \alpha)/(1 + sG)]$ of each clump found by $(1 - s)G$ producers.

Math Box 6.1 (*cont.*)

5. The feeding rate of scrounger is

$$(1 - \alpha)(1 - s)G\lambda/(1 + sG). \tag{6.1.2}$$

6. At ESS, feeding rates of producer and scrounger are equal, and the ESS frequency of scrounger is

$$\hat{s} = (1 - \alpha) - G^{-1} \tag{6.1.3}$$

for $\alpha < (G - 1)/G$; otherwise $\hat{s} = 0$.

7. The feeding rate at the ESS is

$$\lambda(\alpha + G^{-1}) = \lambda(1 - \hat{s}) < \lambda, \text{ for } 0 < \hat{s} < 1. \tag{6.1.4}$$

Group Members Play Producer and Opportunist

1. Frequency of opportunists in group is q; group composition is $(1 - q)G$ individuals playing producer and qG individuals playing opportunist.

2. Individual producer feeds as in group of producer and scrounger. The producer's feeding rate is

$$\alpha\lambda + [(1 - \alpha)\lambda/(1 + qG)] \leq \lambda. \tag{6.1.5}$$

3. Individual playing opportunist finds clumps at rate $\beta\lambda$; fraction $(1 - \alpha)$ of each of these clumps divided among qG opportunists.

4. Individual opportunist takes fraction $[(1 - \alpha)/(1 + qG)]$ of each clump found by $(1 - q)G$ producers.

5. Individual opportunist takes fraction $(1 - \alpha)/qG$ of each clump found by $(qG - 1)$ other opportunists.

6. The feeding rate of opportunist is

$$\alpha\beta\lambda + \frac{(1 - \alpha)\beta\lambda}{qG} + \frac{(1 - \alpha)(1 - q)G\lambda}{1 + qG} + \frac{(1 - \alpha)(qG - 1)\beta\lambda}{qG} = \beta\lambda + \frac{(1 - \alpha)(1 - q)G\lambda}{1 + qG}. \tag{6.1.6}$$

7. At ESS, feeding rates of producer and opportunist will be equal; the ESS opportunist frequency is

$$\hat{q} = [(1 - \alpha)/(1 - \beta)] - G^{-1} \tag{6.1.7}$$

Math Box 6.1 (*cont.*)

for $G^{-1} < (1 - \alpha)/(1 - \beta) < (G + 1)G^{-1}$.

8. The feeding rate at the ESS is

$$\lambda[\alpha + (1 - \beta)G^{-1}] < \lambda, \text{ for } 0 < \hat{q} < 1. \qquad (6.1.8)$$

9. Opportunist ESS frequency in group with individuals playing producer is greater than scrounger ESS frequency in group with individuals playing producer. The ESS feeding rate is lower at producer-opportunist equilibrium than at producer-scrounger.

Group Members Play Opportunist Only

1. Individual playing opportunist takes fraction $\alpha + [(1 - \alpha)/G]$ of clumps it finds and takes fraction $(1 - \alpha)/G$ of clumps found by others.

2. The opportunist's feeding rate is

$$= \alpha\beta\lambda + (1 - \alpha)\,\beta\lambda = \beta\lambda. \qquad (6.1.9)$$

3. An opportunist-only group is sometimes termed an "information-sharing" group. An opportunist-only group is sometimes confused with a producer-scrounger group, but the two are quite different. The feeding rate in an opportunist-only group exceeds the feeding rate of a producer-scrounger group at the ESS whenever $G(\beta - \alpha) > 1$, which is more likely as G increases or as opportunist searching efficiency (β) increases.

Group Members Play Opportunist and Scrounger

1. Opportunist frequency is q ($0 < q \leq 1$). The group composition is qG playing opportunist and $(1 - q)G$ playing scrounger.

2. Individual opportunist takes fraction $\alpha + [(1 - \alpha)/G]$ of clumps it finds and takes fraction $(1 - \alpha)/G$ of clumps found by $(qG - 1)$ other individuals playing opportunist.

3. The opportunist's feeding rate is

$$= \beta\lambda(\alpha + q[1 - \alpha]) < \beta\lambda. \qquad (6.1.10)$$

4. Individual scrounger takes fraction $(1 - \alpha)/G$ of each clump found by the qG individuals playing opportunist.

5. The scrounger's feeding rate is

$$(1 - \alpha)qG\beta\lambda/G = (1 - \alpha)q\beta\lambda. \qquad (6.1.11)$$

Math Box 6.1 (*cont.*)

6. Opportunist feeds faster than scrounger for any $q \in [0, 1]$. However, learning limitation may otherwise constrain the frequency of opportunist below unity.

7 Producer-Scrounger Games in Stochastic Environments

7.1 Introduction

Chapter 6 modeled the producer-scrounger game in a deterministic environment. Patches contained a fixed number of prey and were encountered at constant intervals. Moreover, individuals obtained fixed shares of each patch. In the real world, uncertainty likely affects each element of the process: search times, clump richness, and the way items in a clump are divided between the producer and the scroungers. Therefore, this chapter asks how foraging in a stochastic environment influences the equilibrium number of producers and scroungers in a group.

First we derive a stochastic model that specifies the mean and variance of the producers' and scroungers' food intake. We analyze the game using standard normal approximations of the producers' and scroungers' probability of energetic failure, and then use the exact food-intake means and variances numerically to find Nash equilibrium solutions to the producer-scrounger game. We include a brief survey of empirical results that address the game's predictions.

7.2 A Stochastic Producer-Scrounger Game

Preliminary Considerations

In this section we assume that a group of G ($G \geq 2$) individuals may contain both producers and scroungers. We fix the size of the foraging group, but we identify conditions where effects of scrounging might lead to the group's dissolution (Ranta et al. 1996). Each individual either produces only or scrounges only; we assume complete incompatibility. "Pure" producing and scrounging might be justified when there is strong interference between producer and scrounger efficiencies, when clumps appear frequently enough to keep scroungers occupied, or when success as a scrounger interferes with an individual's learning to produce (Giraldeau and Lefebvre 1987). The plausibility of the latter two conditions may often increase as group size increases.

The models do not incorporate any asymmetry arising from dominance interactions. When the number of scroungers reaches only modest levels, a would-be dominant (or a subgroup of aggressive individuals) could find it difficult to monopolize the food in a clump. In a few species scrounging occurs without aggression, or the economics of producing and scrounging fail to depend on dominance rank (e.g., Giraldeau et al. 1990).

We solve the analytical models in this section for idealized equilibria where each individual incurs the same probability of starvation, and no individual can unilaterally change its resource-acquisition role without increasing its chance of starvation. This solution would qualify as an ESS conditioned on foraging group size, as in the model of Barnard and Sibly (1981). Equilibration of producers' and scroungers' currencies of fitness, whether stable or not, will refer to an entire foraging period where any individual may exploit a number of different food clumps. We do not necessarily assume that within any particular clump the producer and any scrounger(s) can expect to acquire the same number of food items (Giraldeau et al. 1990).

Numerical evaluations of our models indicate that for some group sizes, the discreteness of the applicable combinations of producer and scrounger may allow "fitness" differences at the Nash equilibrium (see Math Box 1.1 at the back of chapter 1), and scroungers "do better" in such cases. Furthermore, our numerical work shows that some Nash equilibrium combinations of producers and scroungers imply that all group members will likely starve. Faced with this situation, the group might simply disperse. But under certain conditions (Axelrod and Dion 1988; Hirshleifer and Rasmussen 1988; Boerlijst et al. 1997), group members might cooperate conditionally over repeated play. Given a Nash equilibrium mix of producers and scroungers, any scrounger that switches to producing decreases each group member's penalty as a consequence. The remaining scroungers, however, benefit more than the producers. As the frequency of scrounging decreases, an individual switching to producing *may* increase its own chance of starvation while decreasing every other group member's penalty (see Boyd and Richerson 1988; Dugatkin 1990). But combinations of food density and population structure promoting repeated interaction might allow conditional cooperation, as in an iterated N-person Prisoner's Dilemma, to reduce scrounging and so increase total food production by the group. Not surprisingly, a group of all producers is the Pareto optimal solution (as defined in Hirshleifer and Rasmussen 1988), where each player in the producer-scrounger game has the same starvation probability. Hence we also consider how the frequency of scrounging might decline as a conditional mutualism.

THE MODEL

The following model is derived by Caraco and Giraldeau (1991). A group of fixed size G contains P producers and S scroungers. The number of producers is a positive integer from one to G (i.e., $1 \leq P \leq G$). The number of scroungers ($S = G - P$) is a nonnegative integer between zero and $G - 1$ (i.e., $0 \leq S \leq G - 1$) so that the group always includes at least one producer, or no food will be discovered.

We shall consider two potential forms of scramble kleptoparasitism. In one case, a producer's expected food intake is diminished by scrounging, but the producer's loss does not depend on the exact number of scroungers (the *producer-priority rule*). In the other, a producer's expected intake declines strictly monotonically as the number of scroungers in the group increases (the *scramble-competition rule*). In either case, clumps are randomly apportioned among the exploiters according to a multinomial outcome.

The total time available to search for food is τ, and the physiologically required food intake is R. The cost of scrounging is ρ ($\rho < 0$ if scrounging saves energy). Both the requirement and the cost of scrounging have units of food items. Then a producer's total physiological requirement is R, and a scrounger's requirement is $(R + \rho)$. We assume that each model forager attempts to minimize the probability that its intake fails to exceed its total energy requirement (Stephens 1981).

Each producer independently discovers food clumps at constant probabilistic rate λ (see Math Box 1.3 at the end of chapter 1). We again make the simplifying assumption that handling times within clumps are negligible, and consequently are independent of the number of foragers exploiting the clump. Given this assumption, the total number of clumps discovered accumulates as a multiple Poisson process with total rate $P\lambda$.

By assumption, scroungers have the advantage of feeding at every clump located, while each producer exploits only the clumps it discovers. But scroungers pay a cost ρ. We take the cost as independent of the number of scroungers, since this seems biologically reasonable. However, in some foraging groups the individual's cost of scrounging may vary with the number of scroungers.

Suppose each clump is composed of c indivisible items. The individual producing the clump and all the scroungers compete for each item. We treat the c items as probabilistically independent multinomial trials (see Math Box 7.1 at the end of this chapter). For any particular item, the probability that the producer obtains and consumes the item is θ ($0 < \theta < 1$). The probability that the item is scrounged is $1 - \theta$. Since we assume competitive equivalence among scroungers, the probability that the jth scrounger ($j = 1, 2, \ldots, S$) acquires the item is $(1 - \theta)/S$.

We shall consider two different forms for θ. Under the producer-priority rule θ is a constant θ_1, where θ_1 can take any value in the open interval (0, 1). θ_1 functions similarly to the producer's share assumed in the deterministic model of chapter 6. In our scramble-competition model we let $\theta = \theta_2$, where

$$\theta_2 = \alpha/(S + 1) \tag{7.1}$$

for $0 < \alpha \leq 1$ (see Mangel 1990).

Under the producer-priority rule ($\theta = \theta_1$) scrounging by one or more group members reduces a producer's expected intake in a clump. However, since θ_1 is a constant, the impact of scrounging on a producer is independent of the number of scroungers as long as one is present. This rule admits the possibility of each producer maintaining a competitive advantage within clumps it discovers. Arriving first at a food clump may allow a forager to position itself advantageously according to any within-clump spatial variability (see Ward 1986). The experimental results of Giraldeau et al. (1990, 1994b) indicate that in avian flocks producers may often consume more of a clump than does the average scrounger. The producer-priority rule offers a simple way to characterize such situations. Although this rule assumes that a producer's expected intake in a clump does not depend on the number of scroungers (once $S \geq 1$), each scrounger expects less food as their number at the clump increases.

The alternative scramble-competition rule ($\theta = \theta_2$), requires that a producer's expected intake in a clump decline with every increase in the number of scroungers. In this situation there is no producer's advantage, and the producer can expect the same share of a clump as each of the scroungers when $\alpha = 1$. Each scrounger's intake within a given clump declines as the number of scroungers increases. Under the scramble-competition rule, all $(S + 1)$ individuals exploiting a clump have the same expected intake in that clump when $\alpha = 1$. When $\alpha < 1$, the producer suffers a competitive disadvantage within clumps it locates.

The ith producer ($i = 1, 2, \ldots, P$) consumes a total of $X_i(\tau)$ food items, all eaten within clumps that producer discovers. The probability function for $X_i(\tau)$, and hence a producer's chance of starving, depends on two sources of random variation; both the number of clumps a producer finds and the number of items consumed within any of these clumps vary randomly.

Producer i discovers $Y_i(\tau)$ food clumps during the available foraging time. $Y_i(\tau)$ has a Poisson distribution with mean $\lambda\tau$. Each of these clumps contains c items. The producer obtains p_y of these c itmes in the yth clump ($y = 1, 2, \ldots, Y_i(\tau)$), where p_y is a binomial random variable with mean θc. Therefore,

$$X_i(\tau) = \sum_{y=0}^{Y_i(\tau)} p_y. \tag{7.2}$$

Our assumptions imply that each $X_i(\tau)$ has an independent Poisson-binomial distribution (Shumway and Gurland 1960) with mean $\theta c \lambda \tau$. See Math Box 7.1 at the back of the chapter for details.

Since $X_i(\tau)$ is a sum of random variables, a normal approximation should be reasonable, provided that the expected number of food items is sufficiently large. A producer's probability of starvation is $\Pr[X_i(\tau) \leq R]$. Taking the mean $E[X_i(\tau)]$ and variance $V[X_i(\tau)]$ from Math Box 7.1, and using a standard normal approximation, this probability becomes $\Phi(z_P)$ where z_P is

$$z_P = (R - \theta c\lambda\tau)/[\theta c\lambda\tau(1 - \theta + \theta c)]^{1/2}. \tag{7.3}$$

Since the normalized probability of starvation ($\Phi(z)$) increases strictly monotonically in its argument, any reduction in the probability of starvation must imply a decrease in z. Further, any two equal starvation probabilities must have the same z-value.

If the group contains no scrounger ($P = G$), the number of items eaten by all the producers is simply the number of clumps encountered times the number of seeds in each; i.e., $X_i(\tau) = cY_i(\tau)$. Since the number of items per clump is a constant, when each individual plays producer we have

$$z_G = (R - c\lambda\tau)/c(\lambda\tau)^{1/2}. \tag{7.4}$$

Of course, as the producer's priority increases, i.e., as θ approaches unity, z_P approaches z_G.

Each of S scroungers attends all of the clumps found by producers. The jth scrounger consumes a total of $X_j(\tau)$ food items. Both the total number of clumps a scrounger exploits and the number of items acquired by a scrounger within any clump vary randomly. These assumptions together imply that a scrounger's total food consumption has a Poisson-binomial distribution with mean $(1 - \theta)cP\lambda\tau/S$; see Math Box 7.1.

$\Pr[X_j(\tau) \leq R + \rho]$ is the probability that a scrounger starves. The associated standardized normal variate is z_S:

$$z_s = \frac{R + \rho - [(1 - \theta)cP\lambda\tau/S]}{(V[X_j(\tau)])^{1/2}}, \tag{7.5}$$

where the variance of a scrounger's food consumption, $V[X_j(\tau)]$, is given by equation (7.1.9) in Math Box 7.1. This last expression shows that a scrounger's probability of starvation increases with the cost of scrounging ρ.

Now that we have the standard normal variate for both producer and scrounger, we can analyze effects of the kleptoparasitism rules we propose. For the producer-priority rule, we simply let $\theta = \theta_1$ in the expression for z_k ($k = P, S$). For scramble competition, we let $\theta = \alpha/(S + 1)$.

Assuming producer priority, neither a producer's expected food-item consumption ($E[X_i(\tau)]$) nor its variance ($V[X_i(\tau)]$) depends on the number of scroungers once scrounging appears in the group. Hence a producer's probability of energetic failure (z_P) increases as soon as one or more individuals first choose to scrounge, but it does not continue to decline as the number of scroungers increases. Both a scrounger's expected food-item intake

(E[$X_j(\tau)$]) and its variance (V[$X_j(\tau)$]) decline as the number of scroungers increases, but the effect on the mean predominates. That is, any increase in the number of scroungers increases the scroungers' probability of energetic failure.

Under the scramble-competition rule, both a producer's and a scrounger's probability of energetic failure necessarily increases as the number of scroungers increases. Figure 7.1 shows how the probabilities of energetic

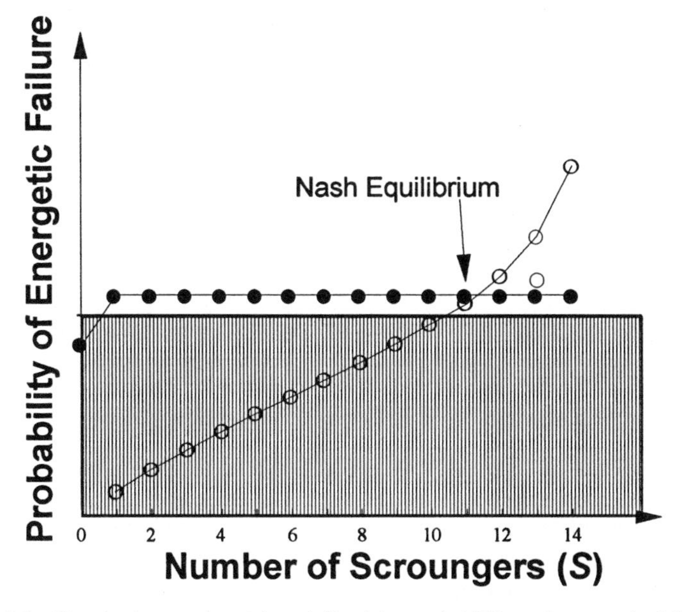

Figure 7.1 Standard normal variates defined by probabilities of energetic failure for producers (z_p; solid circles) and scroungers (z_s; open circles) as the number of scroungers in the group increases using the producer-priority rule. Group size = 15 foragers. The plot is diagrammatic, designed to represent a broad pattern in the numerical results. Note that the first occurrence of scrounging increases z_p but z_p thereafter is independent of the number of scroungers.

As either z-score increases through 0, an individual's expected energy budget changes from positive (sufficient food to meet requirements, shaded area) to negative (insufficient food, unshaded area). If the group contains only producers ($S = 0$), each individual experiences a positive energy budget in this example. But the first producer to choose scrounging decreases its probability of an energetic failure, as does the second, etc. When $S = 11$, a producer increases its probability of an energetic failure by switching to scrounging, and a scrounger increases its probability of failure by switching to producing. Hence ($\hat{P} = 4, \hat{S} = 11$) is the Nash equilibrium. Note that in the example the equilibrium implies that all foragers expect an energetic failure. In some cases, z_s for $S = 1$ exceeds the z-score for each individual when all group members produce. Then ($\hat{P} = G$, $\hat{S} = 0$) is the Nash equilibrium, and no scrounging occurs. Modified from Caraco and Giraldeau (1991).

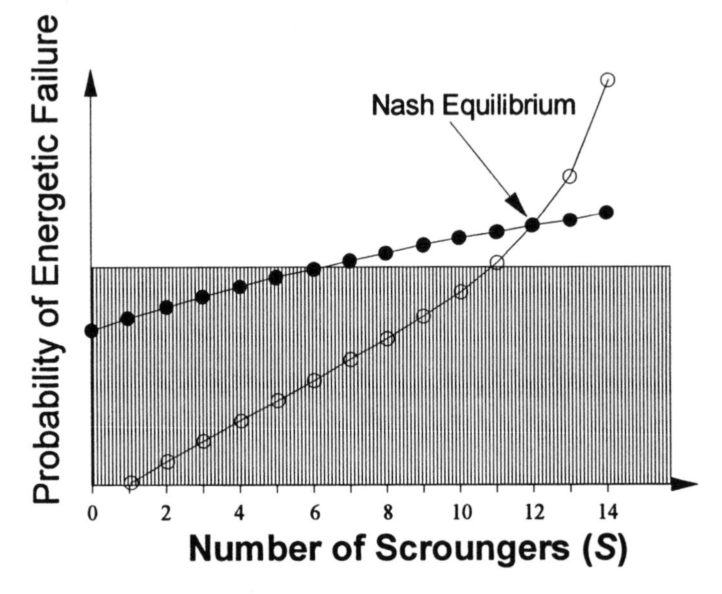

Figure 7.2 Standard normal variates defined by probabilities of energetic failure for producers and scroungers as the number of scroungers in the group increases under the scramble-competition rule. Group size $G = 15$ foragers. Axes and symbols are the same as figure 7.1. For scramble competition, each additional scrounger increases the probability of an energetic failure for all other group members. ($\hat{P} = 3$, $\hat{S} = 12$) is the Nash equilibrium in the diagram and, once again, it implies expected energetic failure.

failure for producers and scroungers vary with the number of scroungers for the producer-priority rule. Figure 7.2 does the same for the scramble-competition rule. The assumptions of the stochastic producer-scrounger game are collected in Summary Box 7.1.

7.3 Analysis of the Stochastic Game

Conditions for Nash Equilibrium Solutions

Our producer-scrounger model always has a stable Nash solution ($\hat{P}$, $\hat{S}$), where $\hat{P} + \hat{S} = G$. Solutions dichotomize conveniently into no scrounging and a mix of producers and scroungers. A group of only producers will be stable if the probability of energetic failure when $\hat{P} = G$ is less than the probability of energetic failure of the first individual switching to the scrounger alternative. So, a group of all producers is stable if

$$\Pr[X_i(\tau) \leq R|P = \mathrm{G}] < \Pr[X_j(\tau) \leq R + \rho|P = G - 1]. \qquad (7.6)$$

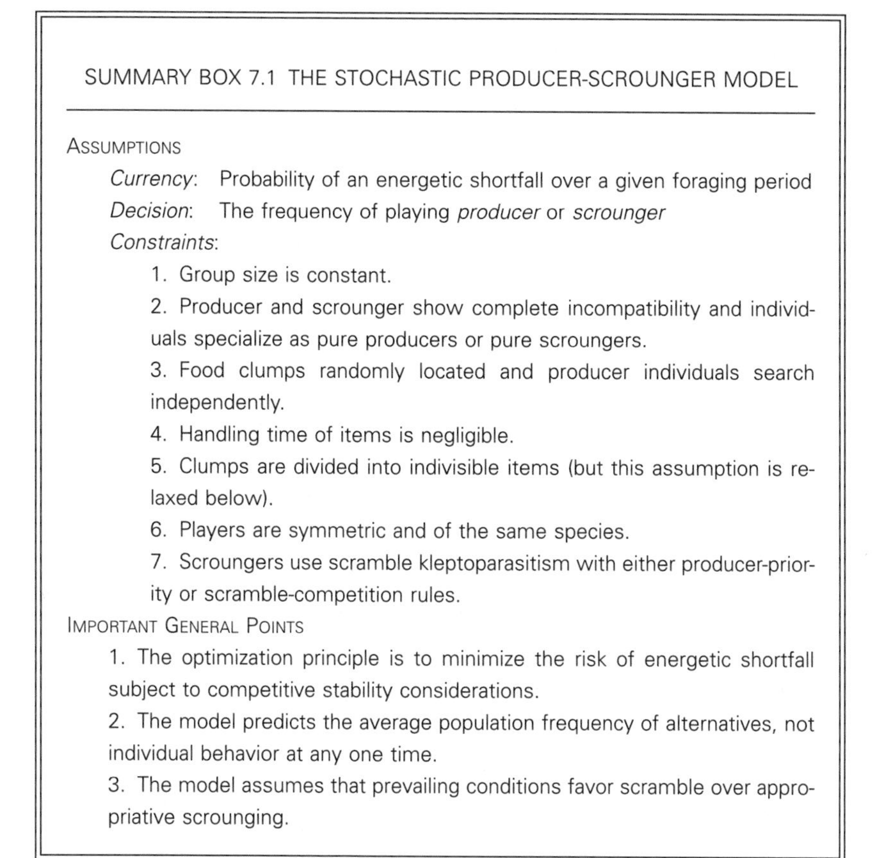

SUMMARY BOX 7.1 THE STOCHASTIC PRODUCER-SCROUNGER MODEL

ASSUMPTIONS

Currency: Probability of an energetic shortfall over a given foraging period

Decision: The frequency of playing *producer* or *scrounger*

Constraints:

1. Group size is constant.
2. Producer and scrounger show complete incompatibility and individuals specialize as pure producers or pure scroungers.
3. Food clumps randomly located and producer individuals search independently.
4. Handling time of items is negligible.
5. Clumps are divided into indivisible items (but this assumption is relaxed below).
6. Players are symmetric and of the same species.
7. Scroungers use scramble kleptoparasitism with either producer-priority or scramble-competition rules.

IMPORTANT GENERAL POINTS

1. The optimization principle is to minimize the risk of energetic shortfall subject to competitive stability considerations.
2. The model predicts the average population frequency of alternatives, not individual behavior at any one time.
3. The model assumes that prevailing conditions favor scramble over appropriative scrounging.

No producer will switch to scrounging since its chance of starvation would then increase. Under the normal approximation, inequality (7.6) implies that $z_G < z_S(S = 1)$ when the Nash equilibrium is $(\hat{P} = G, \hat{S} = 0)$.

Now suppose the solution predicts a group composed of both producers and scroungers. Producers and scroungers have the same probability of energetic failure at equilibrium. Then the condition

$$\Pr[X_i(\tau) \leq R | P = \hat{P}] = \Pr[X_j(\tau) \leq R + \rho | S = \hat{S}] \qquad (7.7)$$

defines the equilibrium where $\hat{P}, \hat{S} > 0$. Under the normal approximation, equation (7.7) requires $z_P(\hat{P}) = z_S(\hat{S})$. In practice, integer values of the number of producers and scroungers will seldom allow exact equality of z-scores. But the implication of the equilibrium is clear. Starting with a group of all producers and no scroungers ($P = G, S = 0$), producers begin switching to scrounging and continue to do so as long as the probability of energetic failure as a producer is greater than the probability of failure after becoming the next scrounger (i.e., $z_P(m) > z_S(G - m + 1)$). At some point this inequality no longer holds; the next producer to switch would increase its probability of an energetic failure. That mix of producers and scroungers where no individual should unilaterally change its resource-acquisition mode constitutes the equilibrium.

Suppose inequality (7.6) holds, so that the equilibrium number of producers and scroungers is G and zero, respectively. Then any individual switching unilaterally to scrounging would incur an increased probability of starvation. To evaluate this case simply, assume that clump size c grows large, so that any advantage to producer is small; that is, assume $\theta \ll (1 - \theta)c$. Then the multinomial allocation of food items within clumps approaches its expectation, and the variance of a scrounger's total food intake (equation 7.1.9 in Math Box 7.1) approaches $P\lambda\tau[(1 - \theta)c/S]^2$. Using this approximation in expression (7.5) for z_s simplifies the stability condition where all group members play producer. When clumps are large, the probability of energetic failure of each of G producers is lower than that of the first scrounger (i.e., $z_G < z_S(S = 1)$) when

$$R[(1 - \theta)P^{1/2} - 1] + (1 - \theta)c\lambda\tau(P - P^{1/2}) < \rho, \qquad (7.8)$$

where the number of producers P is evaluated at $(G - 1)$. Inspection of this inequality suggests several predictions. When clumps contain a large number of items, the incidence of scrounging should be low when

1. The cost of scrounging ρ is high.
2. Producers compete effectively in clumps they discover (i.e., when θ_1 approaches 1 for the producer-priority rule and θ_2 approaches ½ for the scramble competition rule).
3. Group size G is small.
4. The producer expects to find relatively few food clumps during the time available ($\lambda\tau$ is small; below we identify conditions reversing this prediction).
5. The producer's physiological requirement is large if $(1 - \theta) < P^{-1/2}$, or when the requirement is small if $(1 - \theta) > P^{-1/2}$.

Clearly, a high cost of scrounging and a competitive advantage to producers, especially when group size is small, should always limit the incidence of scrounging. The effect of the expected number of clumps ($\lambda\tau$) makes intuitive sense; it implies that the incidence of scrounging is likely to increase as the number of clumps a producer expects to discover increases. However, the more detailed numerical analysis below shows that the prediction may change depending on the ratio of clump size to group size. Prediction (5) suggests that a greater metabolic requirement reduces the incidence of scrounging when a producer has a competitive advantage and group size is not too large. But a smaller energetic requirement is necessary to maintain a low incidence of scrounging when a producer lacks a competitive advantage, especially if group size is large.

Relaxing the Assumption of Indivisible Items: Infinitely Divisible Clumps

Before reporting numerical analyses of the above model, we recall the difference between clumps made up of c indivisible food items and infinitely divisible food clumps (Caraco 1987; Recer and Caraco 1989a; see chapter 5). In the preceding paragraphs, whether we assumed producer-priority or scramble-competition, an individual's food consumption within a clump varied randomly according to a multinomial distribution of indivisible items among foragers. This assumption corresponds to a patch of seeds or any local concentration of discrete items. However, it may not apply when different foragers divide the same large source of food (e.g., lions consuming a zebra or social spiders exploiting a prey item). In the latter case, clumps can be divided according to any set of proportions summing to unity (e.g., Caraco 1981; Clark and Mangel 1986; Ekman and Rosander 1987). So, we compare the indivisible-item and divisible-clump assumptions to ask if they predict different patterns of producing and scrounging.

In this formulation of the model, food intake is counted in units of "clumps." Each producer again encounters food clumps as an independent Poisson process with probabilistic rate λ. The total number of clumps found by P producers through time τ is a Poisson variate with mean $P\lambda\tau$.

Each producer consumes a fraction θ of every clump it discovers. $X_i(\tau)$ is the total food intake of producer i; $\mathrm{E}[X_i(\tau)] = \theta\lambda\tau$, and $\mathrm{V}[X_i(\tau)] = \theta^2\lambda\tau$. As above, $X_j(\tau)$ is the total food intake of the jth scrounger; $\mathrm{E}[X_j(\tau)] = (1 - \theta)P\lambda\tau/S$, and $\mathrm{V}[X_j(\tau)] = [(1 - \theta)/S]\mathrm{E}[X_j(\tau)]$.

We again take a standard normal approximation to the probabilities of energetic failure. If all group members produce, the standard normal approximation to $\Pr[X_i(\tau) \leq R']$ has z-score $z_G = (R' - \lambda\tau)/(\lambda\tau)^{1/2}$, where R' is a producer's requirement in units of clumps. If the group contains both producers and scroungers, the standard normal variate for each producer is $z_P = (R' - \theta\lambda\tau)/\theta(\lambda\tau)^{1/2}$. For the producer-priority rule $(\theta = \theta_1)$ in the expression for z_P. For scramble competition, $(\theta = \theta_2)$ and we have

$$z_P = [R'(S + 1) - \alpha\lambda\tau]/\alpha(\lambda\tau)^{1/2}. \tag{7.9}$$

The corresponding z-scores for scroungers follow easily. For the producer-priority rule:

$$z_S = \frac{(R' + \rho)S - (1 - \theta_1)P\lambda\tau}{(1 - \theta_1)(P\lambda\tau)^{1/2}}, \tag{7.10}$$

where the cost of scrounging (ρ) is scaled in the same units as R'. For the scramble-competition rule,

$$z_S = \frac{(R' + \rho)S(S + 1) - (S + 1 - \alpha)P\lambda\tau}{(S + 1 - \alpha)(P\lambda\tau)^{1/2}}. \tag{7.11}$$

These formulas compare logically with their counterparts in the indivisible-item analysis. Once scaled by clump size (see below) the mean intake levels are the same, but the variances are greater for indivisible items than for divisible clumps. Hence any differences induced in equilibrium numbers of producers and scroungers must follow from effects of different food-intake variances between the two ways of dividing the resources.

Suppose that all group members play producer at the stable solution of the game; i.e., $(\hat{P}, \hat{S}) = (G, 0)$. Then each producer's probability of an energetic shortfall must be smaller than the corresponding probability of the first individual to switch to scrounging. In terms of standard normal variates, $z_G < z_S(S = 1)$. However, if the game's solution calls for a mixture of producers and scroungers, the equilibrium requires that each producer's probability of an energetic shortfall equal the corresponding probability for scroungers. Consequently, $z_P(\hat{P}) = z_S(\hat{S})$ for a Nash equilibrium mix of producers and scroungers.

The inequality $z_G < z_S(S = 1)$ holds where all-producer is stable and reduces to

$$R'[(1 - \theta)P^{1/2} - 1] + (1 - \theta)\lambda\tau(P - P^{1/2}) < \rho, \tag{7.12}$$

where P is evaluated at $= (G - 1)$, and $\theta = \theta_1$ or θ_2. The last inequality recovers expression (7.8), with the physiological requirement and cost of scrounging now scaled in clumps. Hence, the reformulated model leads to similar conclusions concerning the incidence of scrounging. Predictions of the stochastic model under the producer priority rule are listed in Summary Box 7.2. Predictions for the producer-scrounger game under the scramble competition rule are listed in Summary Box 7.3.

7.4 Numerical Evaluation

We calculated z-values and located Nash equilibrium numbers of producers and scroungers across a broad array of group sizes, food availabilities, finder's shares (θ_1 or α as appropriate), scrounging costs, and physiological requirement (Caraco and Giraldeau 1991). Table 7.1 lists parameter values used. All calculations involved the exact means and variances given in Math Boxes 7.1 and 7.2 at the end of this chapter. For divisible clumps we scaled the basic requirement as (R/c) clumps, and the cost of scrounging as (ρ/c)

SUMMARY BOX 7.2 PREDICTIONS: STOCHASTIC PRODUCER-SCROUNGER MODEL FOR THE PRODUCER-PRIORITY RULE

CLUMPS OF INDIVISIBLE ITEMS

Producer priority (θ_1) has strong effect; as θ_1 increases:

1. Stable scrounger frequency decreases.
2. Expected items encountered ($cP\lambda\tau$) increase because of fewer scroungers.
3. Probability of energetic failure declines.

Cost of scrounging (ρ) has weak effect; as ρ increases, $\hat{S}$ decreases.

Clump size (c) is important when it changes sign of energy budget; as c increases, $\hat{S}$ increases.

Group size (G): Has strong effect; as G increases, $\hat{S}$ increases.

Physiological requirement (R): Effect depends on θ_1.

When θ_1 is small, as R increases, $\hat{S}$ increases.

When θ_1 is large, as R increases, $\hat{S}$ decreases.

Producer's expected clump discoveries ($\lambda\tau$): Effect depends on c.

When $c > G/2$, as $\lambda\tau$ increases, $\hat{S}$ also increases.

When $c < G/2$, as $\lambda\tau$ increases, $\hat{S}$ decreases.

Patchiness (decreased $\lambda\tau$ with $c\lambda\tau$ constant): Generally negligible effect.

CONTINUOUSLY DIVISIBLE CLUMPS

All predictions same as above except:

1. When food density is low (small $\lambda\tau$) and foragers expect negative energy budget, $\hat{S}$ is slightly less than for indivisible items.
2. When $\lambda\tau$ is large and expected energy budgets are positive, $\hat{S}$ is slightly larger than for the indivisible-items scenario.

clumps, where c is clump size in the comparable calculation for indivisible items.

USING THE PRODUCER-PRIORITY RULE

Variation in θ_1, the producer's probability of acquiring any given food item it discovers clearly had the strongest influence on the solution to the producer-scrounger game. The equilibrium number of scroungers declined, not surprisingly, as θ_1 increased whether encounter rates with clumps were slow or fast (fig. 7.3). Since fewer scroungers implies more producers, the expected number of items discovered during foraging ($cP\lambda\tau$ items) increases as

SUMMARY BOX 7.3 PREDICTIONS: STOCHASTIC PRODUCER-SCROUNGER MODEL FOR THE SCRAMBLE-COMPETITION RULE

CLUMPS OF INDIVISIBLE ITEMS

Producer's acquisition probability (α):

When α increases, $\hat{S}$ decreases.

Cost of scrounging (ρ); effect stronger than in producer priority:

When ρ increases, $\hat{S}$ decreases.

Clump size (c); effect depends on ρ:

When ρ is small, c has no effect.

When ρ is large, as c increases, $\hat{S}$ increases.

Group size (G):

As G increases, $\hat{S}$ either increases or remains the same (no decrease).

Physiological Requirement (R); effect depends on α:

When α is large ($\alpha \approx 1$), as R increases, $\hat{S}$ decreases.

When α is small, as R increases, $\hat{S}$ increases.

Producer's Food Clump Encounters ($\lambda\tau$); effect depends on ρ:

When ρ is large: as $\lambda\tau$ increases, $\hat{S}$ increases.

When ρ is small: no effect of $\lambda\tau$ on $\hat{S}$.

CONTINUOUSLY DIVISIBLE CLUMPS

All predictions as for clumps of indivisible items except that all foragers more likely to experience a negative energy budget at equilibrium.

θ_1 increases. In turn, this implies that probabilities of energetic failure for both producers and scroungers, at the Nash equilibrium ($\hat{P}$, $\hat{S}$), decline as θ_1 increases. So, as producers leave less per clump for scroungers, fewer individuals should scrounge, and each group member should have a greater chance of meeting its physiological requirement for food.

Table 7.1

Parameter values for numerical calculations of Nash equilibrium frequencies of scroungers in a stochastic producer-scrounger game. For every parameter combination we calculated the standard normal variate for every feasible number of producers and scroungers. Taken from Caraco and Giraldeau (1991). (With permission of Academic Press)

R = 100, 800, 1500	c = 1, 2, 5, 10, 35, 60
G: 3, 5, 8, 15	ρ = −50, 0, 50, 200, 800
$\lambda\tau$: 80, 160, 240, 400, 800	
$\theta_1 = \alpha$ = 0.1, 0.2, 0.25, 0.4, 0.55, 0.7, 0.85 (and α = 1.0)	

Increasing the cost of scrounging had a predictable but rather weak effect on the solution to the game (fig. 7.3). The equilibrium number of scroungers declined as the cost of scrounging increased. When $R = 800$, increasing the cost of scrounging from -50 to 200 food items caused only a small reduction in the equilibrium proportion of scroungers ($\hat{S}/G$), even though a scrounger's total requirement ($R + \rho$) was increased by one-third (fig. 7.3).

Increasing clump size c increased the equilibrium number of scroungers. The effect was most obvious if an increased number of items per clump changed energy budgets at most producer-scrounger combinations from negative to positive (see Caraco and Giraldeau 1991).

Inequality (7.8) predicts that the stable equilibrium proportion of

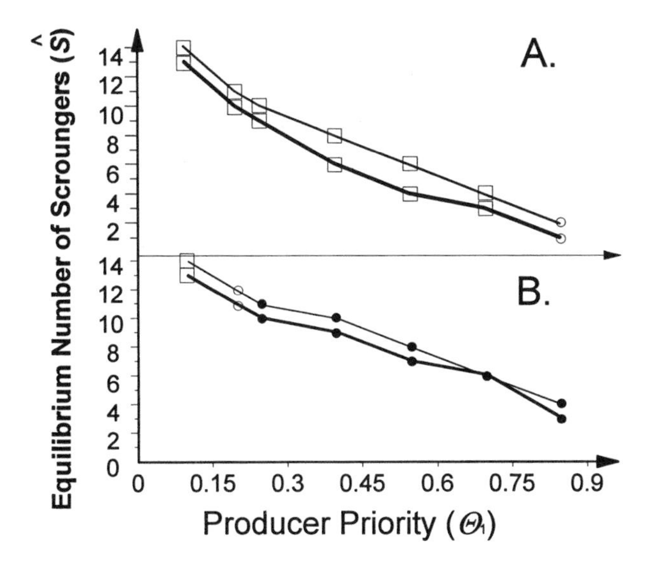

Figure 7.3 Numerical estimates of the Nash equilibrium number of scroungers ($\hat{S}$) in the stochastic, producer-scrounger game as θ_1 increases. Note that the equilibrium number of scroungers ($\hat{S}$) declines as the producer's priority (θ_1) increases. This is so whether a producer's patch encounter rate is low ($\lambda\tau = 80$) or high ($\lambda\tau = 400$). Increasing the cost of scrounging from small ($\rho = -50$; thin line) to large ($\rho = 400$; bold line) has only a weak effect on $\hat{S}$. Group size = 15. Open squares designate equilibria where both producers and scroungers expect a negative energy budget. Open circles designate equilibria where producers expect a negative energy budget and scroungers expect a positive energy budget (simply due to the discrete nature of the number of producers and scroungers). Solid circles designate equilibria where both producers and scroungers expect a positive energy budget. Modified from Caraco and Giraldeau (1991). (With permission of Academic Press)

scroungers should increase as group size increases. Numerical results for the producer-priority rule exhibit this property. Figure 7.4 shows examples where the stable equilibrium proportion of scroungers usually increases with group size. In some cases the proportion of scroungers more than doubles as group size increases from three to fifteen individuals while other parameters are held constant. Vickery et al.'s (1991) model predicts a similar qualitative effect.

Increasing the mean number of clumps a producer finds ($\lambda\tau$) had a surprising effect when the producer-priority rule is used. Based on expression (7.7) one might anticipate that increasing $\lambda\tau$ would lead to increased scrounging. This was true when the number of items per clump (c) was sufficiently large. In our calculations the equilibrium number of scroungers increased as $\lambda\tau$ increased when the number of items in a clump exceeded $G/2$. However, for very small clumps (i.e., $c = 1$ or 2 food items) the equilibrium number of scroungers decreased as the number of clumps a producer expects to find increased; see fig. 7.5. Once clumps were small enough to discourage scrounging, increased food density (λ) actually increased the equilibrium number of producers.

To examine the consequences of increased environmental patchiness, we increased clump size and simultaneously decreased the expected number of encountered patches while holding the product ($c\lambda\tau$) constant. Increments in the patchiness of food either had no effect on the game's solution or, more often, induced a small increase in the equilibrium number of scroungers (cf.

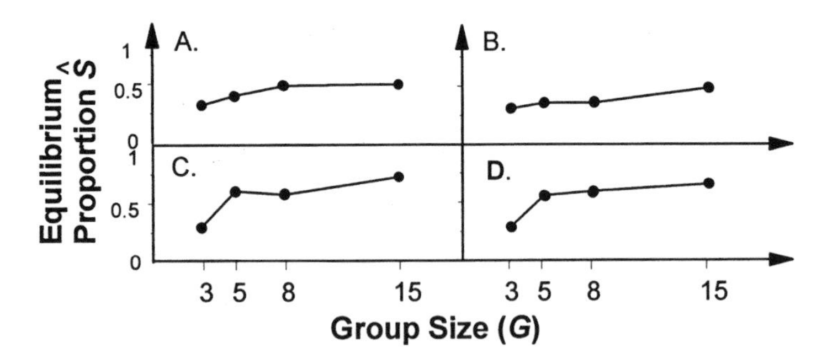

Figure 7.4 The proportion of scroungers in a stochastic producer-scrounger game using the producer-priority rule, based on the numerically estimated Nash-equilibrium number of scroungers. Graphs on the left, (A) and (C), depict low scrounging costs ($\rho = -50$) while those on the right (B and D) depict high scrounging costs ($\rho = 200$). Top graphs (A and B) depict low patch encounter rate ($\lambda\tau = 80$) while the lower graphs (C and D) depict high encounter rate ($\lambda\tau = 400$). Clump size, requirement and producer priority are held constant ($c = 35$, $R = 800$, and $\theta_1 = 0.55$, respectively). Modified from Caraco and Giraldeau (1991). (With permission of Academic Press)

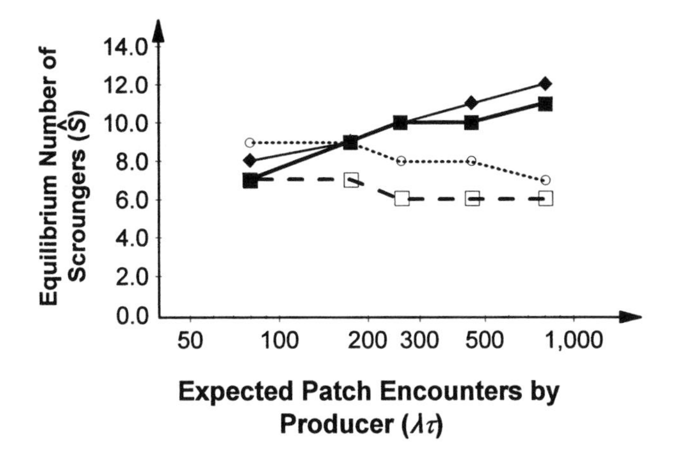

Figure 7.5 The Nash equilibrium number of scroungers ($\hat{S}$) in stochastic producer-scrounger game using the producer-priority rule as a function of the expected number of clumps encountered by producers ($\lambda\tau$). $\hat{S}$ does not always increase with $\lambda\tau$. In a group of fifteen foragers, the effect of $\lambda\tau$ on $\hat{S}$ depends on clump size c, $c = 35$ (full lines); $c = 1$ (dashed lines). Cost of scrounging (ρ) has little effect on direction of effect ($\rho = -50$, thin lines; $\rho = 200$, bold lines). Requirement $R = 800$ in all cases. Modified from Caraco and Giraldeau (1991). (With permission of Academic Press)

Clark and Mangel 1984). Larger, but rarer, clumps can enhance the benefit of scrounging.

Inequality (7.8) indicates that scrounging will decrease as the requirement increases when θ_1 is large, but that scrounging will increase with requirement when θ_1 is small. Our numerical results confirmed these predictions, but the magnitudes of the two effects differed substantially. Table 7.2 gives an example where θ_1 is large; increasing the basic energy requirement causes a strong decline in the equilibrium number of scroungers. Table 7.3 gives an example where θ_1 is small. In this case the few changes noted in the equilibrium number of scroungers are usually increases with increased requirement. But essentially the level of scrounging is independent of requirement (since low θ_1 most often implies a large $\hat{S}/G$).

The pattern exhibited in table 7.3 has an important effect on starvation probabilities. Increasing the physiological requirement decreased the equilibrium number of scroungers, and so producers increased. Increasing the number of producers means more clumps are encountered, so that expected energy budgets at equilibrium often changed from negative to positive. Therefore, increments in the physiologically required food intake have the counter-intuitive result of lowering starvation probabilities through a reduction in the incidence of scrounging.

Rescaling the requirement and cost of scrounging relative to the number

Table 7.2

Results of the numerical calculations of Nash equilibrium number of scroungers ($\hat{S}$) in the stochastic producer-scrounger game under the producer-priority rule when producers hold a large competitive weight ($\theta_1 = 0.85$). $\hat{S}$ decreases as an individual's physiological requirement (R), increases. For given expected rates of encounter with patches by producers ($\lambda\tau$) and cost of scrounging (ρ), $\hat{S}$ decreases across the three values of R. Group size $G = 15$, and number of items per clump $c = 35$ in all cases. Allocation of food items was random. Taken from Caraco and Giraldeau (1991). (With permission of Academic Press)

		$\lambda\tau$:	80	160	240	400
$R = 100$						
	ρ:	−50	10	10	11	11
		0	10	10	10	11
		50	8	9	10	10
		200	6	8	8	9
$R = 800$		$\lambda\tau$:	80	160	240	400
	ρ:	−50	4	6	7	8
		0	4	5	6	8
		50	4	5	6	7
		200	3	5	6	7
$R = 1500$		$\lambda\tau$:	80	160	240	400
	ρ:	−50	3	4	5	6
		0	2	4	5	6
		50	2	4	5	6
		200	2	3	4	6

Table 7.3

Same as table 7.2 except that now producers have weak competitive weight $\theta_2 = 0.1$. The equilibrium number of scroungers ($\hat{S}$) does not change, or may increase, as an individual's physiological requirement (R) increases. Each of the clump-encounter-scrounging cost ($\lambda\tau$, ρ) combinations used in these calculations predicts $\hat{S} = 13$ when $R = 1500$. Taken from Caraco and Giraldeau (1991). (With permission of Academic Press)

		$\lambda\tau$:	80	160	240	400
$R = 100$						
	ρ:	−50	13	13	13	13
		0	13	13	13	13
		50	13	13	13	13
		200	12	12	13	13
$R = 800$		$\lambda\tau$:	80	160	240	400
	ρ:	−50	13	13	13	13
		0	13	13	13	13
		−50	13	13	13	13
		200	12	13	13	13

of items per clump (R/c) and (ρ/c) lets us compare the results for indivisible items to the results of the comparable calculation for divisible clumps. We found no large differences between the stable producer-scrounger combinations. This is the important point of the comparison; it strengthens our confidence in the simpler divisible-clump models (Caraco 1987). When predictions of the two assumptions differed under producer-priority, we noted an obvious pattern. Under low food density and generally negative expected energy budgets, divisible clumps sometimes suggested one or two fewer scroungers than the indivisible-items version did. The pattern was just the opposite under high food density and generally positive energy budgets. Divisible clumps reduce the coefficient of variation of an individual's food intake for given numbers of producers and scroungers when compared to the individual-items case (Caraco 1987). When expected energy budgets are negative, this reduced variance is detrimental and increases in mean intake are necessarily very beneficial. When expected energy budgets are positive, reduced intake variance becomes beneficial, and increases in mean intake are no longer as important. Consequently, the lower intake variability of the divisible-clump version can predict fewer scroungers at low food density and more scroungers at high food density than the indivisible-items version of the producer-priority model does.

Using the Scramble-Competition Rule

Recall that α is the producer's probability of acquiring any given food item in clumps it discovers under scramble competition. When this probability increases, the equilibrium number of scroungers decreases, a sensible result. Effects of varying α in this model are smaller than comparable effects of θ_1 under the producer-priority rule. This follows because the scramble competition rule imposes a greater restriction on a producer's chance of achieving competitive superiority over scroungers.

Increasing the cost of scrounging again reduces the equilibrium number of scroungers. The cost of scrounging is more important under scramble competition than it is for producer priority, since effects of food density and spatial dispersion now depend on the cost of scrounging. If the cost of scrounging is relatively small (e.g., $\rho = -50$ food items), increasing clump size has no effect on the solution to the game. However, if the scrounging cost is relatively large (e.g., $\rho = 800$ items), the equilibrium number of scroungers increases as clump size increases. Figure 7.6 shows how the equilibrium number of scroungers differs between low and high costs of scrounging. Similarly, the equilibrium number of scroungers increases as either the number of clumps encountered or food patchiness (defined above) increases when the cost of scrounging is large.

As suggested by inequality (7.8), the stable equilibrium proportion of

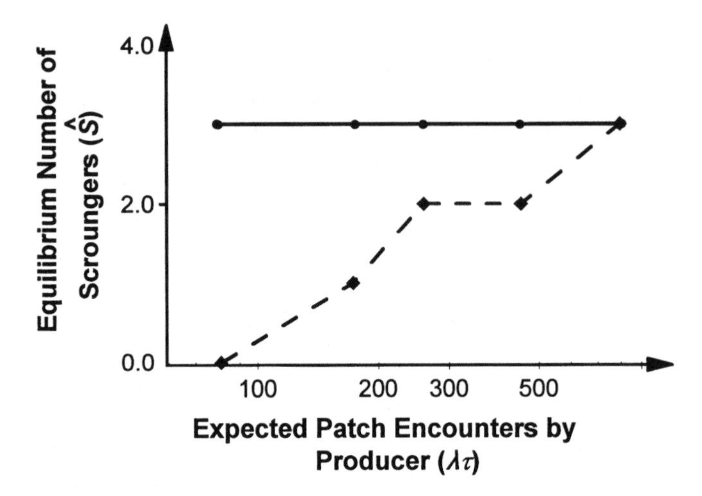

Figure 7.6 The Nash equilibrium number of scroungers ($\hat{S}$) in a stochastic producer-scrounger game using the scramble-competition rule as a function of the expected number of clumps encountered by producers ($\lambda\tau$). The effect of $\lambda\tau$ on $\hat{S}$ now depends on the cost of scrounging ρ. Circles indicate results for small cost of scrounging ($\rho = -50$); diamonds indicate results for large cost of scrounging ($\rho = 200$). In a group of five at low scrounging costs, the equilibrium number of scroungers is independent of the producer's expected number of clumps ($\lambda\tau$). $\hat{S}$ increases with $\lambda\tau$ at a high cost of scrounging. In each case $c = 10$, $R = 800$, and $\alpha = 1$. Modified from Caraco and Giraldeau (1991). (With permission of Academic Press)

scroungers increased with group size (or did not decrease with G). Under scramble competition, the magnitude of this effect depended on the producer's acquisition probability (α) but was essentially insensitive to variation in either the number of clumps a producer expects to find ($\lambda\tau$) or in the cost of scrounging.

As for the producer-priority rule, the results for scramble competition suggest that a greater physiological requirement may lead to reduced scrounging when producers compete as effectively as scroungers. But the equilibrium number of scroungers can increase with the requirement when producers suffer a competitive disadvantage.

We also compared the indivisible-items and divisible-clump versions of the scramble-competition model. Generally, the equilibrium number of scroungers differed little between the two forms of the model. However, the probabilities of energetic failure could differ and sometimes differed considerably even though the solutions were the same.

Under the producer-priority rule, the first individual switching from producing to scrounging increases each remaining producer's probability of an energetic failure. However, further increases in the number of scroungers do not induce additional increments in a producer's chance of failure. Under the scramble-competition rule, each sequential switch from producing to scrounging further increases each remaining producer's chance of an energetic failure. Scrounging penalizes producers more, by design, under the scramble-competition rule than under the producer-priority rule (for sufficiently large θ_1). Hence, it is no surprise that the frequency of scrounging at equilibrium tends to be higher for the scramble-competition rule. What interests us more is the decline in average "fitness" as the increase in the number of scroungers takes the group toward the Nash equilibrium combination of producing and scrounging.

The Nash solution to each game qualifies as an ESS, as defined in Barnard and Sibly's (1981) producer-scrounger model. This is true even though the discrete nature of the producer-scrounger combinations implies that scroungers can do "better" than producers at equilibrium (see fig. 7.1). The interpretation of an ESS suggesting that each individual chooses to produce or scrounge at the beginning of each foraging period with respective probabilities $\hat{P}/G$ and $\hat{S}/G$ (Vickery et al. 1991) will not apply realistically if the foraging period itself is even moderately long or if the consequences of an energetic failure are even moderately severe. Merely suppose that all group members randomly choose to scrounge, an event with probability $(\hat{S}/G)^G$ under this interpretation. Then no group member would search for food during the duration of the foraging period—an outcome we should exclude from consideration if the foraging period is long. A more reasonable interpretation supposes that some individuals consistently produce and some consistently scrounge, as Giraldeau and Lefebvre (1987) found, and the number of individuals in each role varies as influential environmental attributes vary. In any case, learning capacities of many socially foraging species may allow more behavioral flexibility in dealing with intragroup competition than the Nash equilibrium predicts.

In groups of all sizes, the models' Nash solutions *sometimes* exhibit the following property. If there are no scroungers, each producer has a low probability of failing to meet its physiological requirement. However, the absence of scroungers is often unstable to unilateral deviation; the first scrounger reduces its probability of failure (while increasing the others' chance of failure). Individuals continue to switch to scrounging until the number of producers and scroungers reaches the Nash solution (assuming $\hat{S} > 0$). But at equilibrium both the producers and scroungers experience a negative expected energy budget. Every group member will, more likely than not, fail to meet its physiological requirement. By trying to do better, each individual winds up being much worse off than if they all produced food independently. This sort of intragroup competition might force dissolution of the group; otherwise each individual might starve while employing its part of the collectively "noninvadable" strategy set. Alternatively, suppose ecological conditions external to the producer-scrounger game promote group cohesion and repeated interaction. Then conditional cooperation within the group might resolve the problem of insufficient food production sometimes associated with the game's ESS for single play.

Producing, Scrounging, and the N-Person Prisoner's Dilemma

To evaluate the plausibility of conditional mutualism in our producer-scrounger game, we recall the scramble-competition rule. For any feasible $S > 1$, each individual that switches from scrounger to producer necessarily decreases the probability of an energetic failure for every other group member (see fig. 7.2). Then the game qualifies as an N-person Prisoner's Dilemma (NPD), with producing equivalent to "cooperate" and scrounging equivalent to "defect" (see Boyd and Richerson 1988; Motro 1991; Ranta et al. 1996). For a single play of the game, the Pareto optimal solution, subject to the constraint that each group member have the same penalty probability, is that each individual produces. A group of all producers minimizes the average starvation probability within the group and maximizes the expected amount of food discovered.

Given a sufficient likelihood of repeated play, conditional mutualism might reduce the number of scroungers. Different versions of Tit-for-Tat have been hypothesized for the iterated NPD. Each includes initial cooperation, but further cooperation is conditioned on cooperation by all other $(G - 1)$ group members, or on some minimum number less than $(G - 1)$; see Boyd and Richerson (1988) or Dugatkin (1990). The biggest immediate decrease in a single individual's starvation probability (the greatest temptation to defect) accrues to the first individual to switch from producer to scrounger. Although any further scroungers will reduce this temporary advantage, the temptation itself will ordinarily be greater for larger groups. Hence, the likelihood that conditional mutualism can diminish scrounging probably declines in larger groups. As indicated above, individuals that succeed as scroungers may have difficulty learning to produce (Giraldeau and Lefebvre 1987), and "learning to cooperate" may constrain conditional mutualism (Clements and Stephens 1995). The general incidence of scrounging and opportunism among social foragers (see chapter 6) might imply that conditionally cooperative reduction of scrounging occurs only rarely.

7.5 Experimental Evidence of Risk-Sensitive Producer-Scrounger Decisions

Two predictions of the stochastic game have been tested with captive flocks of European starlings (*Sturnus vulgaris*; Koops and Giraldeau 1996). The experiments concern

1. The predicted effect of varying the expected number of clumps encountered by producers ($\lambda\tau$).
2. The predicted effect of varying the physiological requirement (R).

The Effect of Expected Number of Clumps

When the number of food items per clump exceeds half the foraging group size (i.e., $c > G/2$), the stochastic game predicts that increasing the expected number of clumps encountered by a producer ($\lambda\tau$) leads to an increased proportion of scrounging (fig. 7.5). To test this prediction, Koops and Giraldeau (1996) added two test subjects to a core flock of five starlings and recorded whether test subjects obtained their food by producing or scrounging. The procedure was repeated four times with different test pairs. Starlings foraged on a set of fifteen discrete patches that either contained five food items or none. To determine whether the patch contained food items, the starling had to probe the patch with its bill or observe the success of others.

The number of encountered clumps was changed by doubling the number of food-containing patches from 5 of 15 in the low-expectation condition to 10 of 15 in the high-expectation condition. Since the duration of the trials (τ) did not change between conditions, the effect was to double $\lambda\tau$. Although all subjects obtained food both through producer and scrounger alternatives, 8 of 8 subjects increased the proportion with which they used scrounging in the high expectation condition, as the model predicts.

The Effect of Physiological Requirement

Having established that producer individuals in flocks of seven had weak producer-priority (θ_1 was low), Koops and Giraldeau (1996) went on to test the prediction concerning the physiological requirement. When producer-priority is weak, the model predicts that the incidence of scrounging should either be unaffected or increase slightly as the requirement increases (table 7.3). Again, similar procedures were used and eight subjects were added, two by two, to a core flock of five individuals. They found that seven of eight subjects increased their incidence of scrounging in conditions of high daily food requirement, once again as the model predicts.

It is important to point out that despite the model's qualitative success, Koops and Giraldeau's (1996) experimental conditions violated several of the model's assumptions. Namely, individual starlings alternated between both foraging alternatives, food clumps were frequently discovered simultaneously, handling time was not negligible, and the study estimated the average level of scrounging in the group by observing two focal individuals. Moreover, the experimenters did not change physiological requirement directly but instead used food deprivation to alter the rate of food intake required to maintain a constant body mass. Nonetheless, the fact that both experiments provided results that were qualitatively consistent with the model suggests that the stochastic model can realistically depict producer-scrounger economies.

7.6 Concluding Remarks

Implications for Future Work

Our producer-scrounger models predict circumstances likely to advance the frequency of scrounging within foraging groups. Testing any of the model's assumptions or predictions, listed in Summary Boxes 7.1 to 7.3, should prove useful. To date, only Koops and Giraldeau (1996) have tested the model, and their results suggest qualitative support. If producer priority and the physiological requirement for food could be manipulated simultaneously, then one could test for their opposing effects on scrounger frequency in the same animals.

The stochastic model predicts that encounter rate with patches affects the frequency of scrounging (Caraco and Giraldeau 1991; Koops and Giraldeau 1996). All producer-scrounger models must specify how the corporate rate of patch discovery depends on the number of individuals concurrently searching for food, an assumption that was also crucial to risk-sensitive group-membership games (see chapter 4). While all the model's assumptions need to be subjected to empirical scrutiny, it would appear especially important to measure how a producer's patch encounter rate is affected by the addition of scroungers.

Future work should also involve theoretical developments. In particular, a dynamic state-dependent, producer-scrounger game would allow us to see whether diurnal patterns of producing and scrounging are predicted. Koops and Giraldeau (1996) noted a diurnal pattern of scrounging that could not be explained by the model they tested. All else being equal, their food-deprived starlings tended to use scrounging more early in the day, and its frequency declined during later trials. Computational approaches (Houston and McNamara 1987, 1988; McNamara et al. 1997) should provide adequate tools to develop state-dependent dynamic N-person games that may account for such diurnal patterns.

Another important theoretical question concerns player asymmetries. Both the deterministic game of chapter 6 and the stochastic models of the current chapter assume competitive symmetry. Players are considered asymmetric when the payoff they obtain from playing either alternative depends on their own phenotypic characteristics (e.g., age, size, dominance, sex, etc). Solutions to phenotype-limited scrambles have been developed by Parker (1982) and applied to the producer-scrounger game (1984a, but see Gross 1996). However, predictions are generally based on an infinite population assumption, a condition not intended to apply to a foraging group (Barta and Giraldeau 1998).

In chapter 6 we described three forms of kleptoparasitism, yet our models deal exclusively with only one: scramble kleptoparasitism. Although the payoffs obtained from using the two other forms of kleptoparasitism (i.e., aggressive and stealthful) are likely to be frequency-dependent, it is probable that the details of the economic analysis will change. Of special importance is the fact that in both other forms of kleptoparasitism, the scrounger exploits the resource clump singly. Extensions and elaborations of the hawk-dove contest for food (e.g., Houston and McNamara 1988; McNamara and Houston 1989) might predict these forms of kleptoparasitism in foraging groups.

Implications for Scaling Up

Our model fixes group size and applies N-person game theory to predict the number of producers and scroungers in the foraging group (Caraco and Giraldeau 1991). Under this model's assumptions, a substantial range of ecological conditions admits a Nash-equilibrium mix of producers and scroungers and a Pareto optimal solution where all group members produce their own food. The Nash-equilibrium mixture is stable for single and repeated play. The Pareto optimal solution lacks such stability, unless conditional mutualism over probabilistically repeated play (or some other complexity) deters scrounging.

Our analysis of the producer-scrounger interaction concerns the economics of survival probabilities within foraging groups. Since individuals in all-producer (i.e., conditionally cooperative or perhaps kin-altruistic) groups may have greater survivorship than members of groups containing both producers and scroungers, intrademic (within-population) selection at the among-group level may be possible (Wilson 1980; Grafen 1984; Leigh 1991). If reproduction depends only on surviving the within-group foraging interactions, if young disperse prior to forming the next set of foraging groups, and if both producer and scrounger phenotypes (rather than economic choice of resource-acquisition role) are heritable, then foraging groups resemble "trait groups" (Wilson 1980). Groups of producers could collectively contribute more offspring to the next generation than same-sized groups with a mix of producers and scroungers. Selection resulting from this among-group variance may, however, seldom have the strength to overcome selection within groups maintaining scrounging (Leigh 1991). More generally, effects of genetic relatedness could offer an easier way to explain selection promoting the frequency of producers in a population (see Grafen 1984).

Conclusions

Our stochastic models have allowed us to evaluate the effect of reward variance on factors that may govern the frequency of producing and scrounging in small foraging groups. Our models make a number of testable predictions, some of which parallel those made earlier by our deterministic model (chapter 6). Some of the predictions, however, follow directly from risk sensitivity. Specifically, only the stochastic models predict that the number of clumps that producers expect to encounter should affect the frequency of scrounging. Our analyses show that the direction of the effect depends on clump size, for large clumps scrounging should increase with the number of expected clumps, but scrounging should decrease when clump size is small.

Another prediction that follows from risk-sensitivity concerns the effect of the physiological requirement on scrounging. When producer priority is large, increasing requirement should reduce the frequency of scrounging. When producer priority is small, however, the frequency of scrounging is essentially independent of the physiological requirement.

All the producer-scrounger models developed to date assume an environment with a single type of food. They also assume that producers and scroungers exploit each food clump to complete depletion. Both these assumptions are unrealistic. In chapter 11 we allow for two types of food clumps and ask when between-individual searching specialization, sometimes referred to as the "skill pool" (Giraldeau 1984), might prove economically efficient. In the next chapter we explore patch exploitation decisions under competitive conditions. We allow patches to be made up of more than one prey type and analyze dietary decisions under competitive foraging conditions.

Math Box 7.1 Starvation Probabilities for Producers and Scroungers When Clumps Hold a Fixed Number of Items

A producer discovers food clumps as an independent Poisson process. Each clump contains c indivisible items; c is a positive integer.

For Producer

When a clump is discovered, the producer acquires p items and the jth scrounger acquires r_j items, so

$$c = p + \sum_{j=1}^{S} r_j\,, \tag{7.1.1}$$

since clump size constrains the sum of random variables. Each item is allocated independently, so any random $(S + 1)$ vector $(p, r_1, \ldots, r_s)$ has a multinomial probability,

$$\Pr[p,r_1, \ldots, r_s|c] = \binom{c}{p,r_1, \ldots, r_s} \theta^p \prod_{j=1}^{S} \left[\frac{1 - \theta}{S}\right]^{r_j}, \tag{7.1.2}$$

where θ is the probability the producer consumes any given item in the clump. The marginal distribution of p is binomial, since items are independent. The mean and variance of a producer's intake within each clump it discovers are $E[p] = \theta c$, and $V[p] = \theta(1 - \theta)c$.

Similarly, the marginal distribution of r_j is binomial. For $j = 1, 2, \ldots, S$, $E[r_j] = (1 - \theta)c/S$, and

$$V[r_j] = c\left[\frac{1 - \theta}{S} - \left(\frac{1 - \theta}{S}\right)^2\right]. \tag{7.1.3}$$

Producer i discovers $Y_i(\tau)$ food clumps during the available foraging time. $Y_i(\tau)$ has a Poisson probability function with expectation $\lambda\tau$. Producer i consumes a total of $X_i(\tau)$ items within these clumps. Given S scroungers in the group, each $X_i(T)$ has the same distribution, and

$$X_i(\tau) = \sum_{y=0}^{Y_i(\tau)} p_y\,. \tag{7.1.4}$$

Math Box 7.1 (*cont.*)

The distribution of $X_i(\tau)$ is a Poisson "generalized" by a binomial, a randomly stopped sum (Boswell et al. 1979) of independent and identically distributed binomial variates. Then $X_i(\tau)$ follows a Poisson-binomial probability function (Shumway and Gurland 1960):

$$\Pr[X_i(\tau) = x] = \frac{e^{-\lambda\tau} \sum_{y=0}^{\infty} (\lambda\tau)}{y!} \binom{yc}{x} \theta^x (1 - \theta)^{yc-x}, \quad (7.1.5)$$

where $X_i(\tau)$ is defined on the nonnegative integers. The mean and variance are, respectively,

$$E[X_i(\tau)] = \theta c \lambda \tau, \; V[X_i(\tau)] = \theta c \lambda \tau (1 - \theta + \theta c). \quad (7.1.6)$$

Under producer priority $\theta = \theta_1$ in (7.1.6); under scramble competition $\theta = \alpha/(S + 1)$. A producer's probability of starvation is $\Pr[X_i(\tau) \leq R]$. Using a standardized normal approximation, this probability becomes $\Phi(z_p)$. The resulting z-scores are given in the text.

For Scrounger

Each of the S scroungers attempts to feed at every clump discovered. $Y(\tau)$ is the total number of clumps produced; $Y(\tau)$ is a Poisson random variable with expectation $P\lambda\tau$. The jth scrounger consumes $X_j(\tau)$ total food items; $X_j(\tau)$ is another randomly stopped sum of binomial random variables:

$$X_j(\tau) = \sum_{y=0}^{Y(\tau)} r_{jy} \, . \quad (7.1.7)$$

$X_j(\tau)$ follows a Poisson-binomial distribution with mean and variance

$$E[X_j(\tau)] = (1 - \theta)cP\lambda\tau/S \quad (7.1.8)$$

$$V[X_j(\tau)] = \frac{(1 - \theta)cP\lambda\tau}{S}\left[1 - \frac{1 - \theta}{S} + \frac{(1 - \theta)c}{S}\right]. \quad (7.1.9)$$

For producer priority we again let $\theta = \theta_1$ in the last two expressions. For scramble-competition $\theta = \alpha/(S + 1)$. A scrounger's probability of starvation is $\Pr[X_j(\tau) \leq R + \rho]$, where ρ is the cost of scrounging. The associated standardized normal variates z_S are given in the text.

Math Box 7.2 Starvation Probabilities for Continuously Divisible Food Clumps

Each of P producers discovers a Poisson number of clumps with expectation $\lambda\tau$; the total number of clumps then has mean $P\lambda\tau$. In this formulation, intake is counted in units of "clumps." Each producer consumes a fraction θ of every clump it discovers. $X_i(\tau)$ is the total intake of producer i; $E[X_i(\tau)] = \theta\lambda\tau$ and $V[X_i(\tau)] = \theta^2\lambda\tau$. Again, we may apply producer-priority simply be setting $\theta = \theta_1$. Or, we may apply the scramble-competition rule where $\theta = \theta_2$. For the latter case, a producer's mean intake is

$$E[X_i(\tau)] = \alpha\lambda\tau/(S + 1). \tag{7.2.1}$$

The associated variance is

$$V[X_i(\tau)] = \alpha^2\lambda\tau/(S + 1)^2. \tag{7.2.2}$$

$X_j(\tau)$ is the total intake of scrounger j; $E[X_j(\tau)] = (1 - \theta)\, P\lambda\tau/S$, and $V[X_j(\tau)] = [(1 - \theta)/S]\, E[X_j(\tau)]$. $\theta = \theta_1$ under producer priority and under the scramble-competition rule these expressions become, respectively,

$$E[X_j(\tau)] = (S + 1 - \alpha)P\lambda\tau/[S(S + 1)] \tag{7.2.3}$$

and

$$V[X_j(\tau)] = (S + 1 - \alpha)P\lambda\tau/[S(S + 1)]^{1/2} \tag{7.2.4}$$

PART THREE Decisions within Patches

8 Social Patch and Prey Models

8.1 Introduction

Foraging theory divides feeding processes into a number of decisions, each of which has given rise to its own modeling tradition. Models for two such decisions form the core of classical foraging theory: patch and prey models (Stephens and Krebs 1986). Patch models deal specifically with optimal patch exploitation: How long and to what extent should resource parcels be exploited? Most prey models concern the decision to attack or ignore a prey that has just been encountered. Patch and prey models should no doubt be counted among the most active areas of research within behavioral ecology (Krebs et al. 1983; Pyke 1984; Stephens and Krebs 1986; Schoener 1987).

A basic feature of the classic patch and prey models is that they invoke standard optimality theory, a technique that cannot deal easily with the frequency-dependent payoffs that commonly characterize social foraging. Not too surprisingly, therefore, most of classical foraging theory has left social patch and prey decisions unattended. So, for instance, when a few water crickets (*Velia capraï*) capture a large arthropod prey on the surface of water, inject it with digestive fluids, and exploit it by sucking, the question of efficient clump exploitation remains open (Erlandsson 1988). Should the prey item be consumed fully or partially? Should all water crickets leave together, or should they each have a different emigration threshold? Similarly, about half of the predatory marine snails (*Stramonita haemastoma*) observed to be feeding on oysters at two sites off the coast of Louisiana, on the northern shore of the Gulf of Mexico, did so in groups of two or more (Brown and Alexander 1994). Should snails leave the oyster sooner or later when they forage in groups? These questions revolve around the issue of efficient resource clump exploitation under competitive conditions. In this chapter we present some models that have been developed to address the case of social patch exploitation.

Prey selection can also occur under social conditions. Individuals in a flock of feral pigeons (*Columba livia*) presented with a clump of mixed seeds do not select prey randomly. Instead, each individual seems to specialize on a subset of the mixture (Giraldeau and Lefebvre 1985). When given a choice of two seed types under competitive conditions, feral pigeons adjust their prey selection to reduce diet overlap (Inman et al. 1987; Robichaud et al. 1996). The wood pigeon (*Columba palambus*) also appears to choose its seeds on the basis of other pigeons' food choices (Murton 1971). Langen and Rabenold (1994) also found that consumers alter their food choices in the presence of competitors. They showed that subordinate juncos (*Junco hyemalis*) in a flock changed their seed choices, compared to solitary foraging, and reduced competition with dominant flock members. Group foraging promoted specialized seed choice and reduced the diversity of subordinate diets to a greater extent than it did for dominants (Langen and Rabenold 1994). How does competition for prey affect the most efficient prey-exploitation policy? In the penultimate section of this chapter we present two models that address the issue of optimal prey decisions under competitive foraging decisions.

8.2 Models of Social Patch Exploitation

Establishing Where Competition Occurs

Before we can model the effect of competition among consumers on the predictions of the patch model, we must specify the level at which competition is occurring. Broadly speaking, social patch models have considered three kinds of competitive interactions among foragers. The first case assumes that the total number of patches available at any one time is limiting and only a single individual can forage per patch (Parker 1984c; Parker et al. 1993; Yamamura and Tsuji 1987). Competition between consumers regulates access to patches. This would apply, say, to a group of predators who exploit food particles as they flow down a stream (Blanckenhorn and Caraco 1992). Assuming that these predators cannot exploit prey and encounter resources concurrently, they must decide on how long they should exploit each resource before abandoning it in order to wait for the arrival of the next. In the second and third cases, the models apply to situations where a patch is exploited by a group of foragers concurrently. In the second case the resources in the patch decline at some rate that is independent of the forager's exploitation. This assumption may apply when foragers exploit a clump of mobile prey that start to flee once the patch is uncovered. In the third and final case, the patch has a finite quantity of prey that depletes as the foragers consume it. This case may apply to groups of foragers encountering clumps of immobile prey such as seeds or eggs. We present each type of model in turn and collect their assumptions and predictions in Summary Box 8.1.

When Foragers Compete for Patches but Exploit Them Singly

This version of the patch residence problem was first presented by Parker (1984c) and reformulated more generally by Parker et al. (1993). It applies

SUMMARY BOX 8.1 MAIN ASSUMPTIONS AND PREDICTIONS OF SOCIAL PATCH MODELS

MODELS FOR ONE COMPETITOR PER PATCH

Continuous Input of Depletable Patches

Perfect compensation: Explicit solution available. Predictions of stable exploitation time qualitatively similar to non-social patch models. Quantitative differences between social and non-social patch models expected only when exploitation time is long compared to search. Exploitation times predicted to increase with group size and decrease with the input rate of patches to the habitat (Parker et al. 1993).

Imperfect compensation: Qualitatively similar to above but no analytical solution available (Yamamura and Tsuji 1987). Predict that cooperative solutions to the game always gives exploitation times that are longer than competitive solutions.

MODELS FOR *G* COMPETITORS PER PATCH

Ephemeral Patches

All competitors exploit patch to $T_{1\text{crit}}$, at which point sequential emigration of consumers commences until $T_{G\text{crit}}$. $T_{1\text{crit}}$ unaffected by group size, but $T_{G\text{crit}}$ increases with group size. $T_{1\text{crit}}$ and $T_{G\text{crit}}$ both decrease with increasing patch richness, and duration of emigration is longer in poorer patches. $T_{1\text{crit}}$ and $T_{G\text{crit}}$ both increase with increasing travel time, but duration of emigration is unaffected.

Depletable Patches

If consumers are equal and arrive simultaneously at patch, all consumers leave patch synchronously. Predicted exploitation qualitatively similar to nonsocial patch models, but, in addition, foragers leave a patch sooner when group size increases. Level of patch depletion when foragers leave increases with group size.

to socially foraging competitors much as was assumed under the continuous-input habitat models. Patches arrive at some fixed continuous rate k, and each is exploited immediately as it arrives in the habitat. Consequently, patches never accumulate and never disappear without being exploited. The model assumes that the individual's resource intake rate declines during the exploitation of the patch, and only one consumer exploits a patch (e.g., patches are economically defendable, so that no other forager attempts to gain access to one being exploited).

At any particular time, a fixed total number of competitors is divided exhaustively into those exploiting patches and those searching for patches. Foragers cannot search while exploiting a patch. Since the rate of patch arrival is finite, the expected search time elapsing before an individual discovers an unoccupied patch should decline as the number of competitors currently exploiting patches increases. This social-patch model differs principally from the solitary-forager model by incorporating a frequency-dependent search time. Consumers choose patch-exploitation times and in doing so influence the average search time.

Parker et al. (1993) describe how this sort of frequency dependence might arise. A foraging cycle for any individual consists of a period of searching or waiting for a patch (T_S time units), followed by exploitation of a single patch (T_E time units); then the cycle begins again. Parker et al. (1993) first assume that the expected cycle duration D is fixed; that is, $D = (T_S + T_E)$ is constant. Any increase in exploitation time will induce a corresponding decrease in search time.

G consumers have equal competitive abilities. Recall that patches arrive at rate k and are captured by waiting consumers as they appear. Then the assumed fixed cycle duration implies that $D = G/k$. Hence $T_S = (G/k) - T_E$.

The solution to the patch problem for solitary foragers is given by the "tangent" method (fig. 8.1A). The conventional patch model assumes a constant search (travel) time from which a line, tangent to the cumulative exploitation function, is drawn. The point of tangency indicates the optimal exploitation time (T_E^*) that maximizes the long-term rate of energy intake (fig. 8.1A). The tangent method is not entirely satisfactory when search time is a function of exploitation time. For social foragers, the exploitation time defined by the tangent method (T_E^*) gives the exploitation time that will spread most rapidly within a population with the same cycle duration (fig. 8.1B). For this optimal exploitation time to be the evolutionarily stable exploitation time ($\hat{T}_E$), it must also be resistant to invasion from other mutant exploitation times ($T'_E \neq T_E^*$).

Consider three populations with the same cycle duration D (fig. 8.2). In figure 8.2A, a long exploitation time has become fixed in the population. Because of the constraint between exploitation and search times, every population member experiences a short search time. In figure 8.2B, a short exploitation time has become fixed, and all foragers in that population experience a long search time. Suppose a rare mutant exploitation time (T'_E) arises in either population. Because it is rare, it will not influence either population's search time (a standard assumption for large populations). Consequently, the mutant will experience the search time characteristic of the population in which it occurs. Among all feasible T'_E values, the mutant that

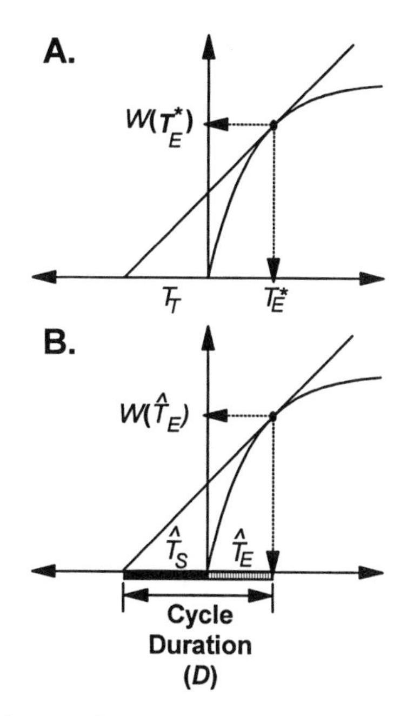

Figure 8.1 A graphical comparison of the conventional and social patch models. In (A) we sketch the well-known tangent method of finding the optimal patch residence time for the marginal value theorem (MVT). Patch time increases to the right of the origin while travel time (or search) increases to the left. The decelerating function gives the cumulative intake of food, and the point of tangency represents the optimal patch time T_E^* that maximizes long-term average intake. The MVT assumes that travel time is not under the consumer's control but is set by the distance between patches in the environment. Similarly, the cumulative gain function is not under the consumer's control but is set by the density of resources in a patch. Given these constraints, the tangent locates the best decision.

The bottom panel (B) gives the equivalent graphical representation for the social patch model. In this case, environmental conditions set cycle duration *D*, the time between successive encounters (the thick section of the abscissa). Any change in patch exploitation time T_E also changes T_S, the time required to search (or wait) for an unexploited patch. The tangent in this case gives the alternative T_E that would likely spread the fastest in the current population. This alternative, however, is not necessarily the ESS.

corresponds to the optimal exploitation time, based on the tangent method, spreads most rapidly in any population. However, because of the frequency dependence, as the mutant strategy increases in abundance it will cause either population's search time to change. The consequence is that as search time varies, opportunities arise for new mutant optimal exploitation times to invade. Ultimately, the process stabilizes at an ESS exploitation time where no mutant can do better, given the population's search time (fig. 8.2C).

Parker et al. (1993) provide an explicit solution for the ESS exploitation

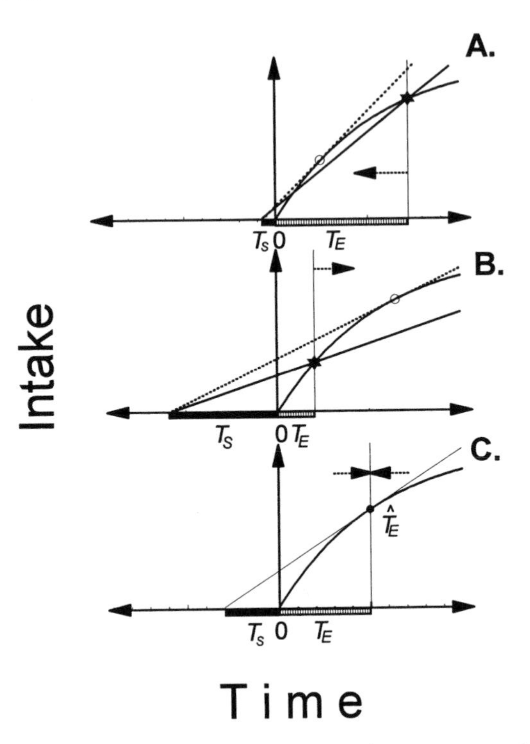

Figure 8.2 The social patch model for a continuous-input system when consumers compete for limited patches. Each panel is the customary graphical representation of the patch exploitation problem for one of three different hypothetical populations. In all cases the abscissa is divided into two regions, search time T_S increases to the left of the origin, while exploitation time T_E increases to the right. The ordinate gives the cumulative resource intake. Each population has the same cycle duration D, indicated by the thick sections of the abscissa. In this case we assume perfect compensation so that any change in T_E is exactly compensated by a change in T_S, keeping D constant. In each population D is divided into different portions of T_E (dashed) and T_S (solid). In (A), a long T_E has gone to fixation in the population, leading to a correspondingly short T_S. In (B) the opposite is true. In both cases, the long-term foraging rate experienced by the population is given by the slope of a line originating from T_S and intersecting the exploitation function (star). An alternative exploitation time is rare, so it experiences the same T_S as the rest of the population in which it originates. In (A), an alternative with shorter exploitation times would do better and would spread. In (B), alternatives with longer exploitation times would do better. Open circles give the alternatives in (A) and (B) that maximize benefits in their respective populations and hence would spread most quickly. The dotted arrows indicates the direction in which D is expected to move along the abscissa as alternative exploitation times become more common and start having an effect on the population's T_S. In (A), T_S decreases from its initial state while in (B) it increases. In (C) the population has reached an ESS of $\hat{T}_E$, where no alternative T_E can do better, given the population's current T_S.

time by assuming that the cumulative exploitation function $W(T_E)$ has an exponential form:

$$W(T_E) = 1 - \exp[-\lambda T_E]. \qquad (8.1)$$

It follows that $dW(T_E)/dT_E = \lambda \exp[-\lambda T_E]$, and so for our purposes patch richness can be defined as $dTW(T_E)/dT_E|_{T_{E=0}} = \lambda$, the instantaneous rate of gain an individual forager experiences as it begins to exploit a patch. Assuming an infinite population, an ESS analysis for this game yields (Parker 1984c):

$$\frac{dW(\hat{T}_E)}{dT_E} = \frac{W(\hat{T}_E)}{D} \qquad (8.2)$$

at the evolutionarily stable strategy. It follows that

$$\hat{T}_E = \ln(D\lambda + 1)/\lambda. \qquad (8.3)$$

In other words, we can locate the stable patch exploitation time $\hat{T}_E$ from the cycle length D and λ, the patch richness.

The qualitative predictions of the effect of resource density are similar to those of the solitary-forager model. As patch richness increases, the model predicts that the stable exploitation time decreases (fig. 8.3A). Note that the effect of patch richness is stronger for longer cycle durations, predicted when there are more competitors or lower rates of patch input.

The social patch model predicts that cycle duration affects the stable exploitation time; this leads to novel predictions (fig. 8.3B). Since cycle duration increases with decreasing rate of patch arrival (k), the model predicts that the stable exploitation time will increase with decreasing rate of patch arrival (fig. 8.3B). Similarly, since cycle duration increases with an increasing number of competitors, the model predicts that the stable exploitation time will increase as group size increases (fig. 8.3B). The effects of both group size and rate of patch arrival are stronger when patches are poorer and, hence, exploited more slowly (fig. 8.3B).

The social patch model makes predictions differing from the conventional model's predictions only when search and exploitation times exhibit frequency dependence. The presence of competitors per se may not be sufficient for this frequency dependence to arise. Applying the social patch model to copulation duration in dung flies (*Scatophaga stercoraria*), Parker et al. (1993) found that the stable exploitation time was 42.0 min, while the optimal exploitation time predicted by the conventional patch model was 42.5 min, a biologically insignificant difference. Congruence of the social and nonsocial models' predictions was attributed to the occurrence of weak frequency dependence when search time is long compared to the stable exploitation time. Hence, the quantitative predictions generated by the social

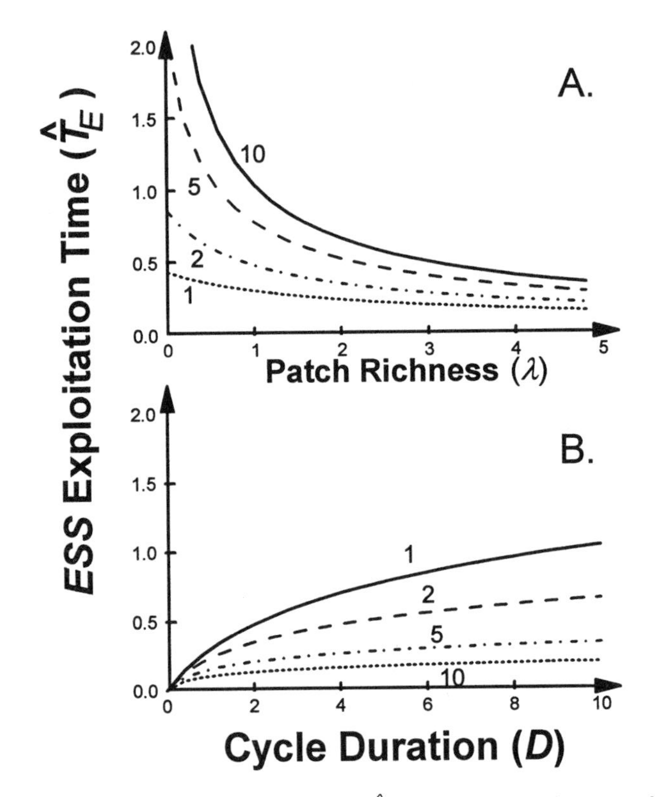

Figure 8.3 The ESS patch exploitation times ($\hat{T}_E$) for continuous input, perfect compensation, and competition for limited patches. $\hat{T}_E$ declines with patch richness (A) but increases with cycle duration (B). In (A), the number next to each curve gives D, the cycle duration. In (B), the numbers next to the curves give λ, the parameter of the exploitation curve that increases with patch richness, the resource density within a patch.

patch model are most useful when exploitation time is an important component of cycle duration, otherwise the conventional patch model is sufficient.

The previous model assumes that all patches are exploited immediately as they appear in the environment, implying *perfect compensation*; that is, search and exploitation times sum to a constant. In some cases, patches may be available for some time before they are exploited, and may be lost to exploitation after a certain time. These conditions lead to a more complex frequency dependence referred to as *imperfect compensation*. The effect of frequency dependence in this case resembles the previous model's scenario, but the relationship between search and exploitation times is now curvilinear.

Curvilinearity of compensation between search and exploitation times arises because the number of undiscovered patches depends on the number of consumers searching. This number, in turn, is a function of exploitation time. Stated in terms of Parker et al.'s (1993) terminology, not only is search time a function of exploitation time, but now the cycle duration itself is a function of exploitation time as well. Since the cycle duration is no longer a constant, Yamamura and Tsuji (1987) suggest that any continuous input system of the type considered here moves toward an equilibrium where the number of undiscovered patches, the number of searching consumers, and expected search time are constant. The equilibrium values should all depend on the stable exploitation time.

To predict the stable exploitation time for a system with imperfect compensation, the following are required: the exploitation function, the group size, the patch input rate, consumers' searching efficiency, the rate at which undiscovered patches are lost, and any mandatory time delay that must be invested once a patch is acquired but before any benefit can be derived. As is true of the patch model for solitaries, there is no explicit solution for stable exploitation time, and the model is solved numerically.

The predictions of the imperfect compensation model are the same as those for perfect compensation in terms of the effect of group size and resource input rate. However, Yamamura and Tsuji (1987) break novel ground when they seek to predict a cooperative solution to the imperfect compensation patch game. To do so, they simply change the currency to be maximized. The objective function becomes

$$\Omega(T_E) = GW(T_E)/[T_S(T_E) + T_D + T_E], \quad (8.4)$$

which is the intake rate for the group as a whole where $W(T_E)$ is any intake function. Unfortunately, there is no explicit solution to the model, but the cooperative (or Pareto optimal) solution can be obtained numerically. The cooperative stable exploitation times predicted are longer than the competitive stable exploitation times. Therefore, foragers that cooperate in exploiting patches should deplete resources in each patch to a greater extent than do strictly competitive foragers.

Group Exploitation of Ephemeral Patches

We next consider the patch exploitation problem when several consumers concurrently occupy the same patch. Models for this problem were first presented by Parker and Stuart (1976) and Parker (1978) to analyze mating systems. Consequently, they do not always translate easily into questions about social foraging. The model we describe assumes that an individual's intake rate decreases as a function of only the time elapsed since the patch became available.

Assume that a patch becomes available or exploitable at some time $T_E = 0$. From this time on, prey arrive within the patch sequentially but at a declining rate. The mechanism for the decline is not specified, but it is entirely *independent* of the forager's behavior. The prey arrival rate depends only on elapsed exploitation time. Hence the arrival rate is independent of the number of consumers. Prey are detected instantaneously as they arrive and are always consumed immediately.

Suppose the intake function $W(T_E)$ defined by equation (8.1) is the cumulative amount of resource that has become available in the patch during T_E time units. By assumption, $TW(T_E)$ is also the total consumption of an individual with no competitors (a solitary). If all G competitively equivalent consumers begin exploiting the patch at $T_E = 0$, then the amount of resource consumed by each competitor over $T_E > 0$ time units is simply $W(T_E)/G$. Given these assumptions, we ask when the competitors should leave the patch, and whether all G consumers should leave at the same time.

Parker and Stuart (1976) and Parker (1978) show that as exploitation time elapses, a first critical point ($T_{1\text{crit}}$) is reached where the benefits of staying are lower than those of emigrating. From this critical point on, individuals leave sequentially until the only remaining competitor departs at the last critical point ($T_{G\text{crit}}$). Only one of the G competitors leaves at the first critical point. So the first and last critical points bound the *emigration-threshold* period. This model is presented graphically in figure 8.4.

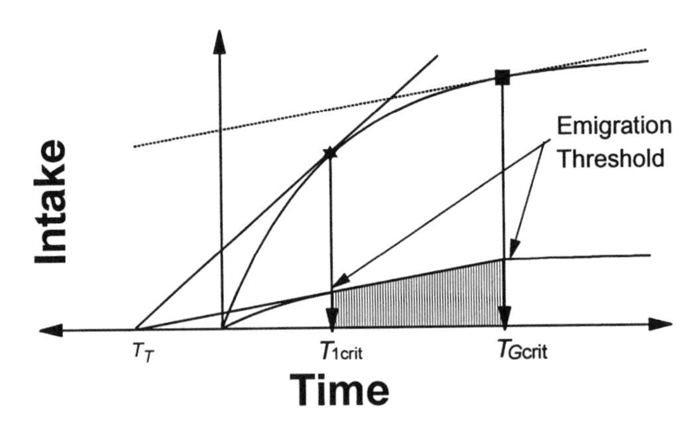

Figure 8.4 Graphical representation of the continuous input patch model when resource input rates decline continuously. Axes are as in figure 8.2. The top curve gives $W(T_E)$, the cumulative intake for a solitary consumer. The lower curve gives $w_G = W(T_E)/G$, an individual's cumulative intake function when G individuals exploit the patch concurrently. The tangent to $W(T_E)$ originating from T_T has slope β_{max}. The tangent to w_G has slope β_{max}/G. Note that both lines intersect their respective cumulative functions at the same exploitation time $T_{1\text{crit}}$, the emigration time for the first competitor. The ESS patch emigration pattern requires that individuals remaining in the patch do at least as well as those leaving (or else they would leave, too). Hence, the emigration pattern from the patch is such that the intake rate during the emigration period remains constant. The last consumer leaves the patch at $T_{G\text{crit}}$ when its rate in the patch falls below β_{max}/G. The emigration period is bounded by $T_{1\text{crit}}$ and $T_{G\text{crit}}$, and its duration depends on the patch's resource input rate.

The ESS for this competitive emigration game is a sequence of departures from the patch such that consumers that remain in the patch continue to do only as well as those that leave. Hence the total foraging rate within the emigration period (T_{1crit}, T_{Gcrit}) will be constant and equal to β_{max}, the rate achieved at the first critical point (fig. 8.4). The number of foragers $G(T_E)$, $1 \leq G(T_E) \leq G - 1$, still present at any given time within the emigration period, is a function of the exploitation time. That is, $G(T_E) = (dW/dt)(1/\beta_{max})$. The last competitor should emigrate when $1 = (dW/dt)(1/\beta_{max})$; see figure 8.4.

The declining continuous input patch model predicts that the emigration thresholds will be affected by the number of competitors in the patch, λ (patch richness) and the time required to travel to the next patch. Increasing the number of competitors does not affect the first critical point (fig. 8.5). However, the position of the last critical point is dramatically affected by increasing the number of competitors (fig. 8.5). As the number of competitors increases, the model predicts that foragers will emigrate at a slower rate from the patch. The model predicts that as λ increases, both the first and last critical points will occur sooner, and the duration of the emigration period will be shorter (fig. 8.6).

Average travel time to the next patch has a somewhat similar effect as λ (fig. 8.7). As travel time increases, both first and last critical points increase. However, travel time will have much less of an effect on the duration of the emigration period, which may increase slightly with increasing travel time.

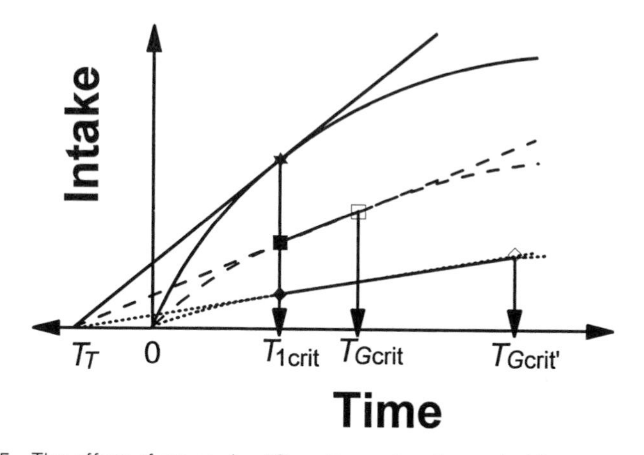

Figure 8.5 The effect of group size (G) on the emigration period for continuous input patches when input rate declines continuously. $W(T_E)$ is given by the top curve. The two lower curves give w_G for a small (dashed) and a large (dotted) group. Note that the size of G has no effect on T_{1crit}, but that T_{Gcrit} increases with G so that the duration of emigration is longer when G increases.

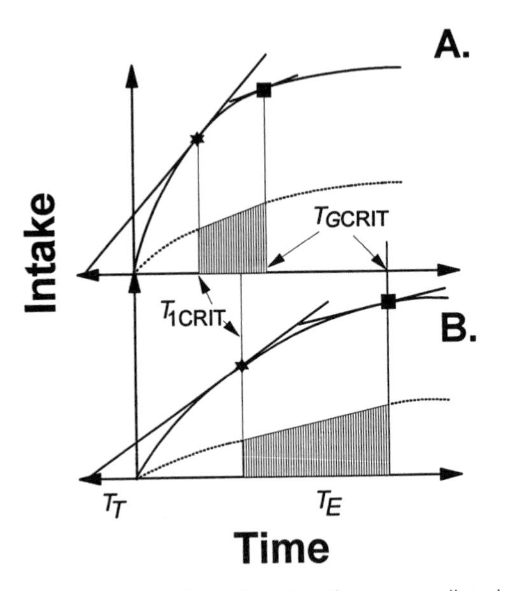

Figure 8.6 Graphical representation of emigration as predicted by the continuous input patch model when input rate declines continuously. In both panels, the solid curve gives $W(T_E)$ but input rate (i.e, patch richness) is higher in (A) than in (B). The dashed curves give $w_G = W(T_E)/G$. The shaded areas give the duration of emigration in both environments. Note that both T_{1crit} and T_{Gcrit} increase with decreasing patch richness, and that the duration of emigration is longer for poorer patches.

Group Exploitation of Depletable Patches

This version of the social patch model probably possesses the most general applicability. It assumes that consumers exploit a nonrenewing resource in each patch. The rate at which a patch is depleted depends on both the number of consumers and consumer strategy. Importantly, this model's predictions differ considerably from those of the previous continuous-input model.

Assume that $W(T_E)$, defined by equation 8.1, is the cumulative resource uptake by a single individual in a patch. If G-searching individuals arrive simultaneously in the patch, $W(GT_E)$, the corporate intake function will be $W(GT_E) = 1 - \exp[-\lambda GT_E]$. Finally, if the G individuals have equal competitive weights, then $w(T_E)$, the cumulative resource uptake of any one individual competing in the patch, will be $w(T_E) = W(GT_E)/G \neq W(T_E)$. Compare these quantities to those in the model for the declining continuous-input case.

This depleting patch model is presented graphically in figure 8.8. Unlike the preceding result, this model predicts that all consumers will leave the patch simultaneously at the *emigration threshold*. Here the patch is depleted

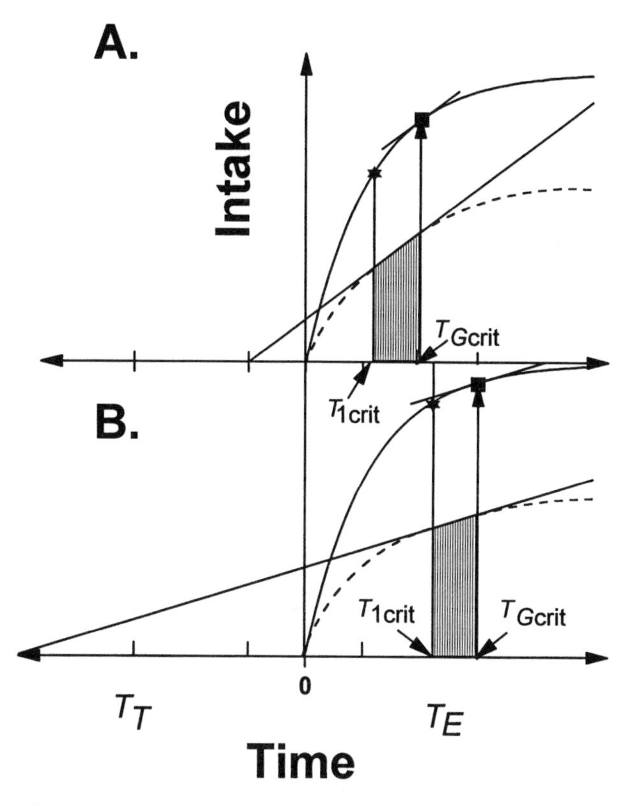

Figure 8.7 Graphical representation of emigration as predicted by the continuous input patch model when input rate declines continuously. Both panels are identical except that travel times are short in (A) and long in (B). $W(T_E)$ (solid) and w_G (dashed) as in Figure 8.6. The emigration periods are represented by the shaded areas. Both T_{1crit} and T_{Gcrit} increase with increasing T_T, but duration of emigration is affected little.

to such an extent that even a solitary consumer would not gain an intake rate superior to the rate expected if it moved to the next patch. In figure 8.8, note that the lines are tangent to $W(GT_E)$ and $w(T_E)$ at the same emigration threshold because $w(T_E)$ is simply $W(GT_E)/G$. Moreover, note that the emigration threshold is much smaller than the optimal patch time predicted by the nonsocial patch model. The model predicts, therefore, that when multiple consumers compete within a depleting patch, individuals will emigrate sooner than if they were alone. But they will have depleted the patch's resources to a greater extent than a solitary consumer. We designate the difference between the two depletion levels ΔW.

Group size has a strong effect on the depleting patch model (fig. 8.9). As group size increases, the model predicts that consumers leave the patch

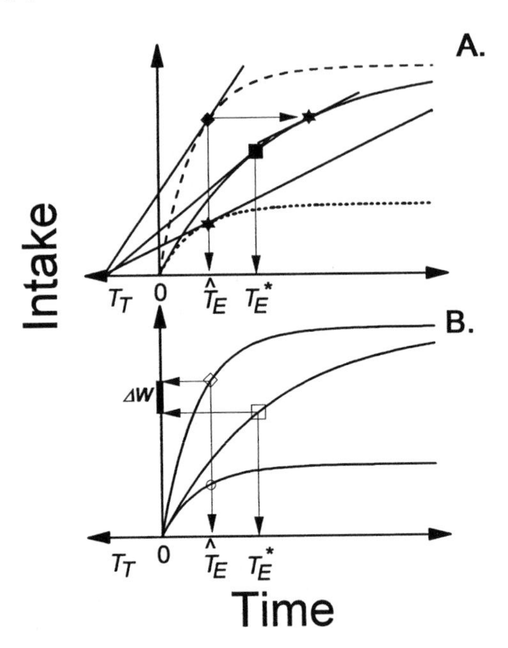

Figure 8.8 Social patch model when consumers deplete a resource within patches. In the top panel (A), the middle curve gives $W(T_E)$, the cumulative intake of a solitary forager as a function of patch exploitation time. The top curve (long dashes) gives $W(GT_E)$, the group's corporate cumulative intake. The lower curve (short dashes) is $w_{\hat{G}}(T_E)$, the cumulative intake of an individual from a group of G consumers. The tangent drawn to $W(T_E)$ gives the optimal patch time (solid square $= T_E^*$) for a solitary individual in the patch. The tangent drawn to $W(GT_E)$ gives the ESS patch time (diamond $= \hat{T}_E$) for the group of foragers. The tangent drawn to $w_G(T_E)$ gives the patch time experienced by each group member (star). Note that a solitary forager in a patch depleted to $W(G\hat{T}_E)$ experiences the same instantaneous patch exploitation rate as a group forager that depletes the patch to the same level (compare the slopes of the tangent to each star). This means that foragers should all leave simultaneously because staying behind, even as a solitary, cannot be profitable for any of them. In (B) the same curves are presented to illustrate that group foragers are predicted to abandon a given patch sooner than a solitary forager ($\hat{T}_E < T_E^*$). At departure, a group of consumers will have extracted more resources from the patch than a solitary forager. The difference between the amount consumed by a group versus a solitary is given by the tick-shaded portion of the ordinate (ΔW).

sooner and deplete it more extensively (so that ΔW increases). The social depleting patch model also makes predictions that are qualitatively similar to the patch model for a solitary forager. As λ, the average patch richness in a habitat declines, the emigration threshold should increase, but ΔW remains relatively unaffected. Similarly, as the average distance among patches in-

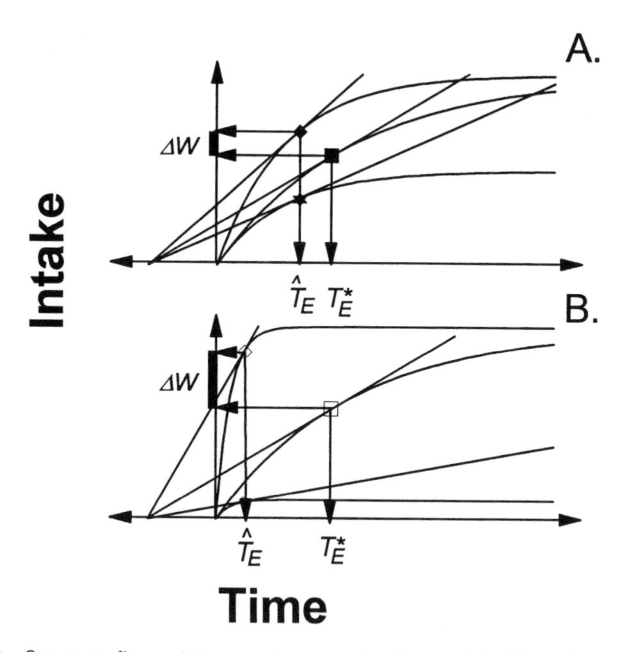

Figure 8.9 Same as figure 8.8 except group size is small in (A) and large in (B) to illustrate the effect of group size on ΔW. The patch quality is the same in both panels; middle curve = $W(T_E)$. When a group is large, its corporate exploitation function rises more quickly; top curves = $W(GT_E)$. Naturally, when group size is large, each individual's intake represents a smaller fraction of the total; bottom curves = w_G). Note that a larger group abandons a patch sooner while having extracted more resources. Hence, ΔW increases with increasing group size.

creases, so does the emigration threshold; however, the value of ΔW, again, is affected little.

Additional complexities, such as the influence of consumer satiation on patch residence time, can be incorporated into the assumptions of a model for social foragers' use of patchy resources. Clark and Mangel (1986) examine a series of biologically realistic extensions of the model just discussed. Social patch exploitation has also been modeled as an *N*-person War of Attrition by Sjerps and Haccou (1994). Most predictions remain qualitatively similar. For instance, in the absence of interference, all players should leave simultaneously. When there is interference, however, a new prediction emerges. Given interference, players should leave asynchronously, using a stochastic departure time within a specified range. Group size, on the other hand, does not have as much of an effect on departure time in Sjerps and Haccou's (1994) War of Attrition model. Whatever the model, however, predictions of asynchronous departures raise problems for group cohesion that will need to be addressed at a different scale.

The depleting patch model assumes that all consumers arrive at patches immediately upon discovery. This assumption is important, since its violation can alter the model's predictions considerably (Rita et al. 1997; Beauchamp and Giraldeau 1997). The next section considers some empirical results that suggest directions for modified theories.

8.3 Tests of Social Patch Models

At least two studies have tested some of the predictions of the social patch model for depletable patches. The first (Livoreil and Giraldeau 1997) has foragers arriving at the patch more or less at the same time, while the other (Beauchamp and Giraldeau 1997) allows the arrival at the patch to be staggered. We briefly review the results of each.

Livoreil and Giraldeau's (1997) study compares patch residence of solitary spice finches (*Lonchura punctulata*) to those of groups of three foraging on identical patches. As predicted, the patch times of solitary finches were longer than those of trios. However, contrary to predictions, trios left more seeds behind than solitary foragers. The results suggest that finches in trios interfered with one another's foraging rate; two individuals within each trio consistently outperformed a third that was most affected by competition. The model's predictions were generally upheld for the two best competitors, but the birds most affected by competition spent too much time in the patch and, at short travel times, they collected more seeds than predicted. Despite the obvious player asymmetries, there was no consistent order of patch departure so that the birds most affected by interference may have been forfeiting foraging efficiency to remain with the other two. Perhaps in nature the predation hazard of traveling singly is too high relative to any extra foraging gain resulting from an early patch departure for birds most seriously affected by interference.

In another study, Beauchamp and Giraldeau (1997) modify the model slightly to allow for asynchronous arrival at the patch (fig. 8.10; see also Rita et al. 1997 for a similar model). Only one of the six birds could find food, and inevitably it was always the first forager to feed from the patch. The five other birds always joined to feed from the discovered patch. When the search time required to encounter the next patch is short, the model predicts that for low levels of interference the patch producer should leave upon arrival of the first scrounger. However, when the search time is long, increasing levels of interference from the scroungers are required to induce the producer to leave upon the arrival of the scrounger (fig. 8.10). Experimental results are qualitatively consistent with the predictions of the model

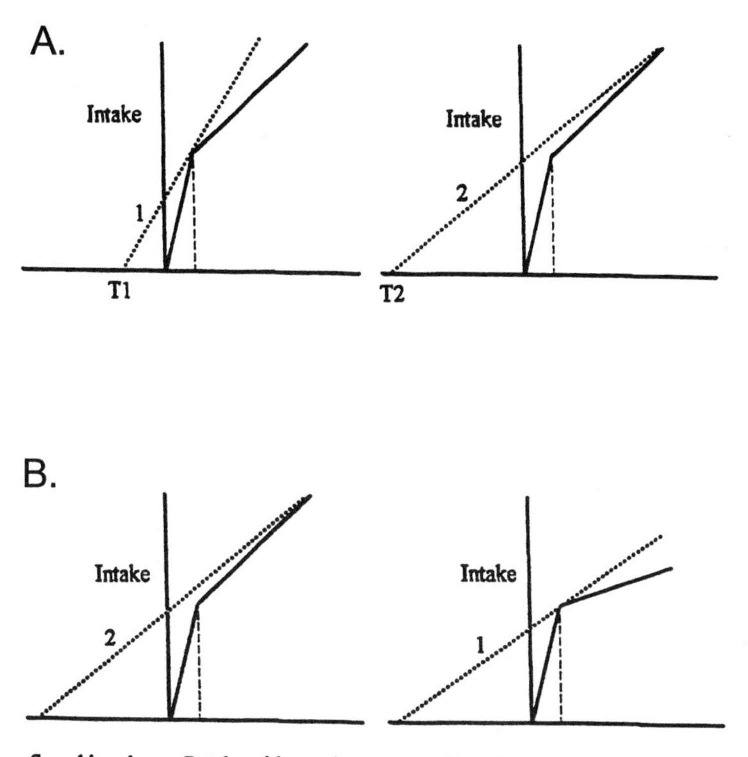

Figure 8.10 Graphical model of patch exploitation for depletable patches when patch arrival is staggered. (A) shows the effect of search time to the next patch while (B) shows the effect of interference caused by the arrival of scroungers. Note that intake rates are linear in this model. Curvilinear intake yields many of the same predictions (see Rita et al. 1997). A patch finder (producer) feeds alone at the patch until the first scrounger arrives, a point indicated by the vertical dashed line. The producer's decision to stay or leave the patch should be based on its long-term expected intake rate given by the dotted line. In (A) for short travel times the interference caused by the arrival of the scrounger is sufficient to make it efficient for the producer to leave. This is not so for long search times where, despite the interference, it is best for the producer to stay, at least until the next scrounger arrives. In (B) the producer stays after the arrival of the first scrounger when interference levels are low; however, when interference is more intense, it may be better for the producer to leave and search for the next patch when the first scrounger arrives. Taken from Beauchamp and Giraldeau (1997). (With permission of Oxford University Press)

(Beauchamp and Giraldeau 1997). The results imply that when search time is long and interference serious, patch finders will tend to leave patches before the scroungers. In a system where individuals switch between playing producer and scrounger, this staggered patch departure would bias an individual that has just played producer successfully toward playing producer again on the next move, if only because playing scrounger would be of low value when all other group members are still feeding at a patch.

8.4 Social Prey Models

While social patch decisions have generated a good number of models, most of which await testing, social prey decisions have been relatively neglected. Only a few models have been described; none has been tested formally.

The conventional optimal diet model predicts a solitary forager's acceptance or rejection of a food item upon encounter with a recognizable prey type. The model assumes that while the solitary searches, each available prey type is encountered at a constant rate, proportional to the density of that type (but see Recer and Caraco 1989b). The model ranks prey types by profitability, the ratio of net energy per item to handling time per item. The conventional model identifies the diet maximizing the expected long-term rate of energy gain. Stephens and Krebs (1986) present a series of useful extensions of the conventional model. Houston and McNamara (1985) present a model for dietary choice where the forager must accumulate a critical level of energy to survive the night; a larger energy reserve can increase the optimal diet breadth (see McNamara and Houston 1992).

If sociality influences foragers' diets, the presence of competitors must affect the individual's encounters with prey items, the profitabilities of some or all prey types, or a combination of these attributes. The most likely competitive effects are a reduction in local prey density and an increase in the individual forager's uncertainty about the local relative abundances of the various prey types. Consequently, the assumption of constant encounter rates may sometimes be unacceptable for social foragers. Instead, encounters with food types are likely to exhibit a frequency dependence generated by the different group members' dietary choices (Inman et al. 1987; Langen and Rabenold 1994).

The economics of social prey models, therefore, can involve dietary choice when prey types are being depleted, perhaps in a nonuniform manner. Few nonsocial prey models have analyzed the effect of prey depletion on optimal diet choice (Heller 1980; Mitchell 1990). Heller (1980) simulated the consequences of different prey choice strategies for a solitary exploiting depletable patches containing two prey types with different profitabilities. When the more profitable prey is common, such that the instantaneous foraging rate is maximized by specializing on that prey, Heller finds that the strategy maximizing long-term intake differs from that maximizing instantaneous gain rate. Heller's model is summarized graphically in figure 8.11; the predictions follow.

Optimal prey choice when food is depleted depends on the time that a consumer spends in a patch. If this time is short, then the best strategy is to specialize exclusively on the more profitable prey. However, if patch residence is long, at some point the consumer will likely have to include some of the less profitable prey in its diet; it will generalize. So, when patch residence times are long, one can ask if the consumer should generalize from the onset (*generalist* strategy) or specialize on the more profitable prey for some time before generalizing (the *expanding specialist* strategy). The consequences of the two strategies, when patch residence is long, appear in figure 8.11. The best strategy for very long patch residence is to generalize from the onset. The cost is a lower intake rate initially (because specializing would have been better early in patch exploitation). But later in the foraging episode the consumer does well because the more profitable prey type is relatively more common than if it had initially specialized on this prey type before generalizing. For intermediate patch residence times, the best strategy is to be an expanding specialist (fig. 8.11).

Heller (1980) deals briefly with the consequences of adding competitors to the depleting prey model. He concludes that the best strategy in any condition is likely to be the expanding specialist, where consumers start off with the best prey type and include prey types of decreasing profitability as depletion proceeds. The treatment of the consequences of competition was addressed more extensively by Mitchell (1990) using a control theory approach.

Mitchell (1990) analyzed the consequences for optimal prey choice when a forager competes with G other consumers, all of which generalize. He argues that consumers' behavior is determined strictly on the basis of the density of the prey items remaining when the consumers leave the patch (this is somewhat equivalent to the duration of patch residence). Mitchell's (1990) analysis concludes that the best strategy is always to start off as a specialist on the more profitable item and switch to a generalist diet at some critical density of profitable items (Heller's [1980] expanding specialist strategy). Increasing the number of competitors decreases the critical density of profitable items at which specialists are predicted to expand to a generalist diet. In other words, more competitors means that foragers should specialize longer on the profitable item and hence deplete it to a greater extent before starting to accept the less profitable item.

The rationale for this effect of increasing competition is intuitively simple. The reason consumers should expand their diet in the first place is to gain long-term foraging rate at the expense of short-term gain. However, when several consumers are competing for the same resources, the long-term gain strategy is invadable by a short-term gain strategy that reaps the benefits of

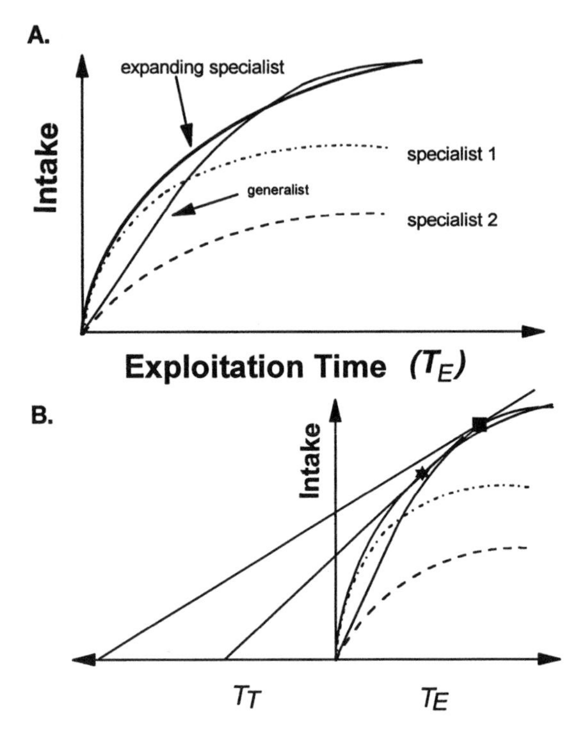

Figure 8.11 Graphical prey model; solitary consumer depletes a patch containing prey of two types. Prey type 1 is more profitable than prey type 2. In (A) we plot the cumulative gross gains of four foraging strategies: specialist on prey 1, specialist on prey 2, generalist taking prey as encountered, and an expanding specialist (specializes on prey 1 initially and then expands). The generalist's cumulative gain is initially below the type 1 specialist because it is accepting lower value prey type 2 that slows it down. However, at some point during exploitation, the intake achieved by the generalist is greater than that of the type 1 specialist because the specialist has depleted its prey and the generalist has not. Note that the expanding specialist strategy initially provides as great an intake as a type 1 specialist. At some point it starts doing better than the type 1 specialist because it can exploit the alternative prey when prey type 1 becomes rarer while the specialist cannot. The expanding specialist does better than the generalist because initially it avoided prey type 2. Consequently, prey type 2 is more abundant later. At longer exploitation times, however, the generalist and expanding specialist curves intersect, and the generalist starts doing better. The intersection occurs because the expanding specialist has reduced the density of prey type 1 more so than the generalist.

In (B) we show how the optimal prey choice policy depends on the optimal patch time. Applying the tangent method shows that the generalist strategy is optimal when long travel times require extensive patch exploitation. At intermediate travel times, the optimal strategy is to be an expanding specialist while specialist 1 is the optimal strategy for short travel times (not on graph). Modified from Heller (1980).

short-term gain while imposing a greater decline in future foraging rate. As Mitchell (1990) realized, the group of consumers involved in prey choice may be faced with a Prisoner's Dilemma. This raises a number of interesting possibilities for the evolution of cooperative prey choice.

The social prey model has never been tested experimentally, so it remains uncertain whether consumers use expanding specialist strategies when faced with competition for depleting prey.

8.5 Concluding Remarks

Implications for Future Work

Both social patch and prey models have generally been ignored by experimentalists. Consider, for instance, the last of the social patch models presented. The model predicts that when consumers arrive at a patch in unison, they should also leave in unison; there is only one ESS patch time for all group members. Moreover, groups should exploit patches more extensively than individuals but abandon them after shorter exploitation times. If this is so, groups of G individuals exploiting a patch should deplete it much more extensively than G solitary foragers exploiting the same patch sequentially (Livoreil and Giraldeau 1997). This difference could have important implications for the use of giving up density as a measure of habitat quality (e.g., J. Brown 1988; Valone and J. Brown 1989). To our knowledge, few of the model's assumptions, let alone its predictions, have begun to be investigated (Beauchamp and Giraldeau 1997; Livoreil and Giraldeau 1997). This neglect is particularly surprising given the remarkable effort that has been directed to studying patch exploitation in a wide variety of solitary organisms (Schoener 1987).

Economic patch exploitation decisions can have important consequences for within-group behavioral diversity (chapter 9). Imagine a situation where rate-maximization requires that the discoverer of a patch be the first one to leave it in search of the next (Beauchamp and Giraldeau 1997; Rita et al. 1997). That means the discoverer can be the only individual searching for the next patch while all others are exploiting the previous one. It is possible that the probability that the finder of one patch also finds the next one increases as a function of the time the individual searches for food alone. If this is so, then the rules for economic patch exploitation can channel individuals into sequences of finder and joiner events, hence specializing individuals into roles.

Both social prey choice models predict similar dietary patterns. When groups of foragers exploit a depleting patch containing a random mix of two prey types, each group member should initially specialize on the more profitable of the two types. Once the better prey has been depleted sufficiently, each forager should expand its diet and accept any item encountered. The critical density of the better prey, where the diet expands, varies inversely with the size of the foraging group.

It is interesting to compare the social prey choice rule, a solution to an *N*-person game, with optimal behavior of a solitary forager. The solitary, attempting to maximize its long-term rate of energy intake, should generalize immediately upon entering the patch. But this tactic, although efficient for a solitary, is susceptible to invasion by competitors that quickly deplete the more profitable prey. Surprisingly, few studies have investigated prey selection under competitive conditions (Inman et al. 1987; Robichaud et al. 1996).

Conclusions

Our review identifies a number of predictions concerning social foragers' patch and prey choices. Many of these predictions remain untested and illustrate the extent to which theories for social foraging offer an array of opportunities for novel research.

Because they were designed in the context of mating systems, only some social patch models may apply generally to foraging. Among the social patch models, those allowing several competitors to exploit a patch concurrently are most likely to interest behavioral researchers. When patches are ephemeral and resources decline independently of the forager's exploitation, the models predict the existence of an emigration period during which foragers leave, one at a time, to search for the next patch. When foragers deplete the patches they exploit, the model predicts simultaneous departure of all foragers at the emigration threshold. Surprisingly, these models have been neglected by foraging research even though they were developed during the years of rapidly increasing interest in foraging theory (Parker 1978). Absence of experimental testing has stunted the refinement of models and has delayed our understanding of behavioral adaptation and constraint.

The study of social prey choice remains the most underdeveloped social foraging decision. Again, lack of experimental data, and perhaps the complicated nature of the models, has deterred progress. However, the problem of social prey choice is likely general and suggests interesting possibilities concerning cooperative foraging.

PART FOUR Models of Phenotypic Diversity

9 Quantifying Phenotypic Diversity

Game-theoretic treatments of phenotypic variation in behavioral ecology often assume that each individual adopts an identical mixed strategy, or that variation arises from a mixture of different pure strategies (see Łomnicki 1988). The first case assumes no variation among individuals; the second assumes no variation within individuals. However, in many circumstances both sources of variation will be significant and biologically important; that is, total variability, or diversity, will have both within and among-individual components. This chapter reviews methods allowing us to partition a foraging group's resource-use variability into its components and suggests predictions concerning patterns in behavioral diversity.

Our analyses in this chapter emphasize statistical models that lead us indirectly to hypotheses about the economics of social foraging. We believe the methods reviewed in this chapter will prove easy to implement and interpret. In particular, hypotheses concerning resource-use variation within and among members of foraging groups should be relatively easy to test experimentally (e.g., Barnard and Sibly 1981).

9.1 Composition of Foraging Groups

Field workers, especially those studying complex social systems, sometimes distinguish foraging groups by their size and composition. For example, Izawa (1976) observed eleven species of New World monkeys in the Amazon Basin and categorized group members according to their sex and age. Beckerman (1983) investigated fishing and hunting groups in the Barí, people inhabiting a river valley along the border between Colombia and Venezuela. He categorized group members according to age, sex, relatedness to other individuals in the group, physical condition, and their hunting and fishing proclivity. Group size and the distribution of individuals among demographic classes according to age, sex, and body size allow convenient, absolute bases for comparing groups. However, in addition to group size, it is often relative attributes like dominance rank and the proportional distribution of aggressive attacks that govern the amounts of food consumed by different members of the same group (e.g., Terborgh 1983; Janson 1985, 1988).

Behavioral ecologists commonly, often of necessity, assume identical individuals and identify a foraging group by its size alone. Assuming identical individuals has obvious advantages. But individuals always differ; the important question is whether or not the variation has ecological significance (see Łomnicki 1988).

Phenotypic variability among group members may result from genetic variation, developmental events, and/or learned differences (e.g., Partridge and Green 1985). Different genotypes may induce different phenotypes, but natural selection may simply not discriminate them; the equally "fit" phenotypes presumably could be foraging strategies. Even with strong optimizing selection on a quantitative trait, the segregation variance (see Slatkin and Lande 1976) inherent to sexual reproduction generates variability among sibs (hence within foraging groups composed of relatives). In some cases, variation among group members may arise as a consequence of a balanced genetic polymorphism at the population level (e.g., D. Wilson and Turelli 1986). Other genetic mechanisms, such as disruptive selection combined with positive assortative mating, also may result in polymorphisms affecting the economics of social foraging.

An individual's genotype may permit a range of possible phenotypes, and the particular phenotype realized may depend on environmental conditions encountered during development (e.g., Lively 1986; Boag 1987). This sort of plasticity may allow a genotypic lineage to track environments that fluctuate with periodicities at least as long as a generation. Even without a great deal of genetic variation, developmental events may generate significant phenotypic variation in a spatially and temporally variable environment.

One of our main interests is learned variation among foraging group members (Werner and Sherry 1987; Diamond 1987). Learning may allow social foragers to track environmental fluctuations shorter than a generation, provided that tracking is both feasible and economically advantageous (Recer et al. 1987; Stephens 1987; Recer and Caraco 1989a). Although different individuals may learn different foraging strategies as a consequence of chance encounters with alternate resources (Partridge and Green 1985), learned variation may more commonly arise as an economic response to the behavior of other group members (e.g., Giraldeau and Lefebvre 1987).

Phenotypic Variation and Foraging Economics

Foraging group members may commonly differ in age (e.g., Dittus 1977; Janson 1985). If different age classes exploit different foods or different parts of a single resource spectrum (see below), an increase in age variation may imply an increase in both the group's total variability and the among-individual variability in diet selection. Polis (1984) develops this concept by modeling an age-structure component of a population's niche width. Application of this approach to species with large differences in size across age classes, or to species with dissimilar larvae and adults, should prove more important than applications to most vertebrate foraging groups.

Variation in age can have economic significance for social foragers if different age classes find, capture, or handle food with different efficiencies (Łomnicki 1988). If juveniles lack the foraging skills of adults, the former may have to spend more energy to locate sufficient food than the latter (Weathers and Sullivan 1989). In turn, the lower foraging efficiency of younger individuals may lead to lower probabilities of survival (e.g., Sullivan 1988).

Most foraging groups contain both males and females. Group members may further differ in reproductive status, physiological condition, and size. Size variation may follow simply from age variation or may result from a more general sexual dimorphism (see Price 1984).

Each of the potential differences mentioned above may imply different energy and nutritional requirements (e.g., Post et al. 1980) and may induce variation in access to resources under competitive conditions. Dominance status often depends on age, sex, and size; status may further be influenced by a parent's social rank and the individual's birth order (e.g., Schulman and Chapais 1980). These demographic variables can have indirect economic significance, since a dominant individual's foraging success often will exceed a subordinate individual's success (e.g., Whitten 1983; Janson 1985; Caraco et al. 1989). Dominant toque monkeys (*Macaca s. sinica*) feed in the parts of trees where fruit is more abundant, and they aggressively force subordinate juveniles to poorer sections; Dittus (1977) documents a remarkable correlate of this spatial segregation. For each unit of physiologically required energy, a juvenile female (the most subordinate class) must forage five times as long as an adult male (the most dominant class). Foraging experience differs between age classes, but dominance appears a far more important influence (Dittus 1977). Fieldwork by Ekman and his colleagues (Ekman and Askenmo 1984; Ekman et al. 1981) reveals a similar phenomenon in winter flocks of willow tits (*Parus montanus*). Older dominants forage higher in pine and spruce trees than do subordinate flock members. The subordinate individuals, as a consequence of their more exposed microhabitat, must spend more time than dominants in vigilance for predators. The time-budgeting difference reduces the subordinate birds' rate of food intake significantly below the dominants' average rate. The younger birds have a lower survivorship, though lack of experience may interact with the cost of subordinance (Ekman and Askenmo 1984), as in the previous example.

The dominance relationship between two individuals may change with change in location, time, season, or reproductive status. But the economic consequences of preferential access to food resources are less variable. In some instances, dominance ranks defined with respect to food may differ in the context of another resource. However, when rank depends strongly on age and size, dominance relationships should usually persist across different resources. Janson (1984) describes an example in the brown capuchin monkey (*Cebus apella*). A group's dominant male controls access to food sources and has nearly exclusive access to fertile females on the most probable days of ovulation.

Dominance status induces economic variation among individuals through aggressive or ritualized interference competition. In the absence of strong dominance, individual differences in foraging ability per se can generate variation in group members' food intake rates and so provide a mechanistic basis for a form of scramble competition. Milinski (1984) reports results where some individuals have a greater probability of capturing prey per attempt than other group members. Similarly, individuals may vary in their rate of discovering food clumps in a patchy environment (e.g., Caraco et al. 1989). Goss-Custard and Durell (1987c) present a useful analysis of the way aggressive interference (a function of group size) and individual foraging ability can simultaneously influence an individual's energy intake rate. Presumably, all of these differences will often be accentuated in multispecies groups (e.g., Barnard and Thompson 1985; Thompson and Lendrem 1985).

The sorts of differences among individuals that we have reviewed will be important to us if the economic correlates influence individuals' propensities to join or leave groups (e.g., Caraco 1987; Fagan 1987b; Ekman and Hake 1988; Giraldeau and Gillis 1988), or to adopt particular behavior within a group (W. Schaffer 1978; E. Smith 1985; Caraco and Brown 1986). The next two sections of this chapter discuss models of intragroup variation in the amount and type of food resources consumed, and consider an example where this variation may reflect competition within a foraging group.

9.2 Quantifying Variability in Foraging Behavior

We next examine some probabilistic methods for studying variability within and among members of a foraging group. Most of the techniques are adopted from niche theory (May 1974; Roughgarden 1979) or from analyses of community diversity (Pielou 1975; Patil and Taillie 1979). The methods developed in this section suggest studies of feeding behavior that complement and extend both "standard" optimality theory (foraging in the absence of social interactions: Stephens and Krebs 1986; Mangel and Clark 1988) and the foraging games we rely on in this book. Rather than starting with an optimization technique, we can look for patterns in observed variability and then interpret the results in terms of ecological benefits and costs (see Hanski 1980; Herbers 1981; Caraco et al. 1989).

Variation within or among individuals may refer to food resources exploited or consumer roles adopted (see Werner and Sherry 1987). We assume that foragers can detect quantitative variation in continuous resource characteristics, or can discriminate among discrete resource types. Consumer roles refer to differences in behavior allowing different individuals to exploit the same unit (clump or patch) of resource simultaneously. Producer and scrounger (see chapter 6) provide examples of common resource-consumer roles within a group. In certain cases (e.g., B. Smith 1987) it can prove interesting to equate consumer roles with different trophic morphologies, but we emphasize modifiable behavior.

We envision a group composed of a fixed number of individuals. The resources exploited by the group may vary quantitatively or qualitatively. Some resources possess a continuous and ordered property useful for describing the availability and consumption of that resource. For example, seed size has proven helpful in describing foraging by granivores (Pulliam 1983). Seed size exhibits continuous random variation. Therefore, any two seed sizes are naturally ordered by the positive real numbers, and seed size variability can be calculated as a variance.

Other resources occur as discrete, unordered classes (Roughgarden 1979). Habitat, patch and prey types (often prey species) are examples. In this case, resources differ qualitatively, and we characterize the distribution of consumption across categories as diversity. Diversity serves as the "variance" of qualitative attributes (Pielou 1975). Some resources allow either a continuous and ordered treatment or a discrete, unordered classification. We might quantify a mixture of seeds as a single probability density with size as the random variable. Alternatively, we might assign the seeds according to species and calculate an index of diversity. We prefer the latter since both food resources and consumer roles usually will qualify as discrete, unordered classes.

Given that resources vary quantitatively or qualitatively, we often must consider availability and some measure of value (energetic profitability, for example). For convenience, we may initially assume uniform resource density and equal fitness value (or preference valence), so that differences among individuals should depend on their interactions within the foraging group.

Phenotypic Variance

Suppose that x represents a continuous, ordered resource attribute. Continuous attributes such as food-item size or distance from a central place have natural numerical orderings. In a group of G individuals we associate the resource-use function $f_g(x) = f(x|\mu_g)$ with the gth group member ($g = 1,2, \ldots, G$). μ_g is the mean value of the resource used by individual g (e.g., average size of the seeds eaten by individual g). In general, individuals' resource-use functions might differ in both mean and variance.

Since the resource attribute x varies continuously, variability within and among group members becomes a matter of partitioning the variance of the group's resource utilization. To do so, we can apply methods employed by community ecologists to calculate niche width (e.g., Roughgarden 1979). First, we define the group's resource-use function $h_G(x)$ as a mixture of the individual $f_g(x)$. We then specify the mean and variance of the group's combined resource utilization. Finally, we partition the group's total resource use variance into

1. The within-individual component, the average resource-use variance of the G individual foragers.
2. The among-individual component, the variance of the individual means μ_g about the group average.

In Math Box 9.1 we assume each $f_g(x)$ has a normal probability density, partition the components of resource-use variance, and calculate examples. Resource-use variances estimated from observation or experimentation can be partitioned by components; Roughgarden (1974) gives an example.

Patterns in resource-use variability can be correlated with social or demographic variables mentioned above, or with "fitness values" dependent on the resource characteristic. We make some economic predictions about behavioral variability within groups of social foragers after considering discrete resources.

9.3 Phenotypic Diversity

Suppose now that resources occur as discrete, unordered categories. For example, we might classify a resource according to its taxon or microhabitat where it is exploited. For resources that differ qualitatively, variability within and among group members' resource consumption becomes a problem of partitioning diversity, rather than variance (e.g., Pielou 1975). To partition diversity, we again turn to methods of community ecology.

Statistical analyses of diversity indices used in community ecology constitute a substantial literature. A number of excellent references are available (e.g., May 1975; Pielou 1975; Grassle et al. 1979; Patil and Rosenzweig 1979). Any diversity index possessing an apparent advantage in one respect will most likely have a countering weakness in another respect. Choosing a diversity index, estimating a confidence interval for the index, and developing significance tests often present problems in community ecology because the total number of species is unknown (see Adams and McCune 1979). In experimental work on group foraging, the investigator can often control the size of the group and the number of different resources available (e.g., Giraldeau and Lefebvre 1985). This sort of control simplifies several statistical aspects of diversity calculations. Our objective, in any case, is describing and interpreting behavioral diversity within and among group members; we do not review the statistical literature on estimating diversity.

Patil and Taillie (1979) point out that diversity may be viewed as average rarity of the elements of a collection or sample. We would reasonably rank a set of a few common "species" as less diverse than a set of many rare "species," and the average rarity per element should be greater in the latter set. We review this idea to motivate use of a diversity index.

Suppose a solitary forager's diet is composed of S unordered prey types. The relative frequency of prey type i in the diet is π_i, where $0 < \pi_i < 1$ for $i = 1, 2, \ldots, S$; and

$$\sum_{i=1}^{S} \pi_i = 1.$$

Let π represent the vector of relative abundances; $\pi = (\pi_1 \; \pi_2 \; \ldots \; \pi_S)$.

The function $\Gamma(\pi_i)$ measures the rarity of the prey type with relative abundance π_i. Rarity must decline as relative abundance increases, so $\partial\Gamma/\partial\pi_i < 0$. Then, taking expected rarity as diversity, the forager's diet has diversity

$$\mathrm{E}[\Gamma] = \sum_{i=1}^{S} \pi_i \Gamma(\pi_i). \tag{9.1}$$

We can quantify rarity in different ways. Patil and Taillie (1979) suggest a general formulation for the rarity function,

$$\Gamma(\pi_i; \beta) = (1 - \pi_i^{\beta})/\beta. \tag{9.2}$$

The associated theoretical diversity is this function's expected value,

$$\mathrm{E}[\Gamma; \beta] = \left(1 - \sum_{i=1}^{S} \pi_i^{\beta+1}\right)\beta^{-1}. \tag{9.3}$$

Requiring $\beta \geq -1$ insures two desired properties: diversity increases as a function of both the number of prey types consumed (S) and the "evenness" of the vector π (see Engen 1979; Patil and Taillie 1979).

Varying β in equation (9.3) yields a family of diversity indices that includes the three formulas most familiar to ecologists (see Baczkowski et al. 1997). When $\beta = -1$, diversity is $(S - 1)$, a count of the resource types. When $\beta = 1$, we have Simpson's index of diversity. We favor the third form. As β approaches 0, $\Gamma(\pi_i)$ becomes $\ln(1/\pi_i)$, and diversity is given by the Shannon index (e.g., Dennis and Patil 1979),

$$\mathrm{E}[\Gamma] = H' = -\sum_{i=1}^{S} \pi_i \ln(\pi_i). \tag{9.4}$$

Suppose a random sample of the solitary's diet yields n_i ($n_i > 0$) items of prey type i ($i = 1, 2, \ldots, S$), and a total of N items. Then $p_i = n_i/N$ is the maximum likelihood estimator of π_i. However, replacing π_i by p_i in equation (9.4) results in a statistically biased estimate. That is, $\mathrm{E}[h'(p_1, p_2, \ldots, p_S) \neq H'$, where h' is the estimator

$$h' = -\sum_{i=1}^{S} p_i \ln(p_i) \tag{9.5}$$

(see Bowman et al. 1971). Fortunately, when the number of prey types (more generally, the number of categories) is known, an excellent estimator of H' is (Adams and McCune 1979)

$$h'_a = h' + [(S - 1)/2N]. \tag{9.6}$$

Expression (9.6) shows that for sufficiently large (N/S), bias should not be a problem when S is known. Recognizing that our method may be subject to a small bias, below we apply h' to a calculation concerning a foraging group's phenotypic diversity.

Partitioning Phenotypic Diversity

The preceding example dealt with a solitary forager's dietary choice, where diversity depended solely on the distribution of the individual's consumption across resources. In a foraging group, total phenotypic diversity depends on variability in resource consumption both within and among individuals. Consequently, we analyze a group's total phenotypic diversity with an h' calculation cross-classified by both individuals and resources.

Suppose again that the foraging group contains a fixed total of G individuals. We index group members by g; $g = 1, 2, \ldots, G$. The foragers collectively exploit a (known) fixed total of S discrete, unordered resources. We index resources by s; $s = 1, 2, \ldots, S$.

We designate the number of items of resource s consumed by individual g as n_{gs}. The n_{gs} are the elements of a $G \times S$ matrix **M**. Following Pielou (1975), we refer to **M** as the occurrence matrix (see Colwell 1979). Each row of **M** specifies a single individual's use of the S different resources. Each column of **M** lists the use of a single resource by the G different consumers.

The total number of items consumed by individual g is

$$n_{g\cdot} = \sum_{s=1}^{S} n_{gs}. \tag{9.7}$$

The total number of items of resource s consumed is

$$n_{\cdot s} = \sum_{g=1}^{G} n_{gs}. \tag{9.8}$$

The total of the elements of **M** is the total number of items consumed by all individuals across all resources:

$$n_{\cdot\cdot} = \sum_{g=1}^{G} \sum_{s=1}^{S} n_{gs} = \sum_{g=1}^{G} n_{g\cdot} = \sum_{s=1}^{S} n_{\cdot s}. \tag{9.9}$$

Different individuals may consume different total amounts of resource. Similarly, different resource types may be exploited in different numbers by both individuals and the group as a whole.

Each $n_{gs} > 0$ defines a sample proportion p_{gs}, where

$$p_{gs} = n_{gs}/n_{\cdot\cdot} \tag{9.10}$$

We use the p_{gs} ($0 < p_{gs} < 1$) to estimate the total, cross-classified diversity $h'(g \times s)$:

$$h'(g \times s) = -\sum_{g=1}^{G} \sum_{s=1}^{S} p_{gs} \ln(p_{gs}). \tag{9.11}$$

We can partition the total, cross-classified diversity in two complementary ways (Patil and Taillie 1979); for a different diversity decomposition, see Roughgarden (1979).

First we calculate the among-resource diversity to ask how much diversity the group's pooled food consumption generates across the S resources. The among-resource diversity is $h'(s)$:

$$h'(s) = -\sum_{s=1}^{S} p_{\cdot s} \ln(p_{\cdot s}). \tag{9.12}$$

$h'(s)$ increases, of course, with S. Given a fixed number of resource types S, the among-resource diversity is relatively large when the group forages in a collectively generalist manner. That is, $h'(s)$ is greatest when the group consumes the same amount of each of the S resources. Note that a group consumes the S resources in a relatively even manner when each individual generalizes, or when different individuals specialize on different resources (see chapter 11). Given that S is fixed, the among-resource diversity $h'(s)$ will be relatively low when the group forages in a collectively specialized manner. If the group specializes as a whole, each individual must have a similar, specialized diet.

Second, we note that diversity within any particular resource is defined by the distribution of the use of that resource across group members. For each resource type s ($s = 1, 2, \ldots, S$) we estimate a conditional phenotypic diversity within that resource:

$$h'(g \mid s) = -\sum_{g=1}^{G} \left(\frac{p_{gs}}{p_{\cdot s}}\right) \ln \left(\frac{p_{gs}}{p_{\cdot s}}\right). \tag{9.13}$$

Then we average the S values of $h'(g \mid s)$, weighted by the group's proportional consumption of each resource, to obtain the average within-resource diversity:

$$\mathrm{E}[h'(g \mid s)] = \sum_{s=1}^{S} p_{\cdot s}\, h'(g \mid s). \tag{9.14}$$

Expression (9.14) averages, across resource types, variability in the group members' use of those resources. To appreciate the average within-resource diversity, suppose each individual forager consumes nearly the same diet; that is, each individual takes about the same number of items of most (or all) resources. Then, independently of whether that diet is specialized or generalized, the total consumption of any given resource will be evenly distributed across the G group members. Consequently, each conditional phenotypic diversity $h'(g \mid s)$ will be relatively large, and the average within-resource diversity must be relatively large.

However, if different individuals consume different amounts of food, the use of most resources will differ substantially across group members. Then the average within-resource diversity $\mathrm{E}[h'(g \mid s)]$ will be low, even if each individual distributes its particular level of consumption across resources in a similar proportional manner. That is, group members may eat the same types of food, but if they eat different amounts of those foods, the mean within-resource diversity will be low.

Now suppose that each group member specializes, but on a different resource. Then the consumption of any given resource will show a low diversity across group members; one or a few consumers specialize on that resource while other foragers tend to avoid it. Then each $h'(g \mid s)$ will be relatively low, and the average within-resource diversity must be relatively low.

The third formulation considers diversity generated by the distribution of the total resource consumption across the G group members. The among-individual diversity (the among-phenotype diversity) is $h'(g)$:

$$h'(g) = -\sum_{g=1}^{G} p_{g\cdot} \ln(p_{g\cdot}). \tag{9.15}$$

$h'(g)$ increases, of course, with group size G. Given that G is fixed, the among-individual diversity is relatively large when each individual consumes about the same number of food items. Physiological or competitive equivalence of group members might imply that resource consumption is not concentrated in a few individuals but is diversely distributed across foragers. A relatively even distribution in the amount of food consumed per individual can arise when each individual generalizes across the various resource types, or when different individuals specialize on different resources but eat about the same number of items. Clearly, the among-individual diversity $h'(g)$ will be relatively low when a few individuals consume large amounts of food and the other group members consume little. Metaphorically, $h'(g)$ considers the view of a single resource item. Its outcome is variable, hence diverse, if every group member has the same chance of consuming the item. But its outcome is far less diverse if one or a few dominants eat most of the available food and are the most likely consumers of the item.

Fourth, we quantify the diversity of each individual's use of resources. For each group member g ($g = 1, 2, \ldots, G$) we estimate a conditional resource-consumption diversity,

$$h'(s \mid g) = -\sum_{s=1}^{S} \left(\frac{p_{gs}}{p_{g\cdot}}\right) \ln\left(\frac{p_{gs}}{p_{g\cdot}}\right). \tag{9.16}$$

We average the G values of $h'(s \mid g)$, weighted by the relative amount of food consumed by each individual, to obtain the average within-individual diversity,

$$\mathrm{E}[h'(s \mid g)] = \sum_{g=1}^{G} p_{g\cdot}\, h'(s \mid g). \tag{9.17}$$

Suppose each individual forager consumes a generalist diet. Then the various $(p_{gs}/p_{g\cdot})$ will have relatively uniform values. The consequent relatively even distribution of consumption across the S resources leads to a large $h'(s \mid g)$ for each group member, and the average within-individual diversity must be relatively large. However, suppose that each individual forager selects a specialized diet. Then, independently of whether different group members specialize on the same or different resources, each $h'(s \mid g)$ will be relatively small, and the average within-individual diversity must be low.

Each of the four diversity estimates responds to different properties of the occurrence matrix **M**. Summary Box 9.1 lists relationships between specialization or generalization in resource use and the diversity measures discussed here.

The total cross-classified diversity $h'(g \times r)$ approaches its maximal value, $\ln(GS)$, when

1. Each individual takes a varied, generalist diet.
2. The total consumption of each resource is evenly distributed across all group members.

We can partition the total diversity in two ways. We first examine diversity within and among individuals, and then within and among resources.

To demonstrate a diversity partitioning, we recall that total diversity is calculated with the proportions p_{gs}. Each of these quantities is the product of individual g's proportional contribution to the group's total consumption of all resources and the proportional contribution of resource s to consumption by individual g. Then we can replace p_{gs} with $p_{g\cdot}\ (p_{gs}/p_{g\cdot})$ in expression (9.11) and write total diversity as

$$h'(g \times s) = -\sum_{g=1}^{G}\sum_{s=1}^{S} p_{g\cdot}\ (p_{gs}/p_{g\cdot})\ \ln(p_{g\cdot}[p_{gs}/p_{g\cdot}]). \tag{9.18}$$

Rewriting the logarithmic term as a sum of logarithms yields

$$h'(g \times s) = -\sum_{g=1}^{G} p_{g\cdot}\ \ln(p_{g\cdot})\left(\frac{\sum\limits_{s=1}^{S} p_{gs}}{p_{g\cdot}}\right) - \sum_{g=1}^{G} p_{g\cdot}\left(\sum_{s=1}^{S}\frac{p_{gs}}{p_{g\cdot}}\ln\frac{p_{gs}}{p_{g\cdot}}\right). \tag{9.19}$$

Since

$$\sum_{s=1}^{S}\left(\frac{p_{gs}}{p_{g\cdot}}\right) = p_{g\cdot}/p_{g\cdot} = 1,$$

the first term on the right-hand side of expression (9.19) is, by equation (9.15), the among-individual diversity $h'(g)$. The second term of (9.19) is the within-individual resource-consumption diversity averaged across group members; see expressions (9.16) and (9.17). That is, the second term is the expected within-individual diversity $\mathrm{E}[h'(s \mid g)]$. Consequently, expression (9.19) becomes

$$h'(g \times s) = h'(g) + \mathrm{E}[h'(s \mid g)]. \tag{9.20}$$

SUMMARY BOX 9.1 DIVERSITY MEASURES FOR SOCIAL FORAGING

A generalized diet includes most or all resource types (or, if applicable, resource-acquisition roles) in roughly equal proportions. A specialized diet includes one or a few resource types at high proportions, and very low proportional levels of the remaining resources. The group's diet refers to the pooled resource consumption of all group members.

AMONG-RESOURCE DIVERSITY $h'(s)$

Low Group specializes because individuals have similar, specialized diets.

High Group generalizes; individuals may generalize or different individuals have different, specialized diets.

AVERAGE WITHIN-RESOURCE DIVERSITY $E[h'(g \mid s)]$

Low Different individuals have different, specialized diets, so group generalizes; similar effect occurs whenever different individuals consume different total amounts of food.

High Individuals have similar diets, whether generalized or similarly specialized; group diet may then be generalized or specialized.

AMONG-INDIVIDUAL DIVERSITY $h'(g)$

Low Individuals differ in amount of food consumed, independently of each individual's specialization or generalization.

High Individuals consume similar amounts of food, independently of each individual's specialization or generalization.

AVERAGE WITHIN-INDIVIDUAL DIVERSITY $E[h'(s \mid g)]$

Low Individuals specialize independently; group may consequently specialize or generalize.

High Individuals generalize; group consequently generalizes.

The total diversity has been partitioned additively into the among-individual diversity and the average within-individual diversity. This partitioning matches the point mentioned in section 9.1; variability in a foraging group's use of resources depends on variability within and among group members.

Proceeding similarly, we can replace p_{gs} with $p_{\cdot s}$ $(p_{gs}/p_{\cdot s})$ in $h'(g \times s)$ and show that

$$h'(g \times s) = h'(s) + E[h'(g \mid s)] \tag{9.21}$$

Written this way, total cross-classified diversity is the sum of the among-resource diversity plus the average within-resource diversity. The two diversity decompositions (9.20) and (9.21) yield complementary information about the biology of social foraging; see Summary Box 9.1. It is perhaps worth noting that the diversity decomposition given by expression (9.20) is not simply the discrete version of the components of phenotypic variance given in Math Box 9.1.

DETERMINANTS OF DIVERSITY

Environmental, demographic, and social variables may directly or indirectly influence diversity within and among foraging group members. Increasing group size (G) ordinarily will increase the among-individual diversity $h'(g)$, since resource consumption will then be dispersed across a greater range of consumers. However, a larger group may imply greater competition for food (e.g., Barton et al. 1996). Variation in competitive success could increase the variability in the amounts of food consumed by different group members and reduce $h'(g)$. Lower food density, dispersed in a manner making aggressive defense economically advantageous, could imply that interference competition lowers the evenness of food consumption, and so lowers both $h'(g)$ and the average within-resource diversity $\mathrm{E}[h'(g \mid s)]$.

A larger group should often result in greater among-resource diversity $h'(s)$. If increased group size increases the intensity of food competition, then individuals may commonly accept less preferred food types, whether competitive interactions are direct or indirect. At the level of the group, the number of resources consumed and/or the evenness of consumption among resources could increase, implying an increased $h'(s)$. When competitive effects are indirect, all group members (rather than only subordinate individuals; see below) might take a broader array of resources, so that the average within-individual diversity $\mathrm{E}[h'(s \mid g)]$ will increase.

Differences among group members in age, sex, size, and physiological condition ordinarily suggest differences in the amount of resources consumed per individual. Decreased evenness in food consumption reduces the among-individual diversity $h'(g)$, since resource use tends to be concentrated in a subset of group members. Similarly, the average within-resource diversity $\mathrm{E}[h'(g \mid s)]$ may be reduced. If variation in group members' nutritional requirements or efficiencies as foragers generates differences in the types of resources individuals select, the group will be composed of contrasting specialists. This effect also tends to decrease $\mathrm{E}[h'(g \mid s)]$, and clearly should decrease the average within-individual diversity $E[h'(s \mid g)]$. Divergent specializations may also increase the range of resources exploited by the group as a whole, and so will increase the among-resource diversity $h'(s)$.

Strong dominance relationships concentrate resource consumption in a subset of group members, so that both among-individual diversity $h'(g)$ and average within-resource diversity $\mathrm{E}[h'(g \mid s)]$ will be relatively low. If dominants sequester higher-quality resources, subordinate individuals will be forced to accept a broader range of resource types. Consequently, the among-resource diversity $h'(s)$ and the average within-individual diversity $E[h'(s \mid g)]$ should be relatively large.

Independently of effects of dominance, an individual's net economic benefit from exploiting a particular resource, or from adopting a particular resource-acquisition role, may decline as the frequency and/or density of group members using that resource increases. As a result, different individuals may specialize on different resources or may forage as a group of similar generalists, depending on the constraints governing the breadth of resources an individual can exploit efficiently. If all group members consume about the same quantity of resources, differing specialists will produce a large among-resource diversity $h'(s)$ and relatively low average within-resource diversity $E[h'(g \mid s)]$ and low average within-individual diversity $E[h'(s \mid g)]$. A group of similar generalists will also exhibit a large $h'(s)$, and average diversities within resources and within individuals will be greater than for a group of differing specialists.

Environmental patterns may also influence the diversity of group foraging. For example, increased productivity could enrich the range of resources available to a group. A greater breadth of acceptable food types could increase the diversity of foraging choice among resources, so that both $h'(s)$ and the average within-individual diversity $E[h'(s \mid g)]$ increase. When individuals do not have to compete for food, an increase in food-quality (or preference-valence) variation should imply convergence in the group members' diets, so that all eat the highest-quality resources. In this case the among-resource diversity $h'(s)$ will decline, while the average within-resource diversity $E[h'(g \mid s)]$ will increase.

The spatial dispersion of resources might influence patterns in dietary diversity. Suppose different resources occur in different patches, and discovery of a patch by one individual alerts the entire group to the location of food. Indefensible, clumped food will then tend to homogenize the diets of different group members. The average within-resource diversity $E[h'(g \mid s)]$ should consequently attain a relatively high value, and the among-resource diversity $h'(s)$ could take a relatively low value. Either a less clumped dispersion of food or increased defensibility of the clumps reduces the economic advantage of scrounging or joining others' discoveries of food. Reduced scrounging can increase the variation among individuals' diets, and thereby decrease $E[h'(g \mid s)]$ while increasing $h'(s)$.

An Example

Given an array of the n_{gs}, or the p_{gs}, we might conduct a statistically rigorous test of an a priori hypothesis. A diversity decomposition offers a complementary approach that can reveal patterns that might otherwise be missed. We present an example, intended simply to demonstrate use of the method.

Giraldeau and Lefebvre (1985) reported strong variation among individuals in urban pigeons' dietary choice. During a program of capturing and banding in Montréal, a total of fifty-seven pigeons from three different flocks succumbed accidentally following narcosis. Giraldeau and Lefebvre counted the crop contents of each individual and calculated a significant heterogeneity statistic, justifying their conclusion about differences among the pigeons' selection of food types.

At each site (designated Campus, Drummond, and Mountain), Giraldeau and Lefebvre provided 2 kg of commercial seed on four consecutive days and then treated the seeds with sedatives on the fifth day. Seven different food types were included in each day's bait: wheat, corn, maple peas, white peas, milo maize, vetch, and oats. The mean number of seeds consumed per individual was 278.7 (SD = 106.6); converted to mass the mean is 15.5 g (SD = 8.81). Here we use our models to reanalyze the number of seeds, by type, consumed by each individual sampled.

Table 9.1 shows the individual dietary choices of the sixteen birds sampled from the Mountain flock. The table is the occurrence matrix. Note that some of the entries are zero; according to our definition of p_{gs}, these entries are not included in our calculations of diversity estimates.

At first glance the data in table 9.1 suggest a preference for wheat and milo maize and an apparent aversion to maple peas. Table 9.2 shows the diversity decompositions for each of the three flocks. The total-diversity estimates do not vary particularly, especially if we consider that the three flocks differed in size. The among-individual diversities are the largest components of total diversity. Total seed consumption in the Mountain flock ranged almost threefold (from 188 to 544 seeds) across individuals, but more than half of the flock took between 200 and 300 seeds. This relative homogeneity in quantity, coupled with the excess of flock size over resource number, explains the large proportional contribution of diversity among individuals, $h'(g)$, to total cross-classified diversity.

Average within-resource diversities $E[h'(g \mid s)]$ appear relatively large, partly due to the large flock sizes and partly a consequence of clear preferences and aversions for different food types in the average behavior of flock members. Despite certain similarities in food choice, we can ask if individuals consumed resources in a less diverse manner than the flock did as a whole. If different individuals adopt different, somewhat specialized diets, then within-individual diversities $h'(s \mid g)$ should be less than the among-resource diversity $h'(s)$. We calculated an approximate 95% confidence interval (see Adams and McCune 1979) for each within-individual diversity in the Mountain flock, and asked whether or not the interval included the flock's $h'(s)$ value. Thirteen of sixteen pigeons took diets more specialized than the flock's collective diet; the other three individuals were more generalized than the flock as a whole. After only four days' experience with the combination of seven resource types, individuals tended to specialize while the group as a whole generalized (Giraldeau and Lefebvre 1985).

Table 9.1

Crop contents of sixteen pigeons of the "Mountain Flock." From Giraldeau and Lefebvre (1985).

Bird	*Wheat*	*Corn*	*Maple Peas*	*White Peas*	*Milo Maize*	*Vetch*	*Oats*
1	144	24	1	9	182	13	2
2	30	45	6	68	25	11	3
3	183	18	1	19	4	0	0
4	118	19	7	54	9	45	0
5	308	16	0	0	112	0	1
6	129	2	1	5	2	7	65
7	131	55	4	4	339	9	2
8	115	39	4	25	7	7	6
9	36	65	0	0	327	4	0
10	91	45	0	1	109	0	7
11	119	12	1	13	91	4	0
12	188	133	1	0	8	0	0
13	379	0	0	0	24	0	3
14	185	17	1	64	9	3	28
15	148	82	4	29	1	7	8
16	16	41	6	15	98	13	8

We cannot claim that this pattern represents a response to food competition. But significant differences among individuals might ameliorate competition within a large group, especially if learning to handle one food type efficiently interferes with efficient consumption of a second type. Presumably, questions concerning social foraging, food competition and diversity could be answered experimentally. Suppose group members respond to one another so that their diets quickly differentiate. Learned dietary divergence might require not only that group membership remain stable, but also that a range of resources remain profitable and temporally predictable. If only a single food type is available, or if a single type has a sufficiently large fitness value, the group's total phenotypic variance will be too small to allow a

Table 9.2

Diversity decompositions for each of three flocks studied by Giraldeau and Lefebvre (1985). Diversity symbols are the same as those used in the text and Summary Box 9.1.

Flock	$h'(g \times s)$	$h'(g)$	$E[h'(s \mid g)]$	$h'(s)$	$E[h'(s \mid g)]$
Campus	3.69	2.92	0.77	1.02	2.67
Drummond	3.86	2.99	0.87	1.38	2.48
Mountain	3.71	2.72	0.99	1.37	2.34

biologically significant and empirically detectable divergence of individuals' use of resources. Sufficiently fast temporal fluctuations in resource availability could constrain individuals' capacity for responding efficiently to both food densities and one anothers' behavior (e.g., Recer and Caraco 1989a), although Inman et al. (1987) found that the diets of paired pigeons diverge rather quickly.

9.4 Concluding Remarks

This chapter begins with the premise that members of a foraging group may often differ in their use of resources or in their modes of acquiring a particular resource. The results of Giraldeau and Lefebvre (1985; see Inman et al. 1987) indicate that dietary variation within groups will at times exceed the homogeneity often assumed in behavioral ecology. The interest in producing and scrounging behavior (see chapters 6 and 7) suggests that variation in resource-acquisition roles may be common and may be ecologically important when role variation extends beyond the asymmetries resulting from social dominance relationships.

Implications for Scaling Up

To address some ecological implications of the concepts discussed in this chapter, suppose that consumers forage socially and impose significant mortality on at least some of the resource populations they exploit. Then the among-resource diversity in consumer groups may predict effects of social foraging on the resource community's structure.

Increases in either the total number of resources exploited (S) or the evenness of their representation in the group's combined resource consumption increases the among-resource diversity $h'(s)$; that is, greater among-resource diversity implies that group members, collectively, have an increasingly generalized diet. If resources have a spatially clumped dispersion, $h'(s)$ will be relatively large when different group members search for different resources (see chapter 11), or when different group members produce the same, generalized set of resource types. But if group members forage as similar specialists, each searching for the same type of resource, $h'(s)$ will be relatively small.

Basic ecological considerations suggest that dietary generalists tend to maintain the species richness of resources exploited (e.g., Ricklefs 1979; Begon et al. 1990). Generalist consumers may regulate densities of the various resources below levels where interspecific competition could exclude one or more resource species. Similar specialization among consumers might lead to a like result, provided that consumers prefer competitively dominant resource species. Otherwise, similar specialists may accelerate competitive exclusion among their resources. Therefore, increasing among-resource production diversity should maintain species richness and promote evenness of abundances among the consumers' resources. Assuming this hypothesis holds, we can ask how social foraging might induce an among-resource diversity different than we would find if consumers foraged solitarily.

In section 9.1 we noted that increasing demographic variation among group members, i.e., differences in age, size or sex, might increase consumers' among-resource diversity. This type of demographic variation can be associated with differences in energy requirements, morphologically based efficiencies, or foraging preferences acquired developmentally (e.g., Partridge and Green 1985). That is, variability among group members may arise without the economic interactions defining social foraging. Individual variation of this sort consequently need not depend on solitary versus social foraging. Under these circumstances, consumers' social foraging will neither amplify nor attenuate effects on the resource community.

Suppose that symmetric or asymmetric food competition in foraging groups may lead to dietary divergence and an increase in among-resource consumption diversity. That is, as consumers try to avoid competition, $h'(s)$ increases, and the group forages collectively as a generalist. Solitary foragers should experience less interference competition than group members; solitaries' diets might then exhibit less collective variability. Under these circumstances, food competition within consumer groups may help maintain species richness of the resource community.

Alternatively, social foraging might homogenize consumer diets. Consider a spatially heterogeneous environment where different resources occur in different habitats. Suppose the formation of social groups localizes the consumer population so that its use of space is more restricted than, say, territorial solitaries. Given these conditions, social foraging could reduce the among-resource diversity and allow competitive exclusion in the resource community.

Final Comment

We hope that greater attention will be directed toward economic variation among social group members. Differences in resource use may imply ecologically significant variation in feeding rates, starvation probabilities, and the energy available for investment in reproduction; that is, we would like to know how phenotypic variation in the use of resources maps onto variation in fitness or reasonable proxy attributes for fitness. Novel insights could emerge if we monitored the development, temporal stability, and economic consequences of differences among group members.

Math Box 9.1 Partitioning Phenotypic Variance

Let x represent a continuous, ordered resource attribute. A foraging group contains G individuals; and we initially assume that each individual g ($g = 1, 2, \ldots, G$) consumes the same amount of resource. $f_g(x \mid \mu_g)$ is the gth individual's resource-use function; μ_g is the mean of the resource attribute levels consumed by individual g. We let each $f_g(x)$ follow a normal probability density with variance $\sigma_g{}^2$; different foragers' resource-use functions may differ in both mean and variance.

To specify the probability density of resource use for the group as a whole, let $k_G(x)$ represent the group's resource-use function. $k_G(x)$ will be a finite mixture (e.g., Ord 1972) of the individual resource curves. Mixing is an averaging process; each probability density $f_g(x)$ is weighted equally in the mixing process when all group members consume the same amount of resource (e.g., the same total biomass of seeds). Since $k_G(x)$ is a probability density, the weighting must assure that

$$\int k_G(x)dx = 1. \tag{9.1.1}$$

Then the weighting is G^{-1}, the inverse of the number of elements in the finite mixture. Therefore,

$$k_G(x) = G^{-1} \sum_{g=1}^{G} f_g(x \mid \mu_g). \tag{9.1.2}$$

The overall mean resource selected (e.g., mean seed size for the group as a whole) is

$$\mathrm{E}[x \mid k_G(x)] = G^{-1} \sum_{g=1}^{G} \mu_g. \tag{9.1.3}$$

The total variance is the variance of the group's collective resource-use function,

$$\mathrm{V}[x \mid k_G(x)] = G^{-1} \sum_{g=1}^{G} \left(\sigma_g{}^2 + \delta_g{}^2\right), \tag{9.1.4}$$

where

$$\delta_g = \mu_g - \mathrm{E}[x \mid k_G(x)]. \tag{9.1.5}$$

Math Box 9.1 (*cont.*)

That is, δ_g is the deviation between the mean resource chosen by individual g and the group's mean resource.

Note that the group's total variance $V[x \mid k_G(x)]$ must exceed the average of the individual resource-use variances. Let $E[V_g]$ represent the average individual variance,

$$E[V_g] = G^{-1} \sum_{g=1}^{G} \sigma_g^{\,2}. \tag{9.1.6}$$

The difference between the total variance and the mean individual variance is

$$V[x \mid k_G(x)] - E[V_g] = G^{-1} \sum_{g=1}^{G} \delta_g^{\,2}. \tag{9.1.7}$$

Substituting for δ_g and rearranging gives

$$V[x \mid k_G(x)] = E[V_g] + G^{-1} \sum_{g=1}^{G} (\mu_g - E[x \mid k_G(x)])^2. \tag{9.1.8}$$

This equality partitions total phenotypic variance as in niche-width calculations (e.g., Roughgarden 1979).

Equation (9.1.8) shows that the phenotypic variance of the group as whole is the sum of two component variances. The within-individual component is $E[V_g]$, the average variance of the G individual resource-use curves. Holding the various μ_g constant, any increase in the average variance of individual resource selection will increase total phenotypic variance. The among-individual component is the variance of the individual means μ_g about the overall average resource selected by the group. Holding the $\sigma_g^{\,2}$ constant, an increase in the dispersion of the individual means μ_g about the group mean $E[x \mid k_G(x)]$ will increase total phenotypic variance.

The assumption of normality for the $f_g(x)$ is important. If the individual resource-use functions lack symmetry about their means, the partitioning of the total phenotypic variance by equation (9.1.8) need not hold. Assuming the individual $f_g(x)$ are normally distributed, the finite mixture $k_G(x)$ for the group may be shaped oddly and highly

Math Box 9.1 (*cont.*)

skewed (especially for small G) without violating the variance partitioning's assumptions.

Examples

Our two examples consider a group of $G = 3$ members. Each individual's selection of the continuous ordered resource is normally distributed with mean μ_g and variance σ_g^2. The first example envisions a relative "generalist" (larger variance) between two relative "specialists" (smaller variance).

For the first example, the individuals' mean-variance pairs are $(\mu_1, \sigma_1^2) = (366, 35)$, $(\mu_2, \sigma_2^2) = (380, 80)$ and $(\mu_3, \sigma_3^2) = (394, 35)$. The group's mean resource use is

$$\mathrm{E}[x \mid k_G(x)] = (366 + 380 + 394)/3 = 380.$$

The deviations of each individual mean about the group mean are

$$\delta_1 = 366 - 380 = -14,\ \delta_2 = 380 - 380 = 0\ \&\ \delta_3 = 394 - 380 = 14.$$

The total phenotypic variance is

$$\mathrm{V}[x \mid k_G(x)] = ([35 + 196] + [80] + [35 + 196])/3 = 180.67.$$

The average within-individual variance is

$$\mathrm{E}[\mathrm{V}_g] = (35 + 80 + 35)/3 = 50.$$

The variance among individuals is

$$\mathrm{V}[x \mid k_G(x)] - \mathrm{E}[\mathrm{V}_g] = 180.67 - 50 = 130.67.$$

Variability among foragers exceeds the average variability within foragers.

For the second example, the mean-variance pairs are $(\mu_1, \sigma_1^2) = (556, 35)$, $(\mu_2, \sigma_2^2) = (580, 70)$, and $(\mu_3, \sigma_3^2) = 622, 150)$. The group's mean resource use is

$$\mathrm{E}[x \mid k_G(x)] = (556 + 580 + 622)/3 = 586.$$

Math Box 9.1 (*cont.*)

The deviations of each individual mean about the group mean are

$$\delta_1 = 556 - 586 = -30,\ \delta_2 = 580 - 586 = -6\ \& \\ \delta_3 = 622 - 586 = 36.$$

The total phenotypic variance is

$$\mathrm{V}[x \mid k_G(x)] = ([35 + 900] + [70 + 36] \\ + [150 + 1296])/3 = 829.$$

The average within-individual variance is

$$\mathrm{E}[\mathrm{V}_g] = (35 + 70 + 150)/3 = 85.$$

The variance among individuals is

$$\mathrm{V}[x \mid k_G(x)] - \mathrm{E}[\mathrm{V}_g] = 744.$$

Asymmetric Food Consumption

The above formulas may require modification when different individuals consume different amounts of food. Let $b(\mu_g)$ represent the biomass (or related quantity) consumed by the individual who selects mean resource μ_g. Then the composite resource-use mixture becomes

$$k_G(x) = \frac{\sum_{g=1}^{G} b(\mu_g) f(x \mid \mu_g)}{\sum_{g=1}^{G} b(\mu_g)}. \qquad (9.1.9)$$

The mean and variance of the composite resource-use function $k_G(x)$ are modified accordingly. In particular,

$$V[x \mid k_G(x)] = \frac{\sum_{g=1}^{G} b(\mu_g)\left(\sigma_g^2 + \delta_g^2\right)}{\sum_{g=1}^{G} b(\mu_g)}. \qquad (9.1.10)$$

Math Box 9.1 (*cont.*)

The model's assumption of a continuous, ordered resource attribute might often apply to groups of generalized frugivores, granivores, or insectivores. For carnivores that share prey carcasses, or herbivores that exploit clumped resources, a diversity analysis might apply more readily.

10 Learning in Foraging Groups

10.1 Introduction

There are two good reasons for ecologists to take an interest in learning, a subject traditionally studied by psychologists. First, as is true of all foraging theory, social foraging models characterize individuals as making decisions on the basis of economic information concerning fitness consequences of alternative courses of action. The mechanism involved in gathering, storing, and recalling this economic information is learning. Knowing something about an animal's learning, therefore, may suggest more realistic and quantitatively predictive foraging models.

The second reason why ecologists should study learning concerns behavioral diversity. Does learning act as a homogenizing force within foraging groups whose members exploit the same resources? Or are there conditions within groups that affect the dynamics of learning and increase variability among individuals? These questions must be addressed if we are to gain a better appreciation of the factors governing the extent of behavioral diversity at the individual and group levels.

In this chapter we address both issues. After proposing a functional classification of different types of learning, which we direct to the interests of behavioral ecologists, we review a number of factors that can operate within foraging groups either to promote or inhibit learning. We also develop learning models that are meant to mimic the learning process as it operates within foraging groups, and we ask whether learning likely enhances or reduces within-group behavioral diversity.

10.2 Some Functional Definitions of Learning

Learning includes a diversified set of phenomena that can be categorized in a number of different ways (Shettleworth 1998). For instance, behavioral ecologists divide learning into two broad functional categories: *learning about* and *learning how* (Krebs et al. 1983). *Learning about* is concerned with the collection of information pertinent to the estimation and assessment of alternatives. So, in order to determine the quality of a patch of *Impatiens* flowers, a bumblebee may need to invest some effort in sampling a number of flowers within the patch. Foraging theory has been drawn to the study of this kind of learning because its habitat, patch, and prey models often require assessment and hence *learning about.*

Learning how, on the other hand, is concerned with the acquisition of behavioral traits involved in search, pursuit, capture, and handling of prey. A forager may have learned, for instance, that to pull a worm from the soil it must hold it in a specific way and adjust the effort to the worm's own response. This kind of learning also has important implications for foraging models. Hughes (1979), for instance, has shown that acquisition of improved handling and capture techniques can alter parameters of the prey model and hence affect its predictions (Croy and Hughes 1990). *Learning how* no doubt also affects many other foraging decisions. The acquisition of increasingly efficient searching behavior will likely alter the pattern of energy gain in a patch and thus alter the time spent exploiting the patch (Green 1987).

To psychologists, learning categories are more mechanistic and often founded on the cognitive processes thought to underlie the phenomenon. One classic example of such mechanistic categories is the distinction between Pavlovian and operant conditioning (Dickinson 1980). Although these types of learning may involve different cognitive processes, the distinctions will not likely affect the outcomes that interest behavioral ecologists. The more useful economic categorization will be whether learning involves merely the acquisition of information necessary for a decision (*learning about*) or the acquisition of a novel motor pattern (*learning how*).

Most research on learning is conducted with subjects tested in isolation, enclosed within an experimental apparatus. These studies do not address influences that sociality may exert on the learning process. From the perspective of those interested in learning's social context, learning can be categorized as *individual* and *social learning*. The information a forager acquires through individual learning is generated solely by the subject's own interaction with available resources. This information can be termed *private* information.

Most attempts to define social learning refer, once again, to the cognitive process involved. Psychologists have suggested a remarkable number of cognitive categories (Galef 1988; Whiten and Ham 1992; Heyes 1993). We propose, instead, a functional definition of social learning that is independent of its cognitive processes and based only on the origin of the information used in the learning process. Whereas individual learning is based on private information, social learning occurs when individuals learn from information generated by the behavior of other individuals. Because this information is available to all group members, Valone (1989) referred to it as *public information*. Hence, we define social learning as any learning that involves the use of public information.

In this chapter we categorize learning in both the *how* versus *about* and the *individual* versus *social* dichotomies. Hence, there can be individual and social ways of *learning how* as well as individual and social ways of *learning about*. We first deal with *learning how*, which, in the context of group foraging, may give rise to phenotypic asymmetries that, for instance, affect the use of producer and scrounger alternatives. We review the effects that group foraging can have on individuals' *learning how*. Then, we present two stochastic models that predict the expected number of skilled food-finding individuals when only individual learning is possible. We turn to exploring the ways in which group foraging can affect social *learning how* and follow this with a stochastic model that allows both individual and social learning to occur. We end by considering both individual and social *learning about*, which in the social foraging framework raises interesting problems relevant to sampling behavior.

10.3 *Learning How*: Individual Learning Only

The Inhibiting Effects of Group Foraging

Individual learning in the present context refers to acquisition of a motor pattern contingent upon the individual initially performing the behavior and then experiencing the resulting beneficial consequences. Imagine that a recognizeable prey type is found under leaves; prey can be discovered only if a leaf is turned over. Two steps are involved in the associated learning:

1. Directing general exploration toward leaves.
2. Experiencing a prey discovery following leaf turning.

To begin, we examine how either step in the individual learning process could be inhibited when individuals forage in groups (Giraldeau 1984).

Time Depletion

The likelihood of acquiring a trait by individual learning depends on the time an animal spends in exploration. Thus, any behavior that reduces naive individuals' exploration time will necessarily reduce their chances of learning a given trait. If a proficient leaf searcher discovers food that can be shared by naive individuals, then these naive individuals will redirect some exploratory time to feeding. As more and more knowledgeable, and hence proficient, leaf searchers generate food that is shared, more and more of the ignorant foragers will spend increasing amounts of time feeding at the expense of exploration. It follows that a greater number of proficient leaf turners should lower the chances that a given naive individual will acquire the trait through individual learning.

SECONDARY ASSOCIATIONS

When a trait such as leaf turning reveals a resource that can be divided, then rewards are available to individuals (joiners) that have not performed the leaf-turning trait. Inevitably, these individuals have performed some other secondary response for which they are rewarded, a response (joining) that is not leaf turning. For instance, Giraldeau and Lefebvre (1987) report that within a captive pigeon flock, those individuals that had not learned to find food had learned, instead, to recognize the food finders and followed them in order to gain rapid access to any discovered food source. The acquisition of such a secondary association can inhibit acquisition of the target association in at least three ways. First, learning one association under a set of conditions can deter from learning a different association under the same set of conditions (Beauchamp and Kacelnik 1991), a phenomenon learning psychologists refer to as *blocking* (Shettleworth 1998). Second, having learned the secondary association may selectively direct an individual's attention away from the stimuli necessary to establish the target association, a phenomenon known to psychologists as *overshadowing* (Shettleworth 1998). Third, and quite independent of any cognitive process, the secondary association can lead to responding that depletes exploration time and hence reduces the likelihood of an individual learning the target behavior (Giraldeau and Lefebvre 1986).

DEPLETION OF REWARD OPPORTUNITIES

The previously mentioned sources of interference with the learning process dealt with the exploratory phase. However, the presence of others can also have effects on the likelihood of gaining a reward for performance of the target behavior. A proficient leaf searcher will locally deplete the number of prey found under leaves. Thus, a naive individual foraging in its vicinity will have a lower chance of experiencing a prey discovery when it eventually turns a leaf during the exploratory phase.

Alternatively, a subordinate naive individual may be foraging close to a despot, so that whenever the naive subordinate happens to turn a leaf and find a prey, the food reward is usurped by the despot. Consequently, the leaf turner may not experience a reward and, if the despot uses aggression, could even experience punishment following leaf turning.

The Enhancing Effects of Group Foraging

The foraging behavior of proficient individuals can modify the environment in a way that increases a naive individual's chances of acquiring a trait through what Galef (1991) calls *exposure*. For instance, chickadees learning to pierce a small container in order to obtain food did so more quickly when they were previously exposed to containers that had already been opened

(Sherry and Galef 1984). These results suggest that the spread of milk-bottle opening behavior to a large number of British birds in the first half of this century (Fisher and Hinde 1949), for instance, could have been promoted by simple exposure of birds to bottles that had already been opened (Lefebvre 1995).

10.4 Models of Individual Learning Only

Giraldeau and Lefebvre (1986, 1987) found that when members of a pigeon flock were required to learn a foraging skill in order to find large patches of seed, only a few individuals acquired the skill. The others obtained their food solely by feeding at patches generated by the knowledgeable birds. This behavioral diversity among individuals persisted for a relatively long time. The observed pattern of finding and joining might have reflected an economic interaction (Barnard and Sibly 1981; Giraldeau et al. 1990; Caraco and Giraldeau 1991; Ranta et al. 1993; see Ruxton et al. 1995). Alternatively, at a purely mechanistic level, skill acquisition might have been constrained because some individuals learned to join and consequently lost the time and opportunity to acquire the skill through individual learning.

In this section we model the dynamics of the frequency of a learned food-finding trait within a foraging group. The model assumes group members are capable of individual learning only. The results indicate how group size, ease of trait acquisition, and the effect of the trait on clump discovery rate influence the equilibrium number of individuals exhibiting the trait. Later, a second model will assume that both individual and social learning operate, and examine how the additional type of learning may affect behavioral diversity within a foraging group. The models of frequency-dependent learning were developed by Giraldeau et al. (1994c).

General Assumptions about the Learning Process

We consider a group searching for clumps of a single type of resource. For simplicity we assume an information-sharing system where when one individual discovers food, all others have no better option than to join. At any given time, group members will ordinarily vary in the rate of discovering clumps (e.g., Caraco et al. 1989); we presume this variation is affected by individual learning.

The models classify each group member as either a slow or a fast clump finder. Slow finders are naive individuals that discover clumps at a constant, lower probabilistic rate s_1. Fast finders have acquired the trait and so discover clumps at a greater, constant probabilistic rate s_2, where

$$s_2 = \alpha s_1; \alpha > 1. \tag{10.1}$$

A slow finder may increase its rate of clump discovery through individual learning. Upon discovering a clump, a slow finder may learn from the experience and become a fast finder, or not learn and remain a slow finder. A larger value of the parameter α implies a greater consequence of learning on finding rate, and a larger increase in each group member's feeding rate when learning occurs. A large effect of learning may be anticipated when, for instance, the spontaneous rate of performance (rate of slow finding) of some behavior is very low but, once acquired, leads to frequent food finding.

Sociological analyses of the intragenerational dynamics of learned traits usually allow forgetting and relearning (e.g., Karmeshu and Pathria 1979; Sharma et al. 1982, 1983). More importantly, empirical results imply that forgetting occurs in both mammalian (D'Amato 1973) and avian foragers (Grant and Roberts 1973; Giraldeau and Lefebvre 1987) when opportunities to use a learned trait are rare, or when tasks are complex (Commons 1981). Consequently, the first model we present assumes that forgetting occurs. Naturally, only fast finders can forget. When they do, they revert to the state of slow finder such that their rate of clump discovery decreases from s_2 to s_1.

The preceding section argues that an increase in the frequency of a learned trait within the group decreases the probability that the next clump is discovered by any given slow finder; that is, the model assumes a simple frequency-dependence of learning opportunity. The probability that a slow finder acquires the trait is inversely proportional to the number of fast finders in the group, because opportunities for learning decline as the number of fast finders increases. The model asks how this frequency dependence affects the equilibrium proportion of individuals that exhibit the trait.

A model with Forgetting

At time t a group of G foragers contains $X(t)$ slow finders and $[G - X(t)]$ fast finders; $X(t) \in \{0, 1, \ldots, G\}$. Individuals discover resource clumps independently at respective probabilistic rates s_1 and s_2. At time t the probability the next clump is discovered by any of the slow finders is $\Theta(t)$

$$\Theta(t) = \frac{s_1 X(t)}{s_1 X(t) + s_2[G - X(t)]} = \frac{X(t)}{X(t) + \alpha[G - X(t)]}. \tag{10.2}$$

When at least one individual in the group is a fast finder (i.e., for $X(t) < G$), the likelihood that any of the slow finders learns to discover resources at the higher rate will be proportional to $\Theta(t)$.

The model assumes that the number of slow finders is the result of a stochastic process. We assume $\{X(t_1), X(t_2), \ldots, X(t_z)\}$ has the same probability distribution as $\{X(t_1 + \tau), X(t_2 + \tau), \ldots, X(t_z + \tau)\}$, so that the stochastic process is stationary (see Kelly 1979). We assume that the process $X(t)$ has the Markov property, so that the present state contains all useful information for predicting future states. The two allowable state transitions are an increase in $X(t)$ by 1 when at least one group member is still a fast finder (i.e., for $X(t) < G$), and a decrease of $X(t)$ by 1 when at least one group member is a slow finder (i.e., for $X(t) > 0$). Therefore, the number of slow finders in a group constitutes a birth-death process on a finite state space (e.g., Kelly 1979). Since the number of slow finders is a stationary birth-death process, its properties ensure that it will have a unique equilibrium probability distribution.

Suppose the number of slow finders increases from r to $(r + 1)$. Then one of the fast finders has forgotten the trait. Assuming that forgetting occurs independently among the fast finders, the probabilistic transition rate from r to $(r + 1)$ slow finders is given by $q(r,r + 1)$:

$$q(r,r + 1) = \mu(G - r); \; r = 0, 1, \ldots, G - 1, \qquad (10.3)$$

where μ is the individual forgetting rate. The forgetting rate may be relatively small but should increase as either task difficulty or the rarity of clump discovery increases.

Now suppose instead that the number of slow finders decreases from r to $(r - 1)$. Then one of the slow finders must have learned to discover resources at the higher rate. The probabilistic transition rate from r to $(r - 1)$ slow learners is given by $q(r,r - 1)$. Given the frequency dependence of learning opportunities, as specified by expression (10.2), $q(r,r - 1)$ is

$$q(r,r - 1) = \beta r/[r + \alpha(G - r)]; \; r = 1, 2, \ldots, G, \qquad (10.4)$$

where β is the learning parameter; larger values of β should be associated with less difficult tasks that are learned reliably when a forager uses the task to discover food. A difficult task implies a lower value of the learning parameter; a difficult task may require a number of repetitions before an individual performs it reliably. Note that the rate at which the task is learned *per naive individual*, $q(r,r - 1)/r$, is greater when there are more slow finders (i.e., larger r). Hence, a slow finder's probability of learning individually decreases as the fast-finding trait's frequency $(1 - r/G)$ increases.

The degree to which a task is difficult to acquire and retain depends on the chance that the forager learns the task when it finds food (β) and on the rate at which the task is forgotten (μ) once the forager has learned. So we take the ratio (μ/β) as an index of task difficulty; a more difficult task implies a greater value of the ratio. Note, however, that the index depends not only on inherent properties of the task, but also on the rate at which opportunities to perform the task occur. For example, an increased density of food clumps could reduce the degree of task difficulty.

Since we defined $X(t)$, the number of slow finders, as a stationary birth-death process, we are assured that $X(t)$ has a unique equilibrium probability distribution. In Math Box 10.1 at the back of this chapter we define the equilibrium and derive the associated distribution. Essentially, as the process continues for a long time, the proportion of time $X(t) = X$ will be given by the equilibrium probability $p(X)$, where

$$p(X) = \frac{\binom{G}{X} [\mu(\alpha - 1)/\beta]^X \, [\Gamma(C)/\Gamma(C - X)]}{\sum_{i=0}^{G} \binom{G}{i} [\mu(\alpha - 1)/\beta]^i \, [\Gamma(C)/\Gamma(C - i)]} \tag{10.5}$$

for $X = 0, 1, \ldots, G$; $C = \alpha G/(\alpha - 1)$, and the gamma function $\Gamma(k) = (k - 1)\,\Gamma(k - 1)$.

We numerically analyzed the equilibrium distribution of slow finders for a range of group sizes, task difficulties (μ/β), and effects of learning on clump-finding rate (α). For each parameter combination, we calculated the mean (E[X]), variance (V[X]), and coefficient of variation (CV[X]) of the equilibrium distribution of slow finders. As the mean number of slow finders (individuals that do not exhibit the trait) increases, the mean of the number of fast finders must decrease. An increase in the variance of the number of slow finders is mirrored in the variance of the number of fast finders:

$$\mathrm{V}[G - X] = \mathrm{V}[G] + \mathrm{V}[X] - 2\,\mathrm{COV}[G, X].$$

But group size is a constant, so variance of the number of fast finders reduces to the variance of slow finders; $\mathrm{V}[G - X] = \mathrm{V}[X]$.

The coefficient of variation in the number of slow finders is $(\mathrm{V}[X])^{1/2}/\mathrm{E}[X]$. The equivalent quantity for the number of fast finders is

$$\begin{aligned} \mathrm{CV}[G - X] &= (\mathrm{V}[G - X])^{1/2}/\mathrm{E}[G - X] \\ &= (\mathrm{V}[X])^{1/2}/(G - \mathrm{E}[X]). \end{aligned} \tag{10.6}$$

An increase in the coefficient of variation in the number of slow finders may imply either an increase or a decrease in the coefficient of variation of the number of fast finders, depending on the magnitude of the changes in the mean and variance of the number of slow finders.

Figure 10.1 shows the mean number of slow finders for twenty different combinations of task difficulty, effect of learning, and group size. For a given group size, the expected number of slow finders increases as either task difficulty or the effect of learning increases. This follows directly from the assumption that the probability that a slow finder discovers a clump (and perhaps learns the task) decreases as the effect of learning on finding rate (α) increases. This implies that skills leading to the greatest enhancement of

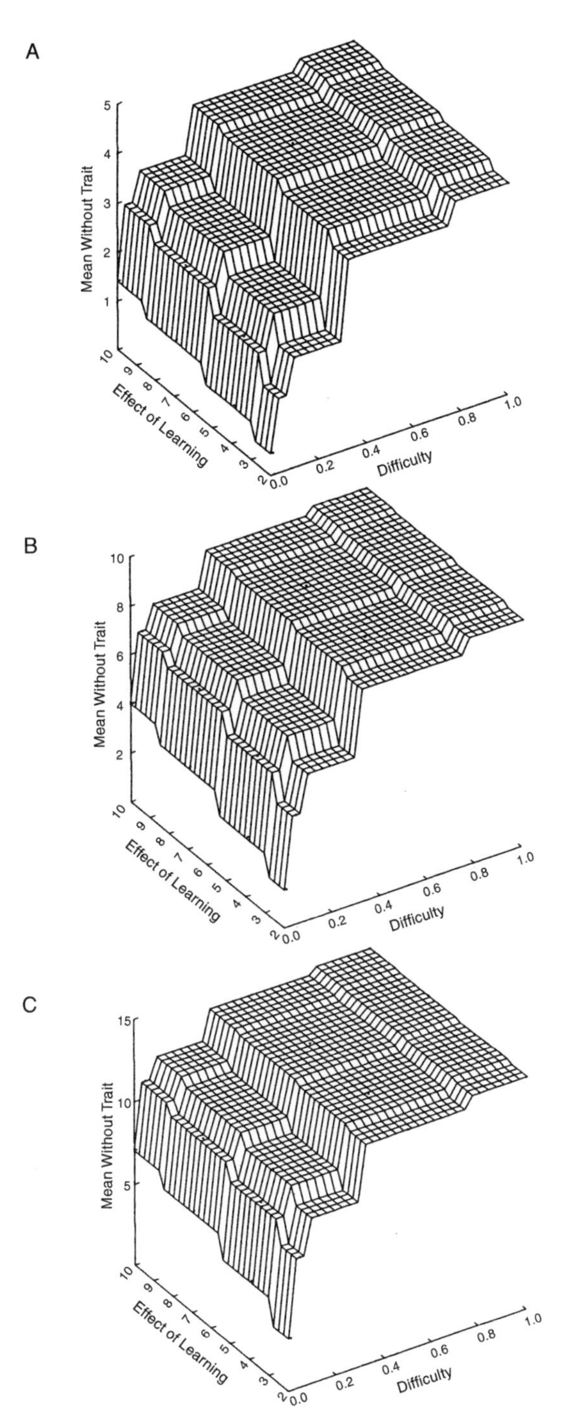

Figure 10.1 The mean number of slow finders (group members that do not exhibit the learned food-finding skill) as a function of the task's difficulty (μ/β) and effect of learning on finding rate (α) when group size is 5 (A), 10 (B), and 15 (C). Surfaces are step functions based on twenty combinations of skill difficulty and effect of learning. Note that the equilibrium number of slow finders increases with skill difficulty as well as the effect of learning on finding rate.

food finding rate ($\alpha \gg 1$) will be observed in fewer group members than those skills that have a lesser effect on food discovery ($\alpha \rightarrow 1$).

The model's most important biological result predicts that the expected number of group foragers exhibiting the fast finding skill will be less than the corresponding expectation had the same individuals foraged solitarily. Solitary foraging implies that each of the individuals forages independently. The expected number of solitary individuals behaving as slow finders at equilibrium is $G\mu/(\beta + \mu) = G(\mu/\beta)/(1 + [\mu/\beta])$. For a given task difficulty (μ/β), this quantity is always less than the group foraging expectation, simply because group members ($G \geq 2$) have, on average, fewer opportunities to learn the foraging skill through their own clump finding. Hence a trait acquired through individual learning can be more common among solitaries than among members of groups with fixed membership.

To demonstrate the comparison of independent and social foragers, we calculated the expected number of slow finders for both cases. We took the ratio of the expected number of slow finders for solitary foraging to the mean for group foraging across a small set of parameter values. Since the ratios never exceed unity, the learned fast-finding trait is always more widespread among independent foragers than among group foragers (table 10.1). When task difficulty is lower and group size is larger, the incidence of fast finders among solitaries increases, relative to the trait's frequency among group members.

Figure 10.2 shows the variance in the number of slow finders estimated from twenty different combinations of task difficulty, effect of learning, and group size. For a given average frequency of the learned trait, an increased V[X] implies greater temporal fluctuation in the number of group members exhibiting the trait. When group size is fixed, the variance in the number of slow finders becomes smaller when the mean approaches either 0 or G (group size). The variance is larger when the mean is about half the group

Table 10.1

Average frequency of slow finders (foragers lacking the food-finding trait) among G solitaries divided by expected frequency among G group members. Results indicate that the food-finding trait is usually rarer among group members.

		G		
α	μ/β	5	10	15
2	0.1	0.24	0.17	0.15
	1.0	0.59	0.55	0.53
5	0.1	0.19	0.15	0.13
	1.0	0.58	0.54	0.53

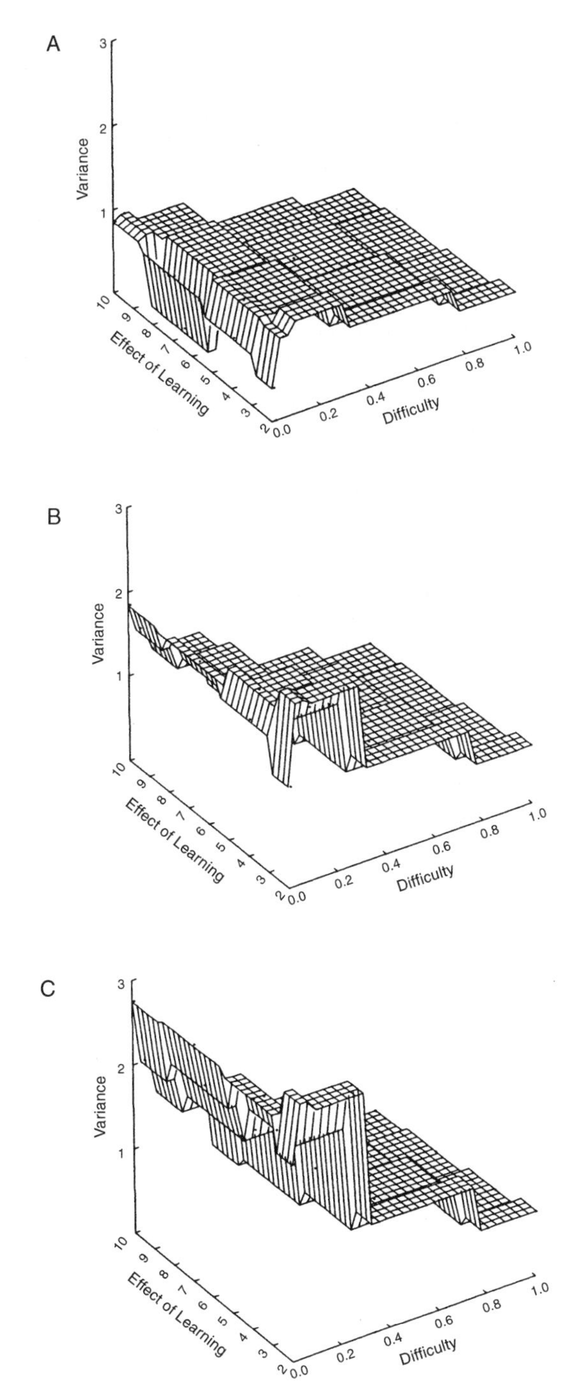

Figure 10.2 The variance in the equilibrium number of individuals that do not exhibit the finding skill as a function of the skill's difficulty (μ/β) and effect of learning (α) when group size is 5 (A), 10 (B), and 15 (C). Surfaces estimated as in figure 10.1.

size. Consequently, the variance of the number of slow finders does not vary monotonically with increases in either skill difficulty or the effect of learning on finding rate. For instance, for all three group sizes, the variance is low for skills of low or high difficulty, but the variance is high when skill difficulty is intermediate (fig. 10.2).

The influence of the effect of learning (α) on the variance of slow finders depends on skill difficulty for all three group sizes. For greater task difficulty ($\mu/\beta \geq 0.4$), the variance in the number of slow finders decreases with increasing effects of learning. However, when task difficulty is low ($\mu/\beta \leq 0.2$) the effect is reversed; variance in the number of slow finders now increases with the effect of learning (fig. 10.2).

Group size and task difficulty apparently have an interactive effect on the variance in the number of slow finders. In groups of five foragers, the lowest variances are observed at low task difficulty. But for groups of 10 or 15 foragers, the variance is lowest at greater task difficulty.

We noted above that the mean number of slow finders generally increases with either task difficulty or the consequences of learning on finding rate. As a consequence, patterns in the coefficients of variation of the number of slow and fast finders depend more on the mean than on the variance. For a given group size, the coefficient of variation in the number of slow finders declines as either task difficulty or the consequence of learning on finding rate increases (fig. 10.3). The effect of task difficulty appears to be much stronger in smaller than in larger groups.

The combined effect of task difficulty and effect of learning gives rise to an interesting pattern when we examine coefficients of variation across group sizes. As group size increases, the coefficient of variation of slow finders clearly decreases at all parameter combinations (fig. 10.3).

At equilibrium, the average proportion of slow finders in the group ($m = E[X]/G$) increases with task difficulty, the effect of learning, and group size (fig. 10.4). Consequently, the expected proportion of fast finders declines in larger groups. The model's predictions are collected in Summary Box 10.1.

A Model without Forgetting

If no forgetting occurs, then it is easy to imagine that all individuals eventually learn the resource-finding task. Furthermore, the effect of forgetting may be neglected if the forager can relearn a forgotten task much more quickly than its initial learning. For these cases, instead of analyzing the number of individuals that do not exhibit the trait at equilibrium, we consider the time elapsing until all group members learn the task. A longer duration of the learning process can imply a greater chance of observing within-group variation. The random variable T represents the total time for all group members

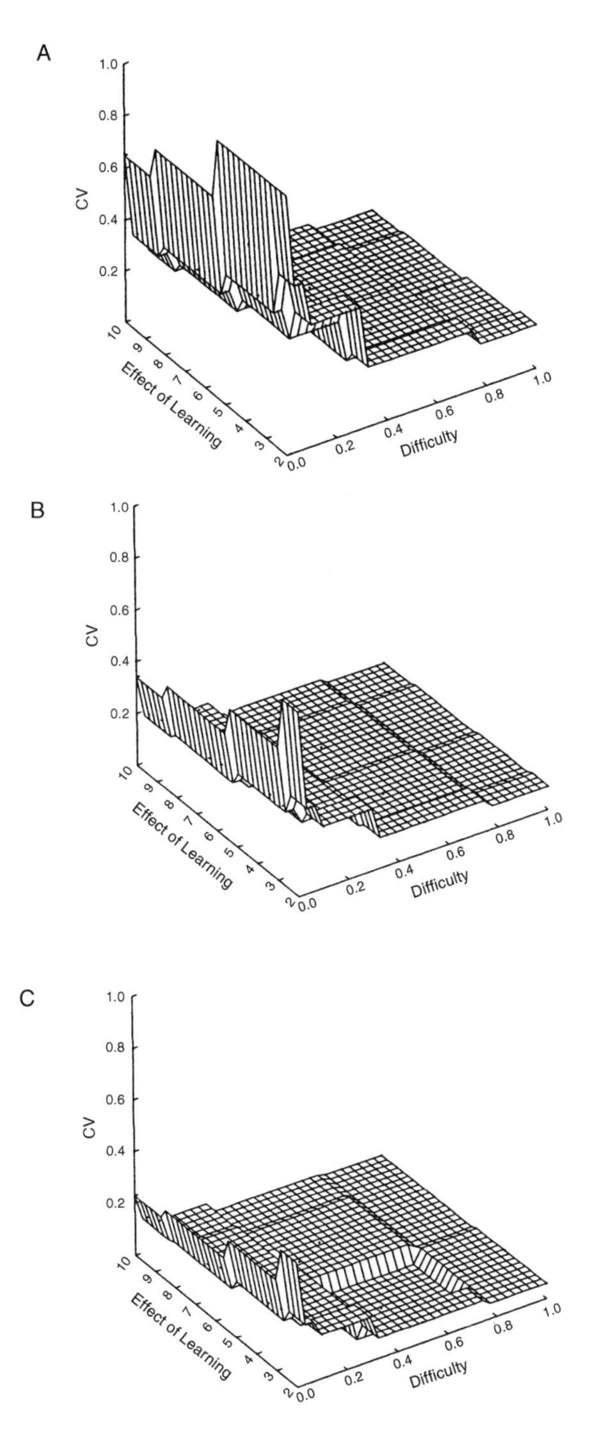

Figure 10.3 The coefficient of variation (CV) in the equilibrium number of slow finders as a function of the skill's difficulty (μ/β) and the effect of learning (α) for group size of 5 (A), 10 (B), and 15 (C). Surfaces estimated as in figure 10.1.

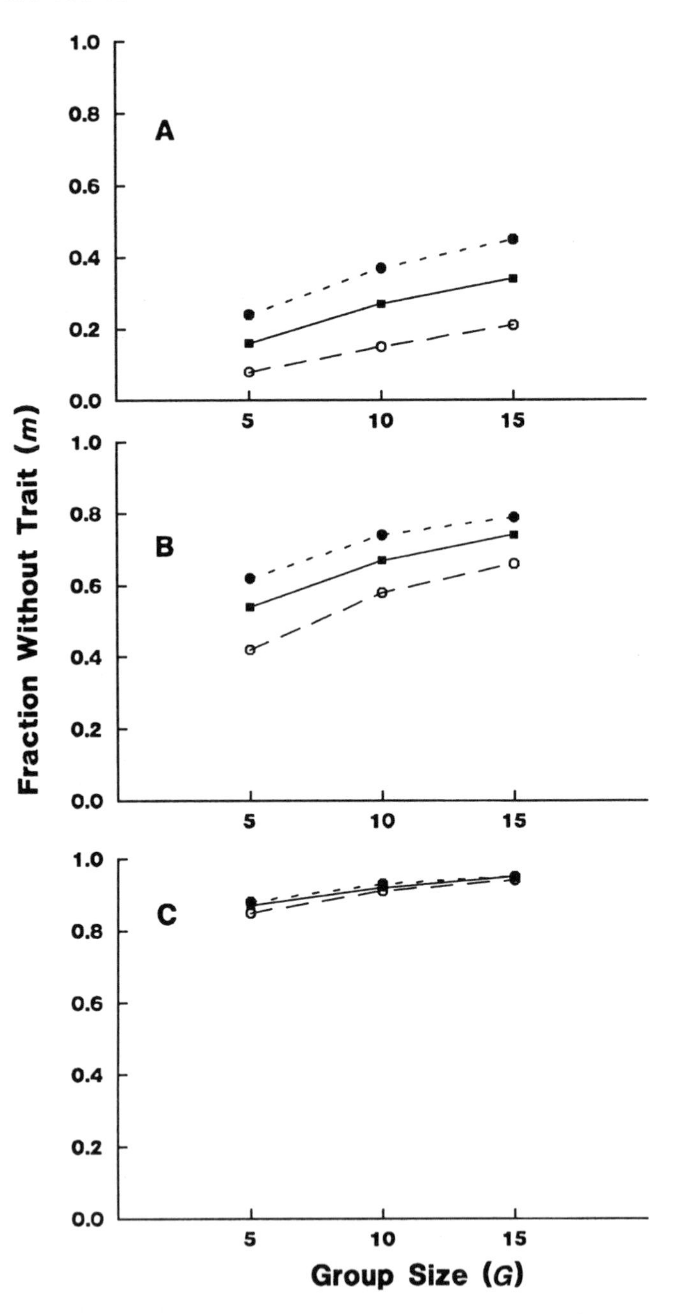

Figure 10.4 The proportion of slow-finding group members (m) as a function of group size (G), when the skill's difficulty is (A) low ($\mu/\beta = 0.01$), (B) intermediate ($\mu/\beta = 0.1$) and (C) high ($\mu/\beta = 1.0$). Dashed line, $\alpha = 2$; solid line, $\alpha = 5$; dotted line, $\alpha = 10$. (Taken from Giraldeau et al. 1994a.) (With permission of Academic Press)

SUMMARY BOX 10.1 ELEMENTS OF A MODEL OF INDIVIDUAL LEARNING IN GROUPS

NOTE:

This model does not predict a decision. The model specifies the way a learned trait increases or decreases among a group's members and identifies the number exhibiting the trait at equilibrium. The dynamics and the equilibrium depend on several properties, listed below as constraints.

CONSTRAINTS

1. A group of G foragers behaves as an "Information-Sharing" system, when one finds food, all $G - 1$ others join and share the food.
2. Learning determines the rate at which each individual can discover food; two states are possible: fast that discovers at rate s_2 and slow that discovers at rate s_1, such that $s_2 = \alpha s_1$; $\alpha > 1$.
3. A slow discoverer turns into a fast one only after individual learning which occurs probabilistically following the discovery of a clump.
4. The probability that a slow finder becomes fast declines as the number of fast finders increases (i.e., negative frequency-dependence of learning).
5. A fast discoverer can turn into a slow discoverer (i.e., forget) probabilistically at rate μ.

PREDICTIONS

1. The equilibrium proportion of individuals that exhibit the learned trait decreases when either task difficulty (μ/β) or the effect of learning on finding rate (α) increases.
2. The proportion of individuals that exhibit the learned trait decreases as group size (G) increases.
3. The coefficient of variation of the number of individuals that do not exhibit the learned trait declines with increased G.
4. The number of individuals that exhibit the trait is always smaller in a group of G than in an equivalent population of G solitary foragers.
5. The coefficient of variation of the number that exhibit the learned trait is independent of G.

to learn the task. In Math Box 10.2 we derive approximations for the mean and variance of T and find:

$$E[T] \approx (G/\beta)[1 - \alpha + \alpha \ln(G + 1)] \tag{10.7}$$

and

$$V[T] \approx \frac{G}{\beta^2}\left[\alpha\left(\frac{\pi^2}{6} - \frac{1}{G + 1}\right) - 2(\alpha - 1)[1 + \alpha \ln(G + 1)]\right] \quad (10.8)$$

Both the mean and the variance of the time until all group members learn increases as group size increases, and decreases as the learning parameter β decreases. $E[T]$ always increases as the effect of learning (α) increases for groups of two or more foragers, but the variance $V[T]$ declines as the effect of learning increases. These results are consistent with the predictions derived when the model allows forgetting (see Summary Box 10.1).

Lessons from the Models

We conclude that group foraging can, given our assumptions, inhibit individual learning. When forgetting occurs, only a fraction of the foraging group shows a food-finding trait at any one time. If forgetting is rare, then all group foragers will eventually learn and exhibit the food-finding trait. But the diffusion of the skill within a foraging group may be so slow that the chance of observing a persistent dichotomy between knowledgeable finders and unskilled joiners remains. These predictions are based on the assumption that the frequency dependence of learning opportunity mimics the essential effects of time depletion, opportunity depletion, and the potential for learned secondary associations.

It is important to consider that the model with $\mu > 0$ predicts an equilibrium frequency of finding and joining that is distinct from an equilibrium based on payoffs (chapter 7). Other factors may interact with individual learning and reduce the persistence of among-individual diversity. Lefebvre (1986) presented a closed aviary flock of pigeons and two open, natural flocks with the same food-finding problem: piercing paper covering a box to gain access to mixed seed. He found that in the closed aviary flock only a small fraction of the birds eventually acquired the skill, a result consistent with predictions of the preceding model. However, in open flocks, where individuals freely joined and left each group, the cumulative number of foragers acquiring the skill steadily increased.

Lefebvre argued that foragers confined to the same group, as were those in the captive aviary flock, are subjected to the constraint of frequency-dependent learning. The extent to which foragers can move between open groups in the field, however, may attenuate within-group inhibition of individual learning. That is, turnover in group membership may eventually allow whole populations to acquire a skill, despite a frequency-dependent decline in learning opportunities within groups.

An alternative view of learning within groups assumes that group foraging promotes trait acquisition. In the next section we consider some of the mechanisms whereby group foraging may advance the frequency of a trait through social learning.

10.5 *Learning How*: Social Learning

Social learning occurs when an individual's acquisition of a behavioral trait depends on information generated when other individuals perform the behavior. There is evidence for the phenomenon in a variety of animals, and a fair share of the examples involve a foraging task (Zentall and Galef 1988; Heyes and Galef 1996). Most models for social learning assume positive frequency dependence. That is, as individuals exhibiting the trait become more common, the rate of learning *per naive individual* increases. Consequently, the total rate at which the trait advances conforms to the assumption of "mass action" (e.g., Giraldeau et al. 1994c; see below). However, the dynamics of social learning may be more complicated (Giraldeau and Templeton 1991; Lefebvre and Giraldeau 1994). This section considers some of these complexities; we model an interaction of individual and social learning in the next section.

Our definitions of individual and social learning are distinct, but acquisition of an ecologically significant trait may involve interaction between direct experience with a resource and observation of other consumers. A forager may attempt to copy another's food-finding behavior but succeed only after the skill is honed through individual learning. For simplicity, we shall treat the two processes as probabilistically independent (Bartholomew 1983).

Food Joining as an Obstacle to Social Learning

Our analysis of individual learning assumed that exploiting clumps found by other group members might inhibit a naive individual's acquisition of the food-finding task. This sort of food-sharing might also interfere with social learning.

Pigeons appear capable of social learning. Palameta and Lefebvre (1985) show that pigeons provided with a demonstrator that pierces a sheet of paper and feeds through the hole learn most quickly how to pierce paper themselves. Individuals provided with incomplete demonstrations learn more slowly, and pigeons with no demonstrator fail to learn.

Despite their potential social learning abilities, when foraging in a group, only a few pigeons learn a new food-finding skill (Giraldeau and Lefebvre 1986; Lefebvre 1986). Giraldeau and Lefebvre (1987) hypothesize that food-sharing (joining) not only prevents performance of the trait (hence individual learning), but also constrains social learning. Fragaszy and Visalberghi (1989) report a similar effect in a primate.

In Giraldeau and Lefebvre (1987) pigeons learned to peck a stopper out of an inverted test tube to obtain food. A naive observer watched a trained demonstrator remove the stopper while always having access to its own stoppered test tube. To mimic food-sharing, the tray between the observer and demonstrator was tilted toward the observer. In the not-sharing treatment, the tray was kept horizontal, so that the demonstrator's food was not available to the observer. Almost all of the food-sharing subjects failed to learn to remove the stopper, whereas almost every not-sharing subject learned the task quickly (fig. 10.5). Hence food-sharing between tutor and observer may inhibit social learning (Giraldeau and Lefebvre 1987; Giraldeau and Templeton 1991).

For individual learning, joining a food discovery merely reduces the opportunity for learning. But joining need not reduce the opportunity for social learning. In fact, joining increases the proximity of observer and model, and may allow a greater opportunity for the observer to see the performance of the relevant skill by the tutor. How joining interferes with social learning in pigeons remains unclear. It is possible that social learning of a food-finding skill is simply inhibited when it is uneconomical for an individual to switch from a food-joining to a food-finding role.

Tutor Dilution

Suppose a group of G individuals contains X naive members and $(G - X)$ members that exhibit a skill. The assumption that social learning proceeds as a mass-action process (e.g., Cavalli-Sforza and Feldman 1981; Giraldeau et al. 1994c) implies that the rate at which the trait's frequency increases is proportional to the product $(G - X)X$. Given mass action, the rate of social learning *per naive individual* will then be $(G - X)$. Hence a focal individual should learn at the same rate as G and X are increased, as long as their difference remains constant.

Suppose a foraging group contains only one member using the target behavior. A focal naive observer in the group choosing randomly to monitor another individual has only a small chance of observing the demonstrator. As more naive individuals join the group, $(G - X)$ remains fixed at unity, but the focal observer's probability of monitoring the demonstrator declines as $(G - 1)^{-1}$. If social learning requires considerable attention to other group members, time constraints may impose a tutor dilution.

Lefebvre and Giraldeau (1994) demonstrate that the effectiveness of a tutor can be reduced by the addition of bystanders providing no relevant information. A single naive pigeon was presented with a single tutor performing a foraging task. The tutor was presented alone, with two, five, or eight noninforming bystanders. The number of observers that eventually learned the skill declined with an increase in the number of bystanders. The result suggests that group size may have an inhibiting effect on social learning, so that mass action might overestimate the advance of a trait through social learning.

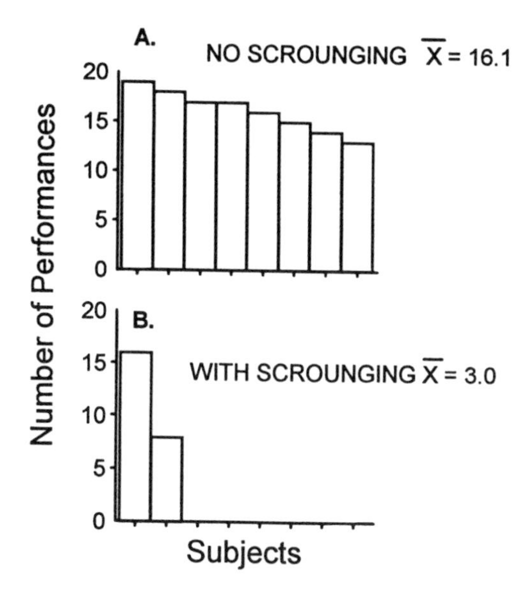

Figure 10.5 Results of the social learning experiment performed by Giraldeau and Lefebvre (1987). The observer had to learn to peck open an inverted test tube that released a standard quantity of millet seed. Histograms give the number of times the eight subjects in each treatment had opened their test tubes by the end of the trials. In (A) the observer pigeons were provided with demonstrations in which they could not have access to the tutor's food and all eight birds learned. In (B), however, the tutor's seeds was available to the observer, allowing it to "join" and, consequently, only two of the eight birds learned. The means are to the numbers of tubes opened by each subject in each treatment.

10.6 Models with Both Individual and Social *Learning How*

Social learning has been surrounded by considerable controversy, due in part to its cognitive-based definitions and the ensuing difficulty of demonstrating the operation of one at the exclusion of others (see Galef 1988; Heyes 1993). One consequence of cognitive-based definitions of social *learning how* is a common assumption that the process could occur in only cognitively sophisticated species such as apes (see Whiten and Ham 1992; Galef 1996; Fragaszy and Visalberghi 1996). However, indisputable demonstrations of social learning have rarely come from studies of apes (Whiten and Ham 1992) but instead from elegant experiments using rats (*Rattus norvegicus*) (Heyes and Dawson 1990), budgerigars, (*Melopsittacus undulatus*) (Dawson and Foss 1965; Galef et al. 1986) and pigeons (Palameta 1989). Social learning, therefore, may be common to a great number of species and may interact with individual learning to influence behavioral diversity in group-living animals.

The model presented earlier in this chapter shows how an increased frequency of the trait within a group may impede individual learning. Here, we add social learning and assume an increased number of fast finders can provide additional opportunities for social transmission of the trait. We assume that social learning occurs as a mass-action process (i.e., we ignore time constraints that may dilute tutor effectiveness). Essentially, the model opposes the inhibitory effect of a trait's frequency on individual learning with an accelerating effect of social learning on the frequency of the trait.

General Properties of the Model

As in section 10.4, $X(t)$ is a stationary stochastic process counting the number of group members not exhibiting the learned trait at time t. $\alpha = (s_2/s_1)$ is the ratio of clump-discovery rates, as in the preceding model. If $X(t)$ increases from r to $(r + 1)$, a fast finder forgets the skill. The probabilistic transition rate $q(r,r + 1)$ again is $\mu(G - r)$ for $r = 0, 1, \ldots, G - 1$. If $X(t)$ decreases from r to $(r - 1)$, a slow finder learns to discover resource clumps at the higher rate. We assume individual learning proceeds exactly as in the first model; the probabilistic rate of transition due to individual learning is $q^*(r,r - 1) = \beta r/[r + \alpha(G - r)]$ for $r = 1, 2, \ldots, G$.

For social learning the probabilistic transition rates are proportional to the product of the numbers of slow and fast finders (mass action). That is, the total rate of social learning should be proportional to the number of demonstrators multiplied by the number of individuals that might learn (e.g., Cavalli-Sforza and Feldman 1981; Bartholomew 1983). So, let the probabilistic rate of transition due to social learning be $\hat{q}(r,r - 1)$, where

$$\hat{q}(r,r - 1) = \Omega r(G - r);\ r = 1, 2, \ldots, G. \qquad (10.9)$$

Ω is the contagion parameter representing the combined effectiveness of the fast finders' demonstrations and the slow finders' social learning. The *total* transition rate due to social learning increases with the number of naive individuals until that number exceeds half the group size, after which the total rate declines as the number of naive individuals continues to increase. The transition rate of social learning *per naive individual*, $\hat{q}(r,r - 1)/r$, decreases as the number of naive individuals increases. Hence the probability of cultural transmission per naive individual increases as the frequency of fast finders increases (see section 10.5), opposite to the frequency-dependent effect assumed for individual learning.

Assuming that individual and social learning are probabilistically independent (see Lefebvre and Giraldeau 1994; Laland et al. 1996), the total transition rate to one less slow finder (i.e., one individual learns the task) is the sum of q^* and $\hat{q}$:

$$q(r,r-1) = r\left[\frac{\beta}{r+\alpha(G-r)} + \Omega(G-r)\right] \tag{10.10}$$

for $r = 1, 2, \ldots, G$.

As in the earlier model, we assume $X(t)$ is a stationary birth-death process. In Math Box 10.3 (end of chapter) we derive the resulting equilibrium probability distribution. At equilibrium, the proportion of time $X(t) = X$ will be given by the probability $p(X)$, where

$$p(X) = \frac{\binom{G}{X}[\mu(\alpha-1)]^X\,[\Gamma(C)/\Gamma(C-X)\,M(X)]}{\sum_{i=0}^{G}\binom{G}{i}[\mu(\alpha-1)]^i\,[\Gamma(C)/\Gamma(C-i)\,M(i)]} \tag{10.11}$$

for $X = 0, 1, \ldots, G$; $C = \alpha G/(\alpha - 1)$; $\Gamma(k) = (k-1)\Gamma(k-1)$, and

$$M(X) = \prod_{r=1}^{X}[\alpha\Omega(G-r)^2 + \Omega r(G-r) + \beta]. \tag{10.12}$$

Consequences of Social Learning

We numerically analyzed the equilibrium distribution of the number of slow finders. We fixed the individual learning parameter β and varied group size G, forgetting rate μ, the effect of learning on finding rate (α), and the contagiousness of the skill (Ω). Since the number of parameters is large, it is possible to assemble a long list of interactive effects. We prefer instead to focus on patterns that examine the generality of the previous model's results. Tables 10.2 and 10.3 show results for groups of 5 and 10 individuals, respectively, and the model's predictions are listed in Summary Box 10.2.

Despite the addition of social learning, group foraging can maintain behavioral diversity. Of course, more individuals will, on average, exhibit the trait when both individual and social learning occur than when individual learning alone is possible.

Not too surprisingly, for a given group size the expected number of slow finders decreases as the contagion parameter increases (tables 10.2 and 10.3). Greater task difficulty, whether interpreted as increased (μ/β) or decreased contagion (Ω), results in greater numbers of slow finders. But the average number of slow finders is less sensitive to variation in the effect of learning (α), compared to results when only individual learning is possible.

When trait acquisition occurred only through individual learning, the frequency of fast finders was lower in a population of group foragers than in a

SUMMARY BOX 10.2 ELEMENTS OF A MODEL FOR INDIVIDUAL AND SOCIAL LEARNING IN GROUPS

NOTE: As in Summary Box 10.1, this model does not predict a decision; it concerns the frequency of a learned trait.

CONSTRAINTS

1. Individual learning proceeds exactly as in Summary Box 10.1.
2. A slow finder can become a fast finder by social learning.
3. Individual learning and social learning are probabilistically independent.
4. Probabilistic transition to fast discovering due to social learning is proportional to the product of the number of slow and fast discoverers and is influenced by the contagion parameter Ω.

PREDICTIONS

1. For a given group size (G), the number that do not exhibit the learned trait increases with task difficulty (μ/β).
2. The number that do not exhibit the learned trait increases with decreasing contagion of social learning (Ω).
3. When social learning is possible, the number that do not exhibit the learned trait is less sensitive to the effect of learning on the finding rate (α) than in cases of individual learning only.
4. The coefficient of variation of the number that do not exhibit the learned trait increases with Ω.
5. The proportion of individuals that do not exhibit the learned trait decreases with G.
6. The number of individuals exhibiting fast finding is larger in a group of G individuals with social learning than for an equivalent population of G solitary foragers without the opportunity of social learning.

population where individuals foraged solitarily (table 10.1). Social learning can produce the opposite result, even at relatively high levels of task complexity. As the contagion parameter increases, the expected number of fast finders in a population of group foragers increases beyond the mean for an equivalent number of solitary foragers, which cannot learn socially. To demonstrate this comparison, we calculated the ratio of the mean number of slow finders for independent foraging to the mean for group foraging with social learning, across a limited set of parameter values. Most entries exceed unity, indicating that the learned trait is more common among group members than among an equivalent number of solitary foragers (table 10.4). Lower task complexity, larger group size, and greater contagion tend to increase the

Table 10.2

Mean (E[X]), variance (V[X]) and coefficient of variation (CV[X]) of the expected number of slow finders (individuals that do not exhibit the trait) as well as the coefficient of variation of the count of fast finders (CV[Y]); where $Y = G - X$). Individual and social learning operate simultaneously; group size (G) is 5, the individual-learning parameter $\beta = 0.6$. μ is the probability of forgetting, Ω the contagion parameter for social learning, and α scales the effect of learning on an individual's clump-discovery rate.

			$\mu = 0.2$		$\mu = 1.2$	
	Ω	α	2	10	2	10
E[X]	0.3		0.83	0.89	3.99	4.1
	0.9		0.28	0.28	2.24	2.32
V[X]	0.3		0.88	0.97	1.44	1.28
	0.9		0.29	0.29	2.74	2.8
CV[X]	0.3		1.13	1.11	0.3	0.28
	0.9		1.92	1.91	0.74	0.72
CV[Y]	0.3		0.22	0.24	1.19	1.26
	0.9		0.11	0.11	0.6	0.62

Table 10.3

Individual and social learning operate simultaneously; group size (G) is 10; all entries are defined in legend for table 10.2.

			$\mu = 0.2$		$\mu = 1.2$	
	Ω	α	2	10	2	10
E[X]	0.3		0.74	0.73	6.26	5.35
	0.9		0.25	0.25	1.51	1.58
V[X]	0.3		0.75	0.69	3.43	6.72
	0.9		0.25	0.25	1.57	1.87
CV[X]	0.3		1.17	1.15	0.3	0.48
	0.9		2.02	2.01	0.83	0.87
CV[Y]	0.3		0.01	0.01	0.5	0.48
	0.9		0.01	0.01	0.15	0.16

Table 10.4

Ratio of expected frequency of slow finders among G solitary foragers to expected frequency of slow finders, in a group of size G.

			G	
α	Ω	μ/β	5	10
2	0.3	0.33	1.51	1.69
		2.0	0.83	1.06
2	0.9	0.33	4.46	5.0
		2.0	1.49	4.41
10	0.3	0.33	1.4	3.4
		2.0	0.81	1.24
10	0.9	0.33	4.46	10.0
		2.0	1.44	4.42

incidence of the trait within groups, relative to its frequency among solitary foragers.

Holding group size constant, the coefficient of variation in the number of slow finders increases as the contagion parameter increases. The coefficient of variation in the number of fast finders correspondingly decreases as the contagion parameter increases, unless the proportion of individuals that learn is close to unity (see table 10.3, where $\mu = 0.2$).

Next, consider patterns among group sizes. In contrast to the model with individual learning only, increasing group size now sometimes results in a reduced expected number of slow finders. It appears that social learning can partially overcome the frequency-dependent restraint on individual learning when groups increase in size.

As group size increases, the coefficient of variation of the number of slow finders increases or remains essentially unchanged. This contrasts with the individual learning model, where the coefficient of variation in the number of slow finders and group size were inversely related. The coefficient of variation in the number of fast finders declines with group size; this quantity was independent of group size in the individual learning model.

The proportion of group members that do not show the trait (m) decreases as group size increases. That is, the learned food-finding trait will be exhibited by a greater proportion of individuals as group size increases when both types of learning occur.

With individual learning only, the frequency of the learned trait decreased as task difficulty, the effect of learning, and group size increased. With both individual and social learning, task difficulty has the same effect as noted in the first model, the effect of learning no longer exerts a strong effect on trait frequency, and group size has the opposite effect of that noted in the first model.

10.7 *Learning About*: Individual Learning

To this point, the discussion has concerned how learning influences the presence or absence of skills and, hence, behavioral diversity among group members. Next, we turn to learning as the way individuals gather information used to make foraging decisions. We consider *learning about:* the acquisition of qualitative or quantitative information about an object, place, or behavior. So, *learning about* may involve determining whether or not food is available in a given area and, if it is, in what quantity. The acquisition of such information requires sampling. We therefore focus on sampling behavior and explore the impact of group foraging on individual sampling.

When individuals need to assess the outcome of alternative courses of actions, the presence of other group members may make the task considerably more complex. Consider foragers that must sample to estimate which of two food locations is more profitable (e.g., Tamm 1987; Shettleworth et al. 1988). Sampling while one is a member of a group may make the task difficult, if group membership constrains the time available for sampling, or if the value of each alternative depends on both resource density and the response of others to the resource (Krebs and Inman 1992). Suppose an individual's payoff for choosing an alternative depends, in part, on the frequencies of the choices among other consumers. Then the reliability of the individual's experience (i.e., how long the sample characterizes the alternative's value) depends on the rate of change in local consumer density. That is, a sample's reliability may vary inversely with the temporal variation in the distribution of consumers across patches. Sampling in the context of social foraging may imply that individuals rely on only the most recent sample.

The economic interactions of social foraging may constrain the utility of sampling information acquired solely through individual experience. However, the successes and failures of other group foragers may indirectly provide information useful in making foraging decisions. So, we briefly review the social sampling information available to many group foragers.

10.8 *Learning About*: Social Learning

Food Availability

There are several ways in which others can provide information concerning food availability. The sight of another individual's feeding activity is a public cue that food is available. Animals commonly respond to such information by engaging in feeding themselves, or, if already feeding, by accelerating their feeding rate. Clark and Mangel (1986) interpret this response in terms of the "milk-shake effect" where consumers feed faster to obtain a greater share of a limited resource.

Beauchamp and Kacelnik (1991) provide experimental evidence that individual foragers attend to this type of information. Their laboratory results suggest that social information concerning the timing of food availability may be so salient as to prevent zebra finches from learning other, possibly superior, direct predictors of food availability.

When resources are clumped, the sight of an individual engaged in feeding also reveals the spatial location of food. McQuoid and Galef (1992) provide an experimental demonstration of this form of social sampling. Hens (*Gallus gallus*) exposed to another hen pecking in a specific place exhibited a strong tendency to peck in the same area for up to 48 hours after exposure.

Stronger experimental evidence of the use of socially generated cues concerning food location requires demonstrating that a change in the meaning of the social cue induces a change in group foragers' responses to them. Templeton (1994; see also Templeton and Giraldeau 1995b) present such evidence using pairs of starlings foraging on artificial patches. They concluded that any cue predicting the occurrence of food, or even lack of food, may provide social sampling information.

Observations of captive flocks of tits (Krebs et al. 1972) and chickadees (Krebs 1973) suggest that unsuccessful individuals were more likely to search patch types that others had explored for food, suggesting that specific location-types may provide social sampling information (see McQuoid and Galef 1992). Hence, foragers may generalize socially sampled information on the spatial location of food, and perhaps on its temporal pattern.

A forager may already know the location of a large number of fruit trees, but may be unaware of which ones currently provide food. Observing another forager at a tree may indicate that its fruit is ripe. Greene (1987) provides a remarkable example of social sampling for food availability. Members of an osprey breeding colony collectively exploited a number of bodies of water located in different directions from the colony. Different foraging localities provided different fish species. Because of the tide, some bodies of water apparently offered fish at high density only during brief, unpredictable periods. Greene reports that some individuals chose the direction of the next foraging trip based on the species of fish carried in the talons of successful individuals recently returning to the colony.

Patch Assessment

Above we listed ways in which individuals can learn whether food is present or not. Patch assessment refers to estimating the amount of resource in a patch being exploited. In a formulation of the patch-residence problem under uncertainty, a forager uses its patch assessment in deciding whether to remain in the current patch or travel to another (e.g., McNamara 1982; Nishimura 1992).

Most models involving patch assessment assume that residence time is a function of individual information, extracted from the forager's direct experience (Green 1984, 1987; McNamara and Houston 1980, 1985). The individual's patch assessment might be based on the number of items encountered in the patch, the time elapsing since the last item was found, or similar estimates of local foraging success. Estimates of patch quality may increase as food is encountered, and decrease as unsuccessful search continues. Clark and Mangel (1984) pointed out that the efficiency of patch assessment might be enhanced if members of a foraging group could pool their respective individual experiences. So, we comment briefly on group foragers' social patch assessment.

If foragers do not rely on social patch sampling, their patch-leaving decisions should depend on their own foraging success, and not on others' success (Valone and Giraldeau 1993; Templeton and Giraldeau 1995b). When two individuals exploit the same patch, the forager that has found the greater number of items likely has the higher estimate of the patch's value. This individual might therefore persist in the patch longer, in the absence of success, than the individual that found fewer items (Templeton 1994; Templeton and Giraldeau 1995b). Hence, the departure order of individuals from a patch might be inversely related to their foraging success if patch assessments strictly reflect individual information.

However, if foragers use social patch assessment, they may have the same estimate of patch quality; their order of departure might then depend less strongly on their individual levels of foraging success. If individually and socially acquired information is equally weighted, both foragers have the same aggregate information concerning the value of the patch.

Valone and Giraldeau (1993) observed captive pairs of seed-eating birds exploit food patches. They found that the departure order of individuals was related to their past success, and concluded that the birds did not use social patch-assessment. However, Templeton (1994; Templeton and Giraldeau 1996) provides field evidence that starlings can and do use social patch sampling information. Templeton (1994) observed the patch departure decisions of two starlings chosen arbitrarily from a flock of individuals foraging in a baited area. Starlings exploited two types of patches. In one, the use of social patch assessment was made difficult; opaque barriers prevented each individual from observing its partner's foraging. In the other patch type, social information could be obtained more easily since no visual barrier was imposed.

Templeton and Giraldeau (1996) found that when the barriers were present, the departure order of the birds was inversely related to their individual success in the patch. Hence, birds used individual sampling information only. However, when no barriers were present and social information could be used, there was no longer any relationship between patch departure order and individual success in a patch. Templeton and Giraldeau (1996) interpret the result as evidence that starlings use social patch sample information when the information can be obtained at little cost (see also Templeton and Giraldeau 1995a).

10.9 Concluding Remarks

Suggestions for Empirical Work

Few studies have examined factors inhibiting gregarious animals' learning of novel behavior. It would be useful to document whether or not food-sharing really does interfere with individual learning (Giraldeau and Lefebvre 1987; Lefebvre and Giraldeau 1994). To answer this question, one could compare foraging groups faced with having to learn the same food-finding skill but differing in the number of individuals that could gain access to and share discovered food. An efficient approach may be to use a core flock design (Giraldeau et al. 1994b; Koops and Giraldeau 1996), where subjects are added to an already existing group for a given number of trials and one counts how many individuals in each type of core flock actually acquire the skill.

Food sharing may reduce the extent of social behavior copying, at least in pigeons (Lefebvre and Giraldeau 1987; Giraldeau and Templeton 1991). Apart from Nicol and Pope's (1994) attempt to document the effect in chickens, few studies have asked whether behavior copying in other species is also inhibited by food sharing. One possibility is that behavior copying of food-finding skills is inhibited when the gains expected from using the food finding skill (i.e., becoming a producer) are lower than not using the skill (i.e., remaining a scrounger).

For many animals, *learning how* is unlikely to be a recurrent daily activity. Instead, it is more probable that it is localized to specific periods in an animal's ontogeny. *Learning about*, on the other hand, appears to be a more common recurring phenomenon. *Learning about*, therefore, may be a subject of more widespread interest to behavioral ecologists. Surprisingly, although some work has been devoted to the inhibitory effects of group membership on individual *learning how*, the equivalent is not true of individual *learning about*. There is considerable scope for empirical investigation of how individual sampling tactics can be affected by the presence of group members.

Somewhat more effort has been devoted to studying social sampling (Valone and Giraldeau 1993; Templeton and Giraldeau 1995, 1996a,b). Pairs of budgerigars apparently do not use social patch assessment, but starlings do use social information, provided that individual sampling is costly (Templeton and Giraldeau 1996a,b). A resulting question is whether social information is given the same weight as personal experience in a forager's decision making.

Conclusions

This chapter addresses issues of learning as it proceeds within groups. We divided learning into two broad categories, *learning how* and *learning about*, and for each category we considered learning based on individual and social information.

Our stochastic individual-learning model suggests that learning can be constrained when it occurs within foraging groups that exploit divisible food clumps. As a result, only a fraction of a foraging group is expected to exhibit a food-finding trait when individual learning alone is operating. We suggest that *learning how* from social information may have arisen as a means to circumvent the shortcomings of individual learning in foraging groups. When social learning is possible, a much greater fraction of the group is expected to exhibit a novel food-finding skill. At the very least, our analyses suggest that the effect that group foraging has on learning is not necessarily always positive. The advanced cognitive abilities possessed by a number of organisms cannot necessarily be attributed to the complexity of their social life. Instead, it may be the result of the problems that social life poses to the efficient operation of individual learning.

In nature, social learning probably homogenizes the types of resources discovered by interacting consumers. Social learning could allow foragers to track spatiotemporal variation in availability of more profitable resources. Consequently, temporal variation in a group member's diet might exceed that of a solitary in the same habitat.

Having investigated how learning can operate within foraging groups either to promote or reduce among-individual diversity, we now turn to a more strategic approach, asking whether some ecological conditions actually make it profitable for individuals to diverge behaviorally and form *skill pools*. That is the subject of the next chapter.

Math Box 10.1 Equilibrium Distribution: Individual Learning Only

The unique equilibrium distribution for the number of slow finders at any time is designated $p(X)$, where

$$p(X) = \lim_{t \to \infty} \Pr[X(t) = X \mid X(0) = k]. \qquad (10.1.1)$$

The proportion of time the process spends at a given number of slow finders (i.e., in state X) converges to $p(X)$ as time increases to infinity. The equilibrium value is independent of the initial number of slow finders. Using the equilibrium state equation (10.5), we derive an equilibrium distribution of slow finders. The equilibrium distribution must satisfy the "detailed" balance conditions (Kelly 1979):

$$\frac{p(r)}{p(r-1)} = \frac{q(r-1,r)}{q(r,r-1)}. \qquad (10.1.2)$$

Starting with $r = 1$, one can derive a general formula for the equilibrium distribution from the balance conditions

$$p(X) = p(0) \prod_{r=1}^{X} [q(r-1,r)/q(r,r-1)] \quad \text{for } X = 1,2,\ldots,G \qquad (10.1.3a)$$

and

$$p(0) = 1 - \sum_{X=1}^{G} p(X). \qquad (10.1.3b)$$

More detailed discussion of the use of birth-death processes as models for social organization can be found in Cohen (1972), Boswell et al. (1979), and Caraco (1980a).

Using equations (10.3) and (10.4), we have $q(r-1,r) = \mu(G-r+1)$, and $q(r,r-1) = \beta r/[r + \alpha(G-r)]$. Then equation (10.1.3a) becomes

$$p(X) = p(0) \prod_{r=1}^{X} \mu(G-r+1)[r + \alpha(G-r)]/\beta r. \qquad (10.1.4)$$

Math Box 10.1 (*cont.*)

Simplification yields

$$p(X) = \binom{G}{X} [\mu/\beta]^X p(0) \prod_{r=1}^{X} [r + \alpha(G - r)]. \qquad (10.1.5)$$

The multiplicand can be written as a ratio of gamma functions (e.g., Caraco 1980a), and substituting the last expression into equation (10.1.3b) gives an expression for $p(0)$.

Then the equilibrium probability function is

$$p(X) = \frac{\binom{G}{X} [\mu(\alpha - 1)/\beta]^X \, [\Gamma(C)/\Gamma(C - X)]}{\sum_{i=0}^{G} \binom{G}{i} [\mu(\alpha - 1)/\beta]^i \, [\Gamma(C)/\Gamma(C - i)]} \qquad (10.1.6)$$

for $X = 0, 1, \ldots, G$; $C = \alpha G/(\alpha - 1)$ and $\Gamma(k) = (k - 1)\,\Gamma(k - 1)$.

Math Box 10.2. No Forgetting: Waiting Time until All Group Members Learn a Skill

At time $t = 0$, all group members are slow finders (i.e., $X(t) = G$). The associated transition rates $q(r, r - 1)$ are given by equation (10.4). We let the random variable t_r represent the duration of the process in the state $X(t) = r$. Each t_r has an exponential probability density with mean $E[t_r] = [q(r, r - 1)]^{-1}$, and variance $V[t_r] = [q(r, r - 1)]^{-2}$. The random variable T represents the total time for all group members to learn the trait, and T is a sum of random variables

$$T = \sum_{r=1}^{G} t_r. \tag{10.2.1}$$

The expected value of T is

$$\begin{aligned} E[T] &= \sum_{r=1}^{G} [q(r,r - 1)]^{-1} \\ &= \sum_{r=1}^{G} [r + \alpha(G - r)]/\beta r. \end{aligned} \tag{10.2.2}$$

This expression becomes

$$E[T] = (G/\beta)(1 - \alpha) + (\alpha G/\beta)\sum_{r=1}^{G}(1/r). \tag{10.2.3}$$

Approximating the partial sum of the harmonic series yields

$$E[T] \approx (G/\beta)[1 - \alpha + \alpha \ln(G + 1)]. \tag{10.2.4}$$

The variance of T is $V[T]$

$$\begin{aligned} V[T] &= \sum_{r=1}^{G} [q(r,r - 1)]^{-2} \\ &= \sum_{r=1}^{G} [r + \alpha(G - r)]^2/(\beta r)^2. \end{aligned} \tag{10.2.5}$$

After expanding, we obtain

$$\begin{aligned} V[T] = {} & (1 - \alpha)(2G/\beta^2) + (1 - \alpha) \\ & (2\alpha G/\beta^2)\sum_{r=1}^{G}(1/r) + (\alpha G/\beta)^2 \sum_{r=1}^{G}(1/r^2). \end{aligned} \tag{10.2.6}$$

Math Box 10.2. (*cont.*)

Applying standard approximations for the partial sum yields the approximate variance of T:

$$V[T] \approx (G/\beta^2)\{\alpha[\frac{\pi^2}{6} - \frac{1}{G + 1}] - 2(\alpha - 1)[1 + \alpha \ln(G + 1)]\}. \tag{10.2.7}$$

Math Box 10.3 Equilibrium Distribution: Individual and Social Learning

Equations (10.3) and (10.8) define a birth-death process with an equilibrium distribution $p(X)$ satisfying equations (10.1.3a) and (10.1.3b) of Math Box 10.1.

In terms of the transition rates, $p(X)$ is

$$p(X) = p(0)\prod_{r=1}^{X} [\mu(G - r + 1)/r]\left[\frac{\beta}{r + \alpha(G - r)} + \Omega(G - r)\right]^{-1} \tag{10.3.1}$$

$$p(X) = \binom{G}{X}\mu^{X}p(0)\prod_{r=1}^{X}\frac{r(1 - \alpha) + \alpha G}{\alpha\Omega(G - r)^2 + \Omega r(G - r) + \beta}.$$

We can simplify the multiplicand as in the previous model and use equation (10.7b) to find the expression for $p(0)$. For simplicity we define $M(X)$ as

$$M(X) = \prod_{r=1}^{X}\alpha\Omega(G - r)^2 + \Omega r(G - r) + \beta. \tag{10.3.2}$$

Then the equilibrium probability function is

$$p(X) = \frac{\binom{G}{X}[\mu(\alpha - 1)]^{X}[\Gamma(C)/\Gamma(C - X)M(X)]}{\sum_{i=0}^{G}\binom{G}{i}[\mu(\alpha - 1)]^{i}\,[\Gamma(C)/\Gamma(C - i)M(i)]} \tag{10.3.3}$$

for $X = 0, 1, \ldots, G$; $C = \alpha G/(\alpha - 1)$.

11 Efficiency of Diversity: The Skill Pool

Several of our models have assumed that food occurs as clumps of a single type of resource. But members of foraging groups must encounter environments where different types of resource, each requiring a particular harvesting behavior, are available simultaneously (Laverty 1994; Beauchamp et al. 1997). Can we predict how this resouce diversity affects behavioral diversity?

Rubenstein et al. (1977) found that granivorous sparrows in mixed-species flocks took more diverse diets than solitary conspecifics. That is, within-individual dietary diversity was greater for group members than for solitary foragers. If seeds of different types were spatially interspersed, scramble competition, mediated through local reduction in food density, might induce group members to accept a greater range of dietary items (see chapter 9). However, suppose different types of seeds occurred as spatially distinct clumps. Then different group members might somehow specialize in searching for particular resource types, while opportunistically feeding at one another's discoveries. This perspective suggests that group foraging can increase individual dietary diversity in accordance with the skill pool hypothesis (Giraldeau 1984).

To model the skill pool, we envision an environment where foragers exploit two types of resources. The skill pool, by definition, requires an independent clumped dispersion for each resource. Each individual searches for food, and all discoveries are shared among group members; there are no scroungers. An individual may specialize and search for only a single type of resource clump, or may generalize and search for two types of clumps. If different individuals direct their specialized searching to different resources, the group fulfills the definition of a skill pool (Giraldeau 1984). That is, different group members develop complementary searching specializations, and each consumes some of the food in any clump discovered, irrespective of resource type. The skill pool might prove advantageous if variation in group members' specializations increment the individual's energy intake or its resource-consumption diversity (Giraldeau and Lefebvre 1986).

We organize this chapter as follows. First, we summarize some points Giraldeau (1984) listed when introducing the notion of the skill pool. Next, we consider a static model where a group of two foragers exploits an environment with two types of divisible food clumps. Each individual may search for both clump types (two generalists), each may search for only the richer clump type (two specialists), or each forager may search for a different clump type (the skill pool). Similar specialists may socially inhibit or enhance one another's rate of clump discovery; the same is true for paired generalists searching for both resources. A generalist's production rate also depends on the way learning/performing different tasks interact; ordinarily, increasing one rate reduces the other if dissimilar skills are required (e.g., Giraldeau and Lefebvre 1986). We quantify these ideas to ask which group composition minimizes the probability of an energetic failure.

The third section examines a stochastic dynamic programming model that extends the static model's predictions. The fourth and last section summarizes our results and compares a skill-pool member with a solitary forager.

11.1 Background

Krebs (1973) hypothesized that an individual might increase its food intake by joining a multispecies group. By assumption, individuals of different species initially differ significantly in the foods they exploit or in the foraging sites they explore. As a member of a multispecies foraging group, each of these individuals might gain exposure to other species' behavior. If food occurs in clumps, variability among individuals' searching behavior might persist. But the among-individual difference in food consumption should decline as each individual learns to exploit discoveries of other species. That is, different individuals' diets become more similar as each species increases its food-consumption diversity through the skill-pool effect.

Giraldeau (1984) restricted attention to conspecific groups in formulating the skill pool hypothesis. He suggested that diversity among group members in the types of resources discovered could be an advantage of social foraging, as long as food was clumped sufficiently to permit mutual exploitation. If a skill pool holds an economic advantage over both similar specialists and a group of generalists, the efficiency differences may involve two distinct causes. First, social interference may inhibit similar foragers' discovery of food clumps (e.g., Ekman and Hake 1988). As more group members search for the same type of resource, each individual's rate of discovering that resource may decline. Specialists on a single clump type are susceptible to social interference. Generalists (by our definition) search for the same set of resource types, and so are also susceptible to social interference. However, skill pool members avoid social interference since different individuals search for different resource types.

The second skill-pool advantage involves comparison with only a group of generalists. Each generalist must learn behavior necessary to discover and process different types of resources. Acquiring skills allowing exploitation of one type of resource may interfere with learning or performing behavior necessary to exploit a second resource (e.g., Pietrewicz and Kamil 1979; Partridge and Green 1985, 1987; Woodward and Laverty 1992; Laverty 1994). Once an individual has acquired a specialist skill, neophobia may inhibit efficient searching for the other resource (Beissinger et al. 1994) and restrict generalism. In an economically similar manner, different types of clumps may occur in different microhabitats so that searching for one type constrains discovery of the other resource. In each case, a generalist may be less efficient at finding a given type of food clump than is the corresponding specialist member of a skill pool (Beauchamp and Kacelnik 1991).

The skill pool hypothesis proposes an advantage for members of foraging groups where the mean within-resource diversity in food production is relatively small (see chapter 9). Equivalently, the hypothesis assumes a disadvantage for members of foraging groups with relatively small among-resource production diversity (i.e., a group of similar specialists), or with relatively large within-individual diversity of resource production (i.e., a group of generalists).

11.2 A Skill Pool: Static Model

We assume a group of two competitively equivalent foragers, identified (where necessary) as 1 and 2. When the two individuals search for the same resource type, we assign them identical discovery rates. For simplicity and tractability, we restrict group size to two, but the concept of a skill pool clearly applies to larger groups.

Food occurs as clumps, and foragers divide all food discovered equally. A clump of resource type 1 yields one unit of energy. A clump of type 2 yields f units, where $0 < f \leq 1$. Foragers have the capacity to search simultaneously for both types of clumps. We ignore handling time within clumps to emphasize effects of searching skills.

The foragers have τ time units to search for food. R is the physiological requirement; a forager suffers mortality or incurs some other fitness penalty if its energy intake fails to exceed R units by time τ. Each forager attempts to minimize its probability of an energetic failure. Since clumps are divided equally, the two foragers either meet or fail to meet their requirement together. This simplifies the model's analysis considerably.

$X_1(\tau)$ is the total number of type-1 clumps the two foragers consume by time τ, and $X_2(\tau)$ is the total number of type-2 clumps the two foragers consume by time τ. We assume that type-1 and type-2 clumps are produced as independent Poisson processes. Let $\chi(\tau)$ represent the total energy intake of the two foragers:

$$\chi(\tau) = X_1(\tau) + fX_2(\tau). \tag{11.1}$$

The associated mean and variance are

$$E[\chi(\tau)] = E[X_1(\tau)] + fE[X_2(\tau)] \tag{11.2}$$

$$V[\chi(\tau)] = V[X_1(\tau)] + f^2V[X_2(\tau)]. \tag{11.3}$$

Note that independence of the $X_j(\tau)$ implies zero covariance. Each individual has cumulative energy intake $\chi(\tau)/2$. Since the two foragers succeed or fail to meet their metabolic requirement together, each individual's objective is equivalent to minimizing $\Pr[\chi(\tau) \leq 2R]$.

Resource Discovery Rates

In the two-resource environment, each individual's feasible clump-discovery rates (one for each clump type) will be influenced by

1. A within-individual (hence between-resource) constraint.
2. A between-forager (within-resource) social interaction.

Within the individual, feasible discovery rates for the two clump types should be governed by the degree of similarity between the respective production skills. Skills might interact so as to enhance total clump discovery, so that increasing the lesser discovery rate may increase the total production of both clump types. Or, the two clump-discovery rates might compensate additively, so that increasing one rate requires an equal decrease in the level of the second (leaving the total rate constant). Most likely, the two clump-discovery rates are governed by an inhibitory constraint. In this case, increasing the lesser rate by learning or practice decreases proficiency at performing the other skill sufficiently to decrease the individual's total rate of clump discovery. Detailed discussion can be found in Pietrewicz and Kamil (1979), Giraldeau and Lefebvre (1986), or Partridge and Green (1987).

Between individuals 1 and 2, feasible discovery rates can be influenced by the social effects discussed in chapter 2. When the two individuals search for the same type of clump, each forager's rate of finding food may be affected positively, negatively, or may be independent of the other's searching behavior.

To quantify these effects, let a forager searching for type-j ($j = 1, 2$) clumps encounter them at constant probabilistic rate λ_j. We assume λ_j is a product of two functions:

$$\lambda_j = s_j(\gamma, D)\ \theta_j(\alpha, G_j). \tag{11.4}$$

$s_j(\gamma, D)$ quantifies the within-individual constraint on the capacity to search for both resources simultaneously. We shall refer to γ as skill compatibility. $\theta_j(\alpha, G_j)$ quantifies the between-forager social effect on the individual's rate of discovering resource j. Below we refer to α as social interference, as a convenience. We develop the two factors of λ_j in turn.

Within-Forager Skill Compatibility

The within-individual constraint depends on the similarity of learned skills required to produce the two resources. We assume that both group foragers are subject to the same constraint.

Let s_j represent the constant probabilistic rate at which either individual (1 or 2) discovers type-j ($j = 1$ and 2) clumps when foraging solitarily. The forager's two clump discovery rates are subject to the scalar-valued equality constraint:

$$s_1{}^{\gamma} + s_2{}^{\gamma} = D^{\gamma}, \tag{11.5}$$

where $s_1, s_2 \geq 0$, and $\gamma, D > 0$. The equality constraint defines the function $s_j(\gamma, D)$ by specifying feasible (s_1, s_2) combinations; **S** represents the set of feasible rates. The identical constraint applies when the animal forages in a group, since skill compatibility reflects properties inherent to the individual and resources.

Figure 11.1 shows the three different forms of the skill-compatibility constraint (cf. Vickery et al. 1991). If $\gamma > 1$, searching skills have an enhancing quality. In this case, $(s_1 + s_2) \geq D$, so that two generalists might discover clumps at the greatest rate. If $\gamma = 1$, $(s_1 + s_2) = D$, and the constraint is linear. If $\gamma < 1$, $(s_1 + s_2) \leq D$; searching skills are sufficiently dissimilar that learning or exercising one skill can reduce not just the other skill, but the sum of the clump-discovery rates. The shape of the constraint set **S** depends on skill compatibility γ.

To examine the within-individual constraint in more detail, let m represent the marginal rate of substitution of s_1 for s_2 on **S**:

$$m = -\frac{ds_1}{ds_2} = \frac{(\partial/\partial s_2)(s_1{}^{\gamma} + s_2{}^{\gamma})}{(\partial/\partial s_1)(s_1{}^{\gamma} + s_2{}^{\gamma})} = \left(\frac{s_2}{s_1}\right)^{\gamma-1}. \tag{11.6}$$

The interpretation of m is simple. Given that the forager's discovery-rate pairs must belong to the set **S**, if the animal chooses to increase its rate of discovering type-2 clumps by one unit, it must reduce its rate of discovering type-1 clumps by m units. Then m defines the allowable trade-off between s_1 and s_2. Clearly, the marginal rate of substitution can depend on the particular (s_1, s_2) pair where m is evaluated (see Keeny and Raiffa 1975; Rachlin et al. 1976, 1980; Houston and McNamara 1988; McNamara and Houston 1994).

The degree of skill compatibility affects the trade-off between discovery rates. If $\gamma = 1$, then $m = 1$. The marginal rate of substitution is constant; s_1 and s_2 are interchangeable with respect to clump discovery when the equality constraint is linear. However, the difference in energy value of the two clump types remains important when $f < 1$.

Suppose the two skills are highly compatible, so that $\gamma > 1$. Then $m < 1$ for $s_2 < s_1$. If f, the energy available in type-2 clumps is sufficiently close to

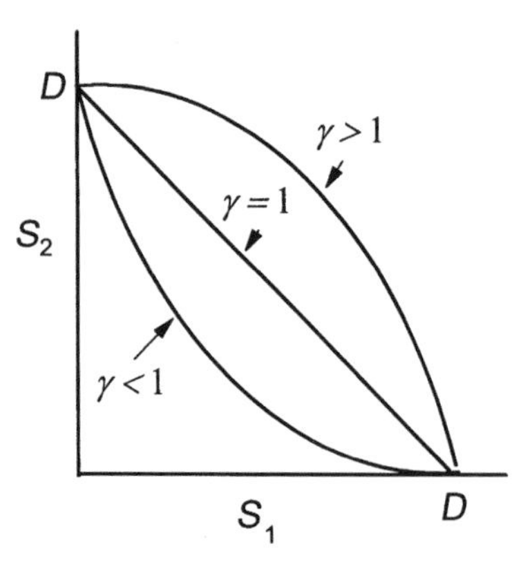

Figure 11.1 Solitary forager's feasible production rates s_j. In the absence of any social effects, a forager finds type-j clumps at constant probabilistic rate s_j; j = 1, 2. Feasible (s_1, s_2) combinations must satisfy equality constraint (11.5).

1, the forager might advantageously reduce s_1 to gain an increase in s_2, thereby reducing the difference between the two rates. However, if $\gamma > 1$, and $s_2 > s_1$, then $m > 1$. Now, decreasing s_2 one unit to gain an m-unit increase in s_1 could be advantageous. In both cases, increasing the lesser s_j might be favored. Highly compatible skills ($\gamma > 1$) imply a relaxed within-individual constraint and consequently tend to favor a generalist forager, unless the value of the poorer resource is too low (e.g., Rapport 1980).

Now consider incompatible skills, where $\gamma < 1$. Then $m < 1$ for $s_1 < s_2$. If f is sufficiently close to 1, the forager might advantageously reduce s_1 to gain an increase in s_2, increasing the difference between discovery rates. However, if $\gamma < 1$, and $s_1 > s_2$, then $m > 1$. Now, decreasing s_2 to gain an increase in s_1 would ordinarily be advantageous. In both cases, increasing the greater s_j might be favored. Relatively incompatible clump-discovery skills ($\gamma < 1$) consequently tend to favor a specialist forager (Rapport 1980).

Between-Forager Social Effect

Recall that λ_j, the rate at which a searching forager encounters type-j clumps, is the product of $s_j(\gamma, D)$ and the function allowing a between-individual interaction $\theta_j(\alpha, G_j)$. To model the between-individual social effect, let G_j count the foragers searching for type-j clumps; $G_j \in \{0, 1, 2\}$. As in chapter 2, we assume $\alpha > 0$ and let $\theta_j(\alpha, G_j) = G_j^{\alpha - 1}$, so that α scales the effect of social interference on clump discovery. Then an individual encounters type-j clumps at constant probabilistic rate,

$$\lambda_j = s_j G_j^{\alpha - 1}, \tag{11.7}$$

subject to $s_1{}^{\gamma} + s_2{}^{\gamma} = D^{\gamma}$.

Then,

$$\lambda_j \in \{0, s_j, s_j 2^{\alpha - 1}\}$$

according to the value of G_j.

If $\alpha < 1$, foragers searching for the same type of clumps interfere with one another's efforts. Then,

$$\lambda_j(G_j = 2) = s_j 2^{\alpha - 1} < s_j = \lambda_j(G_j = 1). \tag{11.8}$$

Each of two identical foragers attempting to find type-j clumps does so at a rate lower than would a solitary in the same environment. However, the two foragers' combined rate, $s_j 2^{\alpha}$, exceeds the solitary's rate.

If $\alpha = 1$, social interference disappears, and

$$\lambda_j(G_j = 2) = s_j = \lambda_j(G_j = 1). \tag{11.9}$$

$\alpha = 1$ implies social independence. An individual producing type-j clumps does so at a rate unaffected by the other forager's searching behavior.

If $\alpha > 1$, social interference becomes social enhancement, and we have

$$\lambda_j(G_j = 2) = s_j 2^{\alpha - 1} > s_j = \lambda(G_j = 1). \tag{11.10}$$

Each of two identical foragers searching for type-j clumps does so at a rate greater than would a solitary in the same environment. In general, we expect $\alpha \leq 1$, so that α will ordinarily indicate the level of social interference. As interference increases, α declines. As indicated above, social interference can affect similar specialists and generalists, but not members of a skill pool.

Failure Probabilities

The G_j identify the classes of strategies. Let $\mathbf{G} = \{G_1, G_2\}$. The elements of $\mathbf{G}$ count the number of foragers searching for the respective resource types, i.e., the two G_j values. We consider three G_j pairs:

1. $\mathbf{G} = \{2, 0\}$. Both foragers specialize on searching for type-1 clumps, the richer resource.
2. $\mathbf{G} = \{2, 2\}$. Both foragers generalize.
3. $\mathbf{G} = \{1, 1\}$. One forager specializes on finding type-1 clumps, and the other forager specializes on finding type-2 clumps; this is the skill pool.

We exclude the two possibilities where $G_2 > G_1$, since type-2 clumps provide less energy. We also exclude $\mathbf{G} = \{2, 1\}$, since it is intermediate to the second and third options.

If an individual is a type-j specialist, $s_j = D$ for that individual, by expression (11.5), and the forager does not attempt to find the other type of clump. That individual's clump-discovery rate then depends on whether

$G_j = 1$ or 2 (i.e., on the social effect). When both foragers generalize, we assume each has the same $(s_1, s_2) \in \mathbf{S}$, and we select the "best" generalist according to the efficiency criterion considered (see below).

Recall that $\chi(\tau)$ is the total energy found by the group during the available time. We list the expectation of $\chi(\tau)$, its variance, and the z-score associated with $\Pr[\chi(\tau) \leq 2R]$ for each group composition. First, we let $\mathbf{G} = \{2, 0\}$; both foragers specialize on type-1 clumps. For each of the identical individuals,

$$\lambda_1 = D2^{\alpha-1} \text{ and } \lambda_2 = 0. \tag{11.11}$$

Type-1 clumps are produced at combined rate $D2^{\alpha}$, and type-2 clumps are not exploited. The total number of clumps consumed by time τ has a Poisson probability function with expectation $D\tau 2^{\alpha}$. Since each clump yields a unit of energy, we have

$$\mathrm{E}[\chi(\tau)] = D\tau 2^{\alpha} = \mathrm{V}[\chi(\tau)]. \tag{11.12}$$

The standard normal approximation yields the value of our fitness criterion for similar specialists:

$$z(2R \mid \mathbf{G} = \{2,0\}) = (2R - D\tau 2^{\alpha})/(D\tau 2^{\alpha})^{1/2}. \tag{11.13}$$

Second, we let $\mathbf{G} = (2, 2)$; each forager generalizes. For each of the two group members,

$$\lambda_j = s_j 2^{\alpha-1};\ s_j < D,\ \mathrm{j} = 1, 2. \tag{11.14}$$

Each of the two clump types is found at total rate $s_j 2^{\alpha}$, with the s_j constrained by expression (11.5). For any given $(s_1, s_2) \in \mathbf{S}$, $\chi(\tau)$ has mean and variance

$$\mathrm{E}[\chi(\tau)] = \tau 2^{\alpha}(s_1 + f s_2) \tag{11.15}$$

$$\mathrm{V}[\chi(\tau)] = \tau 2^{\alpha}(s_1 + f^2 s_2). \tag{11.16}$$

Therefore, the z-score for the generalists is

$$z(2R \mid \mathbf{G} = \{2, 2\}) = \frac{2R - \tau 2^{\alpha}(s_1 + f s_2)}{[\tau 2^{\alpha}(s_1 + f^2 s_2)]^{1/2}}. \tag{11.17}$$

As $s_1 \rightarrow D$, the generalists' z-score approaches the value for similar specialists.

When we compare mean energy-consumption levels, we shall first solve for $(s_1{}^*, s_2{}^*)$, the feasible (s_1, s_2) pair maximizing $\mathrm{E}[\chi(\tau)]$ for generalists. When we compare z-scores similarly, we evaluate expression (11.17) at $(s_1, s_2) = (s_1{}^\circ, s_2{}^\circ)$, the element of $\mathbf{S}$ minimizing the z-score and (hence) minimizing the generalists' $\Pr[\chi(\tau) \leq 2R]$.

For the skill pool, $\mathbf{G} = \{1, 1\}$, and each type of clump is discovered at probabilistic rate D. For the skill pool, $\chi(\tau)$ has mean and variance

$$\mathrm{E}[\chi(\tau)] = D\tau(1 + f) \tag{11.18}$$

$$\mathrm{V}[\chi(\tau)] = D\tau(1 + f^2). \tag{11.19}$$

The z-score for the skill pool is

$$z(2R \mid \mathbf{G} = \{1, 1\}) = \frac{2R - D\tau(1 + f)}{[D\tau(1 + f^2)]^{1/2}}. \tag{11.20}$$

Maximizing Expected Energy Intake

Table 11.1 lists the expected total energy consumption by time τ, $\mathrm{E}[\chi(\tau)]$, for the type-1 specialists, the generalists, and for the skill pool. To begin, we compare the skill pool and similar specialists. A member of the skill pool will have a greater expected energy intake than either of paired type-1 specialists when

$$D\tau(1 + f) > D\tau 2^{\alpha},$$

which reduces to

$$\alpha < \ln(1 + f)/\ln 2. \tag{11.21}$$

As $\alpha \to 0$ and/or $f \to 1$, the skill pool will more likely have the greater mean. If $\alpha > 1$, inequality (11.21) cannot hold. Therefore, if the skill pool's mean intake is to exceed the type-1 specialists' mean,

1. There must be some social interference between specialist foragers.
2. The energetic value of type-2 clumps cannot be too small.

The result makes intuitive sense. The skill pool's use of poorer clumps must be more than compensated for by a reduction in the production of type-1 clumps by the specialists, or else the skill pool will have the lower mean intake.

Table 11.1

Expected energy consumption and z-scores for different searching strategies. $\chi(\tau)$ is total energy consumed by two foragers. z-scores are obtained from $\Pr[\chi(\tau) \leq 2R]$, where R is the individual's physiological requirement. $\mathbf{G} = \{G_1, G_2\}$ identifies the strategies (see text). For generalists, the numerical values of s_1 and s_2 depend on the selected optimization criterion.

G	$E[\chi(\tau)]$	$z(2R \mid \{G_1, G_2\})$
{2, 0}	$D\tau 2^{\alpha}$	$(2R - D\tau 2^{\alpha})/(D\tau 2^{\alpha})^{1/2}$
{1, 1}	$D\tau(1+f)$	$[2R - D\tau(1+f)]/[D\tau(1+f^2)]^{1/2}$
{2, 2}	$\tau 2^{\alpha}(s_1 + fs_2)$	$[2R - \tau 2^{\alpha}(s_1 + fs_2)]/[\tau 2^{\alpha}(s_1 + f^2 s_2)]^{1/2}$

To compare generalists to the alternatives, we first maximize $E[\chi(\tau)]$ for the generalist strategy. The optimization, subject to equality constraint (11.5), is presented in Math Box 11.1 at the end of the chapter. The results show that if searching skills are sufficiently compatible (i.e., $\gamma > 1$), paired generalists' mean energy intake is maximal when each adopts (s_1^*, s_2^*), where

$$s_1^* = \frac{D}{(1 + f^{\gamma/\gamma-1})^{1/\gamma}} \qquad (11.22)$$

$$s_2^* = f^{1/\gamma-1}\, s_1^*. \qquad (11.23)$$

When skills are incompatible within the individual ($\gamma \leq 1$), generalists cannot have a mean energy intake exceeding specialists on type-1 clumps (see below).

From equation (11.23) we note that

$$f = (s_2^*/s_1^*)^{\gamma-1} < 1.$$

For $\gamma > 1$, paired generalists maximize their expected intake when the marginal rate of substitution m equals f, the energy value of type-2 clumps. That is, a unit increase in discovering type-2 clumps is worth a decrease of f (≤ 1) units in discovering type-1 clumps. But scaled in terms of expected energy (the objective being maximized under constraint), the increase and decrease just balance at the optimum (s_1^*, s_2^*).

Suppose $f = 0$, so that only type-1 clumps provide energy. Then, from expressions (11.22) and (11.23), $(s_1^*, s_2^*) = (D, 0)$. That is, the generalist strategy maximizing expected energy intake collapses to specialization when type-2 clumps yield no energy. But $f > 0$ by definition, so that both $s_j^* > 0$ by their definitions. From these observations, a standard exercise in optimization under constraint shows that when $\gamma > 1$, generalists' mean intake always exceeds the mean for specialists. Generalists, on average, do not consume as many type-1 clumps as do specialists. But the energy generalists acquire from type-2 clumps more than compensates when searching skills are sufficiently compatible, since generalists expect to find more clumps. That is, $(s_1^* + s_2^*) > D$, a specialist's discovery rate, when $\gamma > 1$. Now suppose $\gamma < 1$, so that skills are incompatible. A similar argument shows that the specialists' mean energy intake always exceeds the mean for generalists [i.e., for any $s_1, s_2 > 0$, where $(s_1, s_2) \in \mathbf{S}$]. When $\gamma < 1$, incompatible skills imply that $(s_1 + s_2) < D$. So, the comparison of expected energy intake for specialists and generalists depends on the form of the within-individual skill-compatibility constraint. For $\gamma > 1$, generalists have the greater mean. But for $\gamma \leq 1$, specialists cannot have the lesser mean.

The remaining comparison of expected energy-intake levels concerns generalists and the skill pool. Assume $\gamma > 1$, where generalists may succeed. If the skill pool's mean intake exceeds that of the paired generalists, then

$$D\tau(1 + f) > \tau 2^{\alpha}(s_1 + fs_2).$$

Substituting the equality constraint yields

$$\frac{\left(s_1{}^{\gamma} + s_2{}^{\gamma}\right)^{1/\gamma}}{s_1 + fs_2} > 2^{\alpha}/(1 + f). \tag{11.24}$$

Members of the skill pool may have an advantage over each of the paired generalists when

1. Generalists experience social interference ($\alpha < 1$).
2. Food searching skills are not so compatible that they strongly enhance one another (i.e., γ is not too large).

To compare expected energy-intake levels numerically, we set $D\tau = 1$ without loss of generality. Figure 11.2A shows that specialists have the greatest $E[\chi(\tau)]$ only when $\gamma < 1$ and f is sufficiently small that inequality (11.21) does not hold. The skill pool has the greatest mean energy intake when f is sufficiently large and γ is not too large. Generalists have the greatest mean energy intake at any level of f when γ is adequately large. Note that under social interference ($\alpha = 0.4$ in fig. 11.2A), the skill pool can achieve a greater mean intake than generalists even when $\gamma > 1$, provided that the energy found in type-2 clumps (f) is large enough.

Figure 11.2B shows (for $\gamma = 1.3$) how varying the level of social interference α affects the comparison of the skill pool's expected intake with the mean for generalists. As f increases, a greater value of α (reduced social interference) is required for the generalists' expected intake to exceed the skill pool's mean. Some of these predictions recur in the next section where we compare z-scores.

Minimizing the Probability of an Energetic Failure

Table 11.1 lists the parameter values used in the z-score formulas for specialists, generalists, and the skill pool. We compared the three behavioral options numerically by locating parameter combinations where each achieved the minimal $\Pr[\chi(\tau) \leq 2R]$. We set $D = \tau = 1$, and calculated z-scores across levels of R, α, f and γ; we let $D\tau = 2$ in some supporting calculations. Table 11.2 lists the range and granularity for each parameter. At any given parameter combination, we increased s_1 from 0.05 to 0.95 in increments of 0.05. For each s_1, expression (11.5), with $D = 1$, gives the corresponding s_2 as

$$s_2 = (1 - s_1{}^{\gamma})^{1/\gamma}. \tag{11.25}$$

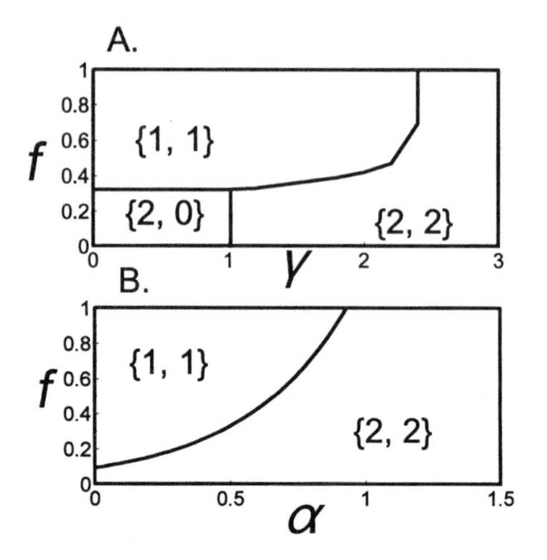

Figure 11.2 Strategy maximizing expected energy intake. f is the energy available in type-2 clumps; γ determines the shape of the constraint on each forager's two production rates. In (A) $\alpha = 0.4$, in (B) $\alpha = 1.3$. The skill pool, **G** = {1, 1}, can achieve the highest mean intake when social interference affects the other strategies (i.e., $\alpha < 1$), f is sufficiently large, and γ is sufficiently small. In the absence of social interference ($\alpha \geq 1$), the skill pool cannot maximize expected intake. Similar specialists, **G** = {2, 0}, can achieve the greatest mean when γ is small; for $\alpha < 1$, similar specialists are favored when f is small. Generalists, **G** = {2, 2}, achieve the greatest mean intake when γ is sufficiently large.

The best generalist, (s_1°, s_2°), minimized the z-score (among generalists) for that parameter combination. We then used that z-score to compare generalists with the other two behavioral options.

The results again fall naturally into two categories, according to whether $\gamma \leq 1$ or $\gamma > 1$. Above we noted that generalists cannot have the greatest mean energy intake when searching skills lack compatibility ($\gamma \leq 1$). In parallel, two generalists never have the lowest $\Pr[\chi(\tau) \leq 2R]$ when $\gamma \leq 1$. Either the skill pool or the paired-specialist group has the least chance of an energetic failure when searching skills are incompatible. The skill pool is the better of these options when $z(2R \mid \mathbf{G} = \{1, 1\}) < z(2R \mid \mathbf{G} = \{2, 0\})$, which implies, via expressions (11.20) and (11.13),

Table 11.2

Parameter values used in z-score calculations.

R: 0.25, 0.5, 1.0, 1.5		α: 0.4, 0.7, 1.0, 1.3
f: 0.1, 0.2, . . . , 0.9		γ: 0.1, 0.2, . . . , 3.0
	$D\tau = 1, 2$	

$$\frac{2R}{D\tau}\left[1 - \frac{(1 + f^2)^{1/2}}{2^{\alpha/2}}\right] < 1 + f - 2^{\alpha/2}(1 + f^2)^{1/2}. \tag{11.26}$$

Two predictions that parallel the analysis of expected intake levels are clear from expression (11.26). The likelihood that the skill pool will minimize the chance of an energetic failure increases when

1. Similar specialists experience social interference ($\alpha < 1$).
2. The energy value of type-2 clumps (f) is relatively large.

The z-score comparison produces a third prediction that cannot be deduced by comparing mean levels of energy consumption. Suppose $(1 + f^2) < 2^\alpha$, so that the left-hand-side of inequality (11.26) is positive. For given α, an increase in the ratio $(R/D\tau)$ can require a greater f to maintain the inequality. Similarly, an increase in α, given $(R/D\tau)$, requires a greater f if the skill pool is to be favored over similar specialists. The quantity $(R/D\tau)$ is simply the ratio of the individual physiological requirement to the expected number of food clumps produced by a member of a skill-pool, and we have the additional prediction:

3. As the individual's required energy intake increases relative to food-clump density, the minimal level of energy in type-2 clumps favoring the skill pool increases.

Figure 11.3A shows that (for fixed α) the skill pool's advantage over similar specialists requires a greater f as the physiological requirement R increases, but a smaller f is sufficient as $D\tau$ increases. Figure 11.3B shows qualitatively similar results, but for an increased α. The least f implying an advantage for the skill pool, for given R and $D\tau$, increases at the greater level of α (i.e., at a reduced level of social interference).

For the other category of z-score comparisons, $\gamma > 1$, so that skill compatibility within the individual is not so strong a constraint. Numerical calculations indicate that the skill pool can imply a lower probability of an energetic failure than a specialist or a generalist group when the energy value of type-2 clumps (f) is sufficiently large, and γ is not too large. When type-2 clumps provide little energy, similar specialists can be favored, especially at lower levels of γ. A sufficiently large γ, except at the lowest values of f, assures that paired generalists achieve the lowest $\Pr[\chi(\tau) \leq 2R]$. As γ increases, generalists must, at some point, gain an economic advantage.

Figure 11.4, A through C, shows that reduced social interference (i.e., increased α, for $\alpha \leq 1$) decreases the number of (γ, f) combinations where the skill pool is favored, and increases the parameter range where similar specialists do best. Figure 11.5, A through C, shows a qualitatively similar effect of increasing α. Comparison of figures 11.4 and 11.5 indicates that increasing the individual's physiological requirement R decreases the num-

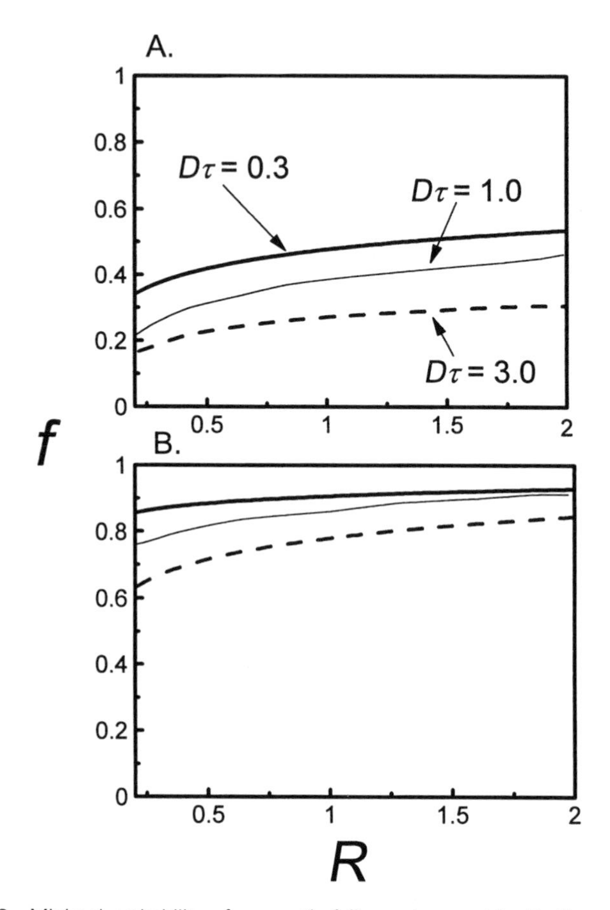

Figure 11.3 Minimal probability of energetic failure when $\gamma < 1$. f is the energy value of type-2 clumps; R is the individual's physiological requirement. $D\tau$ is the expected number of clumps a skill-pool member will produce. In (A) $\alpha = 0.4$, and in (B) $\alpha = 0.9$. Each curve, for its associated $D\tau$, divides the (R, f)-space into two regions. Above the curve, the skill pool induces a lower probability of an energetic failure than does the similar-specialist option. Below the curve, similar specialists have a lower chance of energetic failure. For a given α, the region where the skill pool is the better option increases as f increases, as $D\tau$ increases, and as R decreases. Increasing α reduces the (R, f) combinations where the skill pool is favored.

ber of (γ, f) combinations where the skill pool is favored, and increases the parameter range where generalists do best.

As indicated in this chapter's introductory comments, social interference may reduce the searching efficiency of similar specialists. Dissimilarity of searching skills may further constrain the efficiencies of generalists. The combination of these circumstances can render the skill pool's complemen-

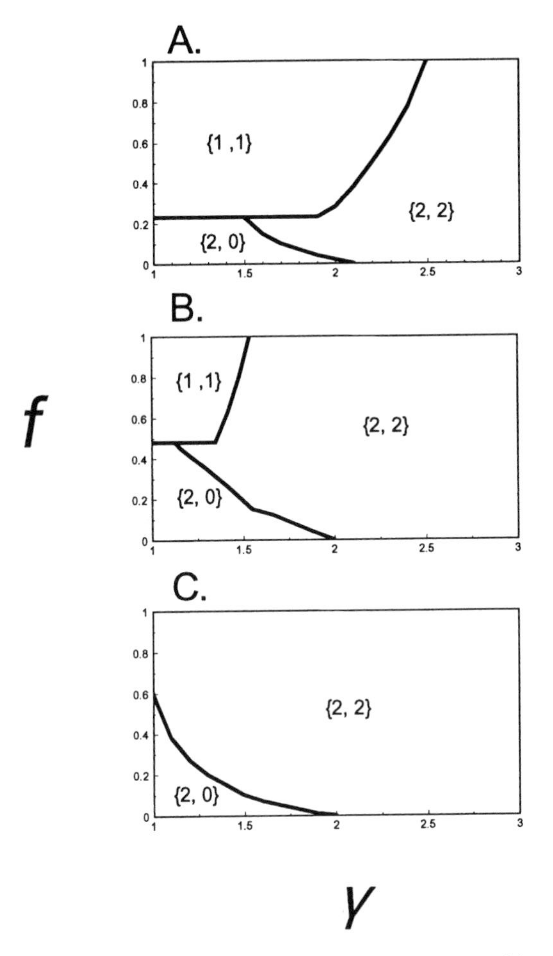

Figure 11.4 Minimal probability of an energetic failure; $\gamma > 1$ and low physiological requirement. f is the energy value of a type-2 clump, and γ determines the shape of the equality constraint on generalists' clump-discovery rates. In (A) through (C), $D = \tau = 1$, and $R = 0.25$. In (A), $\alpha = 0.4$; in (B), $\alpha = 0.7$; and in (C), $\alpha = 1.0$. For $\alpha < 1$ the skill pool, $\mathbf{G} = \{1, 1\}$, minimizes the probability of an energetic failure for sufficiently large f and sufficiently small γ. As α increases, the region where the skill pool is the best strategy diminishes. For $\alpha \geq 1$, the skill pool is not favored. Similar specialists can be favored when both f and γ are sufficiently small. Increasing γ eventually must favor generalists.

tary specializations the best foraging combination, especially when all food clumps are valuable and metabolic requirements are not too demanding (see Summary Box 11.1).

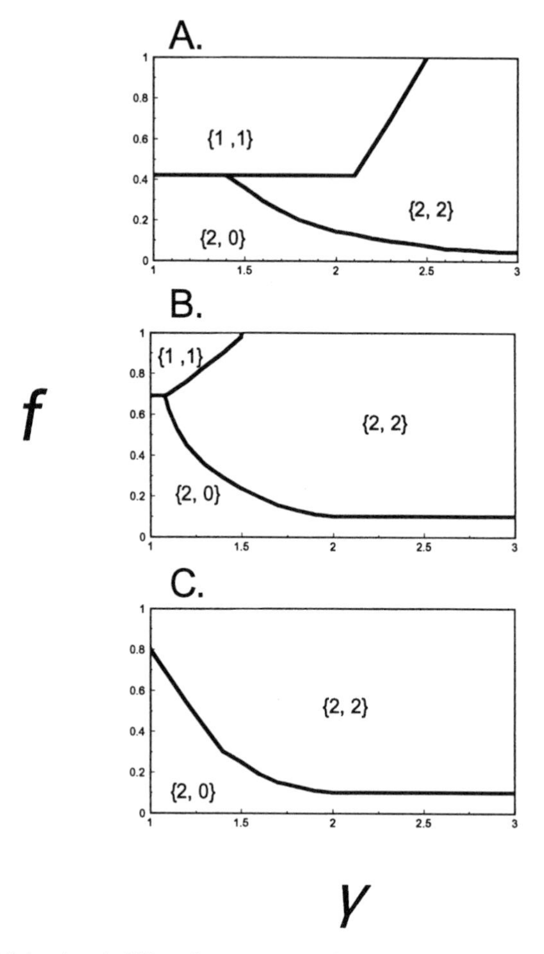

Figure 11.5 Minimal probability of an energetic failure; $\gamma > 1$ and high physiological requirement. In (A) through (C), $D = \tau = 1$, and $R = 1.5$. In (A), $\alpha = 0.4$; in (B), $\alpha = 0.7$; and in (C), $\alpha = 1.0$. Comparison with the previous figure shows that increasing the physiological requirement can decrease (and never increases) the fraction of the (γ, f) space where the skill pool minimizes the chance of an energetic failure.

11.3 A Skill Pool: Stochastic Dynamic Model

In this section, we apply stochastic dynamic programming to the comparison of a skill pool with both similar specialists and generalist foragers. Mangel and Clark (1986, 1988) and also Houston and McNamara (1986, 1988; see Houston et al. 1988) discuss the general significance of stochastic dynamic

SUMMARY BOX 11.1 THE STATIC, TWO-RESOURCE SKILL POOL MODEL

ASSUMPTIONS

Decision: Specialize on one resource, generalize or skill pool

Currency: Probability of an energetic failure

Constraints:

1. Each of two foragers searches for food; all clumps discovered are shared equally.
2. Only two clump types; type 1 is more profitable than type 2.
3. When searching for the same resource type, foragers find food at the same probabilistic rate.
4. A generalist's searching is subject to a compatibility (within-individual) constraint. A generalist's total rate of clump-discovery may be less than ($\gamma < 1$), equal to ($\gamma = 1$), or greater than ($\gamma > 1$) a specialist's rate of finding food clumps.
5. When foragers search for the same resource type, each individual's rate of clump discovery may be reduced by social interference ($\alpha < 1$), a between-individual constraint. Alternatively, each individual's searching may be independent of the other's efforts to find the same resource type ($\alpha = 1$), or their efforts might interact positively ($\alpha > 1$).

Predictions: Foraging as a skill pool can be favored when

1. Similar specialists and generalists experience sufficient social interference ($\alpha < 1$).
2. Energy in type-2 clumps is not too small.
3. Physiologically required energy intake is not too large, compared to energy in type-2 clumps.

optimization in evolutionary ecology. This type of model, which ordinarily requires numerical solution, predicts properties of sequences of behavior (rather than single actions). Dynamic models increase the potential for biological realism because an action at one time may influence future behavior through that action's effect on a physiological state variable correlated with fitness. For example, a foraging decision at one time may cause a change in the animal's energy-reserve level. The new physiological state, as well as the advance of time, might alter the individual's foraging economy, so that the animal adopts a different behavior at the next decision point. Hence, dynamic models stress feedback of one action on the next, with the dependence driven by a state variable governing the individual's survival or reproductive success.

As in section 11.2, the environment contains two types of clumps; $j = 1,2$. When a group of two foragers exploits a type-j clump, they divide the food equally, so that each forager's energy reserve increases by $Y_j/2$ units. Y_j varies randomly with mean $E[Y_j]$ and variance $V[Y_j]$. We assume $E[Y_1] > E[Y_2]$, following the static model's assumption that type-1 clumps are larger.

Time (t) advances discretely. The foragers have a total of τ time units to exploit food clumps. At each time $t = 0, 1, \ldots, (\tau - 1)$, the foragers decide to search as similar specialists, as generalists, or as a skill pool. The state variable of interest is χ(t), the two foragers' combined energy reserves (see McNamara et al. 1994). We assume the two individuals' reserves are equal at $t = 0$. Since food is divided equally thereafter, each individual's reserve is $\chi(t)/2$ at time t, given that the foragers have survived to that time.

The dynamic programming model requires that we define how the state variable $\chi(t + 1)$ depends on the preceding level χ(t). To do so, we first consider how within-individual and between-individual constraints govern clump-discovery probabilities. Then we specify how foraging choices and resource-clump characteristics affect the dynamics of the state variable. We assume a relationship between the state variable and the animals' starvation versus survival and analyze the model's properties numerically. Certain symbols used in the static model have an altered definition in the dynamic model; this results from the discrete-time structure of the latter analysis.

Resource Discovery Probabilities

During each time interval a group of two foragers either finds a single food clump or fails to discover food. We could directly convert the probabilistic rates assumed in section 11.2 to probabilities (see, e.g., Durrett and Levin 1994). However, we adopt a simpler approach that allows easier numerical computation.

$\lambda_j(G)$ is the probability a type-j clump is produced during one time interval by a group with composition $\mathbf{G} = \{G_1, G_2\}$; $0 \leq \lambda_j(G) < 1$ for $j = 1, 2$. As above, $\mathbf{G} = \{2, 0\}, \{1, 1\}$ or $\{2, 2\}$, and G_j counts the foragers searching for type-j clumps. The probability the group fails to produce a clump during a single time interval is $1 - [\lambda_1(G) + \lambda_2(G)]$. Note that $\lambda_2(G)$ is always zero when $\mathbf{G} = \{2, 0\}$; similar type-1 specialists never produce type-2 food clumps.

Searching-skill compatibility and social interaction can influence the $\lambda_j(G)$. To address these effects, we again begin by considering a solitary forager's between-resource constraint. Let s_j represent the probability that the individual finds a type-j clump during a single time interval. Obvious constraints apply: $0 \leq s_j < 1$, and $(s_1 + s_2) \leq 1$. The individual's discovery probabilities s_j also are constrained by the scalar-valued equality,

$$s_1{}^{\gamma} + s_2{}^{\gamma} = D^{\gamma};\ \gamma > 0.$$

Requiring $D \leq 1/2$ assures that the s_j can be taken as probabilities for any $\gamma > 0$.

As in the preceding model, skill compatibility increases with γ. Recall that $\gamma > 1$ implies $(s_1 + s_2) > D$, perhaps favoring generalists. But $(s_1 + s_2) < D$ when $\gamma < 1$, so that specialists or the skill pool are more likely favored when clump-discovery skills are incompatible. In our computations we set $D = 0.25$; we let $\gamma = 0.3$, and then $\gamma = 1.3$. As in the static model, the identical between-resource (within-individual) constraint applies when the animal forages socially.

Clump-discovery probabilities for similar specialists and for generalists (but not for the skill pool) are further influenced by social interference between individuals. G_j, $G_j \in \{0, 1, 2\}$, individuals search for a type-j clump. Given G_j, each individual's probability of discovering a type-j clump in a single time interval will be $s_j G_j^{\alpha-1}$; $0 < \alpha \leq 1$, and $i = 1, 2$. $\alpha < 1$ imposes social interference; social effects disappear when $\alpha = 1$. Since at most one clump is produced per time interval, the probability a type-j clump is discovered (i.e., by either forager) is

$$\lambda_j(G) = s_j G_j^{\alpha}, \tag{11.27}$$

whether $G_j = 0$, 1 or 2. We use two levels of α, 0.5 and 1.0, in our calculations.

Collecting assumptions, we have the resource-production probabilities $[\lambda_1(G)\ \lambda_2(G)]$ for the various strategies. The probabilities are independent of both time and the state variable $\chi(t)$.

When both foragers specialize on searching for type-1 clumps, $\mathbf{G} = \{2, 0\}$ and the $\lambda_j(G)$ are

$$[D2^{\alpha}\ 0] = [(0.25)2^{\alpha}\ 0]. \tag{11.28}$$

For the skill pool $\mathbf{G} = \{1, 1\}$, and the $\lambda_j(G)$ are

$$[D\ D] = [0.25\ 0.25]. \tag{11.29}$$

Specialists cannot expect to find more clumps than the skill pool. But specialists may have an economic advantage since we assume $\mathrm{E}[Y_1] > \mathrm{E}[Y_2]$.

When both foragers generalize, $\mathbf{G} = \{2, 2\}$ and the $\lambda_j(G)$ are

$$[s_1{}^{\gamma}2^{\alpha}\ s_2{}^{\gamma}2^{\alpha}] = [s_1{}^{\gamma}2^{\alpha}\ (0.25^{\gamma} - s_1{}^{\gamma})2^{\alpha}]. \tag{11.30}$$

In the static model we located the best generalist for each parameter combination before comparing the various z-scores. To limit the computational effort spent analyzing the dynamic model, we define two generalist strategies. When $\gamma < 1$, we consider a single generalist strategy, where

$$s_2/s_1 = \mathrm{E}[Y_2]/\mathrm{E}[Y_1], \tag{11.31}$$

with the s_j subject to applicable constraints. We refer to this case as "matching generalists," since each forager's clump-discovery ratio (s_2/s_1) matches the ratio of expected rewards (see Houston and McNamara 1981; Staddon and Ettinger 1989).

When $\gamma > 1$, we also apply the marginal rate of substitution of s_1 for s_2 and formulate a second generalist strategy. We redefine f as $\mathrm{E}[Y_2]/\mathrm{E}[Y_1]$; equation (11.27) then implies

$$s_2/s_1 = (\mathrm{E}[Y_2]/\mathrm{E}[Y_1])^{(1/\gamma-1)} = (\mathrm{E}[Y_2]/E[Y_1])^{3.33}, \tag{11.32}$$

where the exponent takes its value because $\gamma = 1.3$ in our calculations. We refer to this case as "marginal-rate generalists," noting that expression (11.32) requires $\gamma > 1$, where the skill-compatibility constraint is not too severe.

Dynamics of the State Variable

The state variable $\chi(t)$ sums the two foragers' (equal) energy reserves. We could define the state variable as the individual's level of reserves, but our assumptions assure that the results would be identical. Otherwise, our state-variable dynamics resemble the energy-reserve models of Houston and McNamara (1986), Mangel and Clark (1986, 1988), and Newman (1991).

We constrain values of the state variable by assuming a critical lower level and a maximal energy reserve

$$0 \leq \chi(t) \leq \chi_c. \tag{11.33}$$

Suppose that $\chi(t)$, for $t < \tau$, can fall to a nonpositive value. Then we set $\chi(t) = 0$, and both foragers starve since their energy reserve has reached the lower critical level. If $\chi(t)$ should exceed χ_c for any t, we set $\chi(t) = \chi_c$. We might choose the maximal reserve χ_c sufficiently large that it cannot be attained during the τ foraging periods; doing so assures that all food produced is consumed (Mangel and Clark 1988). When we write the dynamics of $\chi(t)$, we combine these two constraints with a chop function (Mangel and Clark 1986; Newman 1991):

$$\mathrm{chop}(\chi(t);0,\chi_c) = \begin{matrix} \chi_c & \text{if } \chi(t) > \chi_c \\ \chi(t) & \text{if } 0 \leq \chi(t) \leq \chi_c \\ 0 & \text{if } \chi(t) < 0. \end{matrix} \tag{11.34}$$

Essentially, neither forager's energy reserve can exceed a common, maximal physiological limit. Nor can energy reserves fall below a lower lethal boundary (Houston and McNamara 1986).

At the termination of the foraging process, the animals must have accumulated enough energy to survive the subsequent nonforaging period. The individual physiological requirement is R, so that both individuals starve if the state variable fails to exceed $2R$ at the end of the foraging period. So, starvation occurs if $\chi(t) = 0$ for $t < \tau$. Given that the foragers survive to time τ, starvation occurs if $\chi(\tau) \leq 2R$.

For each strategy we have identified the probability of producing a type-j clump and, by complementarity, the probability of not finding any food. The next step is specifying the consequences of these events.

Independently of whether or not a food clump is produced during a given interval, each forager incurs a metabolic cost ρ_m. The metabolic cost is scaled in the same units as food energy. ρ_m depends on neither choice of strategy nor energy-reserve level (see Mangel 1990). Therefore, the dynamics of the state variable include a decrement of $2\rho_m$ during each interval.

Production of a type-j clump increases $\chi(t)$ by Y_j units; Y_j has realizations y_j. In our calculations we restrict Y_j to the integers 2, 3, and 4. As indicated above, we hold $E[Y_1] > E[Y_2]$.

At this point we can write a general statement describing the dynamics of the state variable. Suppose $\chi(t) = X$ for any $t < \tau$. Then

$$\chi(t + 1) = \text{chop}(X - \rho_m + y_1; 0,\chi_c), \text{ with Pr: } \lambda_1(G)\, Pr[Y_1 = y_1]$$

$$\chi(t + 1) = \text{chop}(X - \rho_m + y_2; 0,\chi_c), \text{ with Pr: } \lambda_2(G)\, Pr[Y_2 = y_2]$$

$$\chi(t + 1) = \text{chop}(X - \rho_m; 0,\chi_c), \text{ with Pr: } [1 - \lambda_1(G) - \lambda_2(G)].$$

Next we connect the dynamics of the state variable with our currency of fitness (survival). Let $\Omega(X,t,\tau)$ represent the maximal attainable probability that both foragers survive from time t to the final time τ and meet their physiological requirement, given that they have survived to the beginning of the tth and the energy reserve is $\chi(t) = X$. That is, $\Omega(\cdot)$ is the survival probability, given the current state and time, under the assumption of optimal behavior. Consider Ω at the final time $t = \tau$. From above,

$$\Omega(X,\tau,\tau) = \begin{cases} 0 \text{ if } X \leq 2R \\ 1 \text{ if } X > 2R, \end{cases} \tag{11.35}$$

so that the physiological requirement defines $\Omega(X,\tau,\tau)$. One interval earlier the likelihood of survival, given that $\chi(\tau - 1) > 0$, depends on the probability that $\chi(\tau) > 2R$. This probability depends, in turn, on ρ_m, the various $\lambda_j(G)$, and the $\Pr[Y_j = y_j]$, as well as the strategy G at $t = (\tau - 1)$. Then, for $\chi(\tau - 1) > 0$,

$$\begin{aligned} \Omega(X,\tau - 1,\tau) = \max_G \{&\lambda_1(G)\, \Sigma_y \Pr[Y_1 = y_1]\, \Omega(X_1,\tau,\tau) \\ &+ \lambda_2(G)\, \Sigma_y \Pr[Y_2 = y_2]\, \Omega(X_2,\tau,\tau) \\ &+ [1 - \lambda_1(G) - \lambda_2(G)]\, \Omega(X_0,\tau,\tau)\} \end{aligned} \tag{11.36a}$$

and

$$\Omega(0,\tau - 1,\tau) = 0, \tag{11.36b}$$

where

$$X_1 = \text{chop}(X - \rho_m + y_1; 0,\chi_c) \tag{11.37a}$$

$$X_2 = \text{chop}(X - \rho_m + y_2; 0,\chi_c) \tag{11.37b}$$

and

$$X_0 = \text{chop}(X - \rho_m; 0,\chi_c). \tag{11.37c}$$

Once the $\Pr[Y_j = y_j]$ are specified, $\Omega(X,\tau - 1,\tau)$ can be calculated for each feasible energy-reserve level. The resulting values immediately allow calculation of $\Omega(X,\tau - 2,\tau)$, and so forth. In a general form, the functional equation for our dynamic programming problem is

$$\begin{aligned}\Omega(X,t,\tau) = \max_G \{&\lambda_1(G) \Sigma_y \Pr[Y_1 = y_1]\, \Omega(X_1,t + 1,\tau) \\ &+ \lambda_2(G) \Sigma_y \Pr[Y_2 = y_2]\, \Omega(X_2,t + 1,\tau) \\ &+ [1 - \lambda_1(G) - \lambda_2(G)]\, \Omega(X_0,t + 1,\tau)\}\end{aligned} \tag{11.38a}$$

and

$$\Omega(0,t,\tau) = 0. \tag{11.38b}$$

Maximization, of course, occurs over strategies $\mathbf{G} = \{G_1, G_2\}$.

Maximizing the Probability of Survival

We numerically evaluated our dynamic model to assess the generality of the static model's predictions, and to identify new predictions. In our dynamic model, surviving the foraging process and the subsequent nonforaging period requires meeting separate criteria for averting starvation (e.g., Stephens 1981). Energy reserves must remain above the lethal boundary for $t < \tau$, and energy accumulation by the final time τ must exceed the physiological requirement R (see Houston and McNamara 1986).

In our calculations we fixed the final time τ at 10, the maximal individual energy reserve χ_c at 10, and we kept the individual metabolic expenditure per period at 0.5. In half our calculations we let the physiological requirement $R = 2$, and in the other half we set $R = 6$. We used the same reward probabilities for type-1 clumps in all calculations. Given production of a type-1 clump, the foragers' combined energy intake Y_1 advances according to

$$\Pr[Y_1 = 3] = 0.5, \Pr[Y_1 = 4] = 0.5.$$

Consequently, $E[Y_1] = 3.5$ and $V[Y_1] = 0.25$ in all calculations.

In most calculations we set $E[Y_2] = 3.0$. However, to examine effects of varying f (the relative value of type-2 clumps), we sometimes set $E[Y_2] = 2.0$. For either level of $E[Y_2]$, type-2 clumps were on average less rewarding than type-1 clumps.

For $E[Y_2] = 3.0$, we examined two levels of variance in Y_2. When $V[Y_2] > V[Y_1]$, Y_2 was distributed according to

$$\Pr[Y_2 = 2] = 0.5, \Pr[Y_2 = 4] = 0.5,$$

so that $V[Y_2] = 1.0 > V[Y_1]$. When type-2 clumps were the less variable type, the reward in a type-2 clump was 3.0 with certainty ($V[Y_2] = 0$). In calculations where we reduced $E[Y_2]$ to 2.0, we set $Y_2 = 2.0$ with certainty.

Social Interference and Incompatible Skills

When $\alpha = 0.5$, social interference inhibits both the type-1 specialists' and generalists' clump-production probabilities, but does not affect the skill pool. Setting $\gamma = 0.3$ further limits generalists since increasingly effective use of one skill can sharply decrease the chance of finding food via the other skill. Therefore, in this series of calculations (α and $\gamma < 1$), the skill pool should have an economic advantage.

Figure 11.6 (A and B) shows results from this series of calculations at two levels of the requirement R. For the parameter values selected, no alternative strategy is ever superior to the skill pool. Either the skill pool alone is the best choice, or all four strategies are equivalent. Increasing the physiological requirement (fig. 11.6B) increases the range of $(\chi(t),t)$ combinations where the skill pool is optimal. But further increments in R reverse this trend, but only because survival becomes impossible under any strategy.

Figure 11.7 (A and B) allows two comparisons within the first series of calculations. Comparing 11.7A with 11.6B indicates that increasing the variance of type-2 resource clumps produces only a minimal effect. Specialization is favored at a few $(\chi(t),t)$ combinations at the lower variance level; increasing the variance favors the skill pool in place of specialization. Second, comparing 11.7A with 11.7B indicates that decreasing the expected reward in type-2 clumps (hence decreasing f) moderately increases the number of $(\chi(t),t)$ combinations where specialization will be favored over the skill pool. Lower reserve levels, implying lower survival probabilities, sometimes can favor increased specialization on type-1 clumps as $(E[Y_2]/E[Y_1])$ decreases.

Results of the first series of calculations resemble predictions of the static model. Social interference between foragers ($\alpha < 1$) and incompatible skills ($\gamma < 1$) combine to favor the skill pool. Increasing the expected reward in type-2 clumps tends to favor the skill pool over type-1 specialists. In the dynamic model, varying the final requirement R does not induce a simple

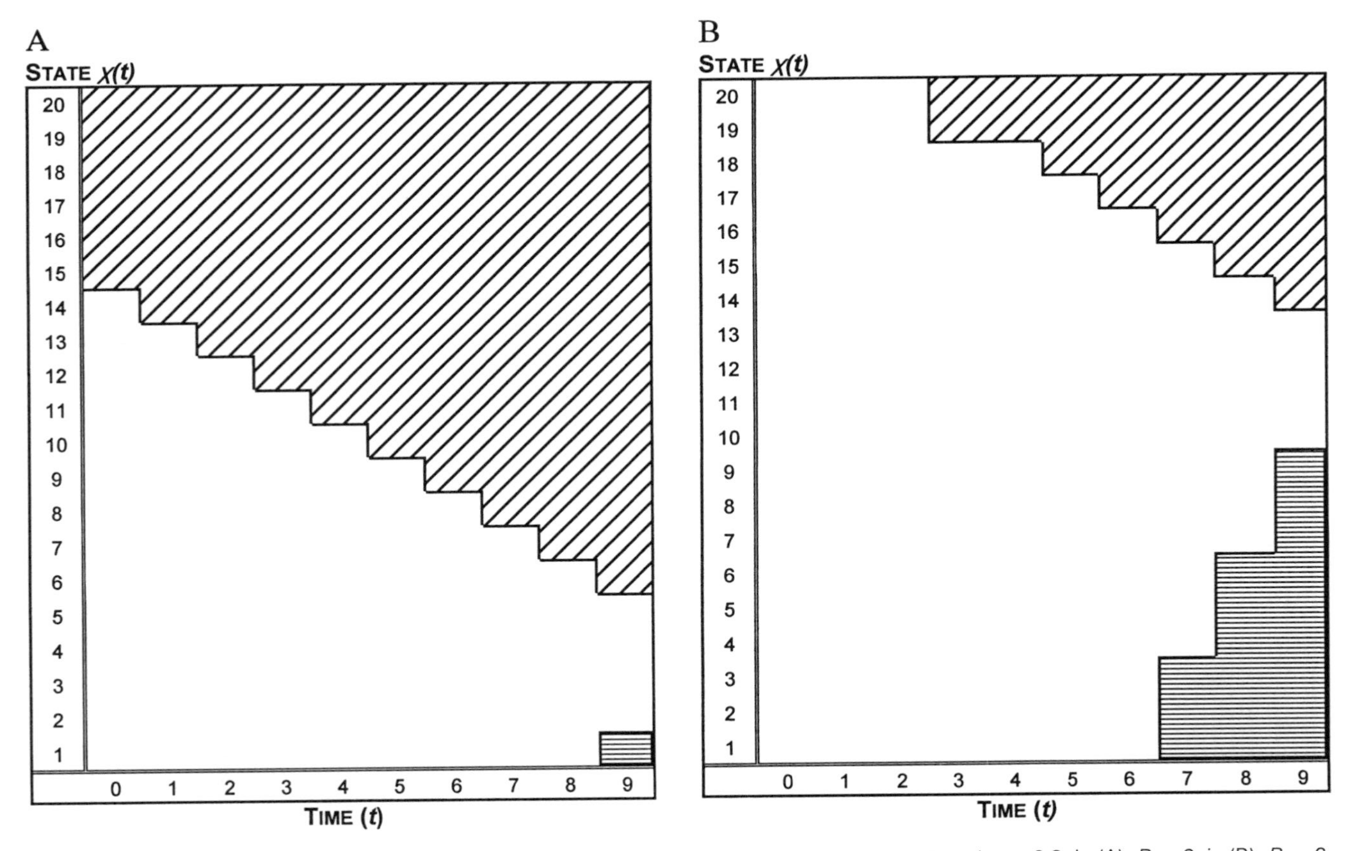

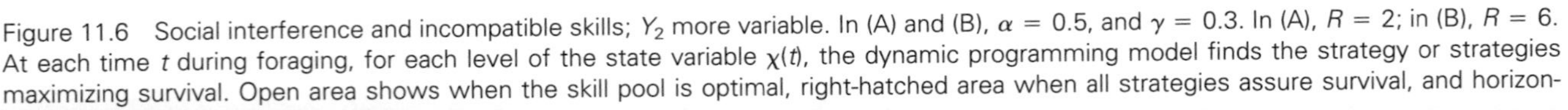
Figure 11.6 Social interference and incompatible skills; Y_2 more variable. In (A) and (B), $\alpha = 0.5$, and $\gamma = 0.3$. In (A), $R = 2$; in (B), $R = 6$. At each time t during foraging, for each level of the state variable $\chi(t)$, the dynamic programming model finds the strategy or strategies maximizing survival. Open area shows when the skill pool is optimal, right-hatched area when all strategies assure survival, and horizon-

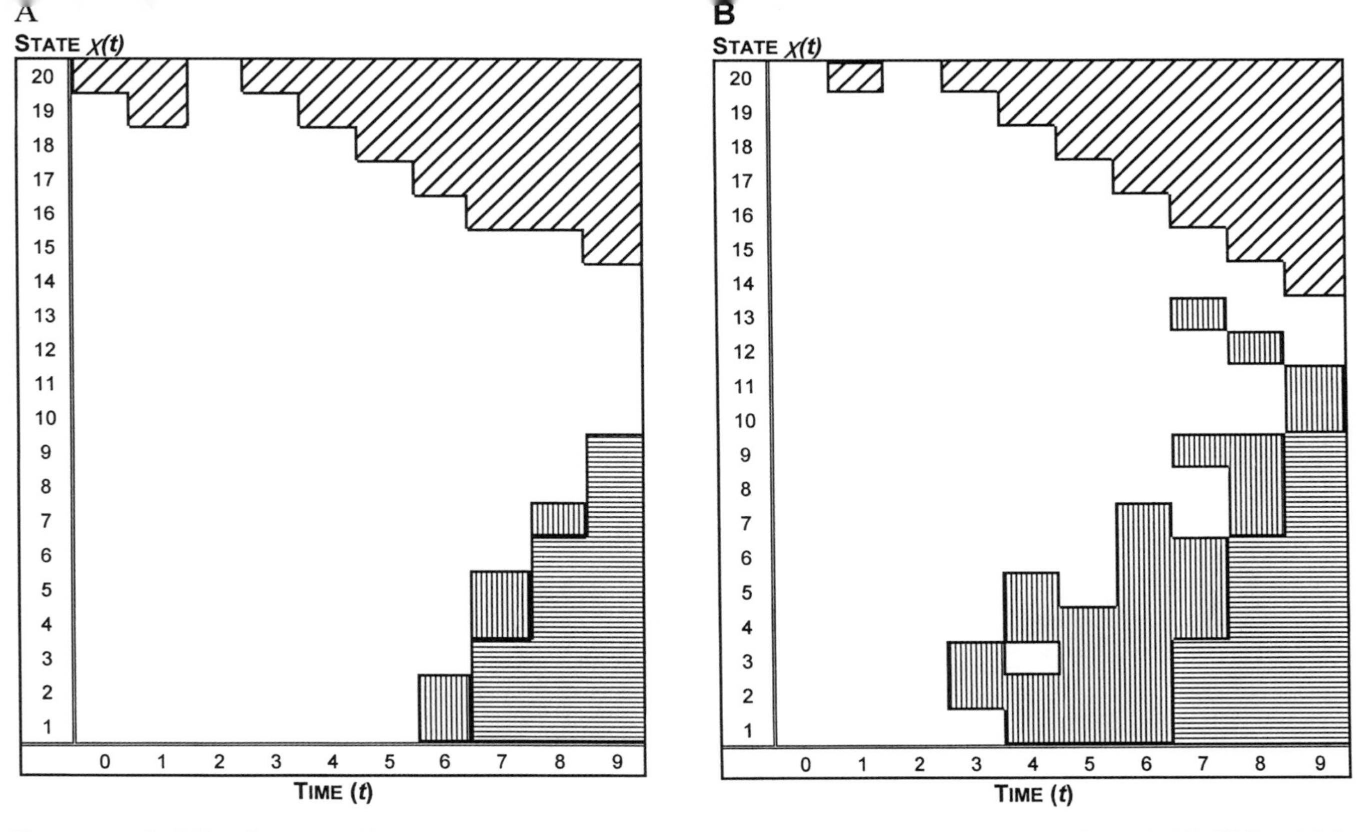

Figure 11.7 Social interference and incompatible skills; Y_2 less variable. In (A) and (B), $\alpha = 0.5$, $\gamma = 0.3$, and $R = 6$. In (A), $E[Y_2] = 3.0$; in (B), $E[Y_2] = 2.0$. Vertically hatched areas show when specialization is optimal. Other areas as in figure 11.6.

effect, since values of the expected terminal reward $\Omega(\chi,t,\tau)$ decrease rapidly for all strategies at lower reserve levels.

Incompatible Skills without Social Interference

When $\alpha = 1.0$, a forager's probability of finding a type-j clump is independent of whether or not the other individual searches for the same type of resource. Setting $\gamma = 0.3$ retains the within-individual skill incompatibility assumed above. We used these parameter values in the second series of calculations. Since $E[Y_1] > E[Y_2]$, type-1 specialists should have an economic advantage when skills are incompatible, but clump discovery is not affected by social interference.

Figure 11.8 (A and B) shows results from the second series of calculations. At lower reserve levels, progressively lower as time advances, specializing on type-1 clumps is the sole best strategy. Increasing the requirement R (fig. 11.8B) increases the range of $(\chi(t),t)$ combinations where specialization is superior to any alternative considered. In both figure 11.8A and B, the skill pool and specialization are sometimes equivalent and superior to generalization (at higher reserve levels). The equivalence results from the simple $\Pr[Y_i]$ we employed. But the dynamic model, in contrast to the static model, does suggest that the skill pool may remain an efficient (although suboptimal) strategy when foragers' clump-discovery probabilities are not affected by social interference (i.e., when $\alpha = 1$).

Figure 11.9 (A and B) indicates that decreasing the expected reward in type-2 clumps increases the number of $(\chi(t),t)$ combinations where specialization is favored. Even at the lower level of f (fig. 11.9B), however, the skill pool is not always inferior to specialization.

Results of the second series of calculations resemble predictions of the static model. Social independence ($\alpha = 1$) and incompatible skills ($\gamma < 1$) combine to favor specialization on type-1 clumps, especially when the expected reward in type-2 clumps is reduced.

Social Interference and Compatible Skills

In the third series of calculations we let $\alpha = 0.5$, so that social interference inhibits specialists' and generalists' clump production. We set $\gamma = 1.3$ to ask if generalists would benefit, comparatively, from enhancement of their searching skills. We found that the effect of social interference dominated the relaxation of the within-individual constraint. When $V[Y_2] = 1.0$, the optimal strategies are identical to those in the first series, for both $R = 2$ and $R = 6$ (as in fig. 11.6). Furthermore, when $V[Y_2] = 0$ and $R = 6$, the third series' results were nearly identical to those in the first series (as in fig. 11.7). Hence, for the parameter values we examined, reducing the within-

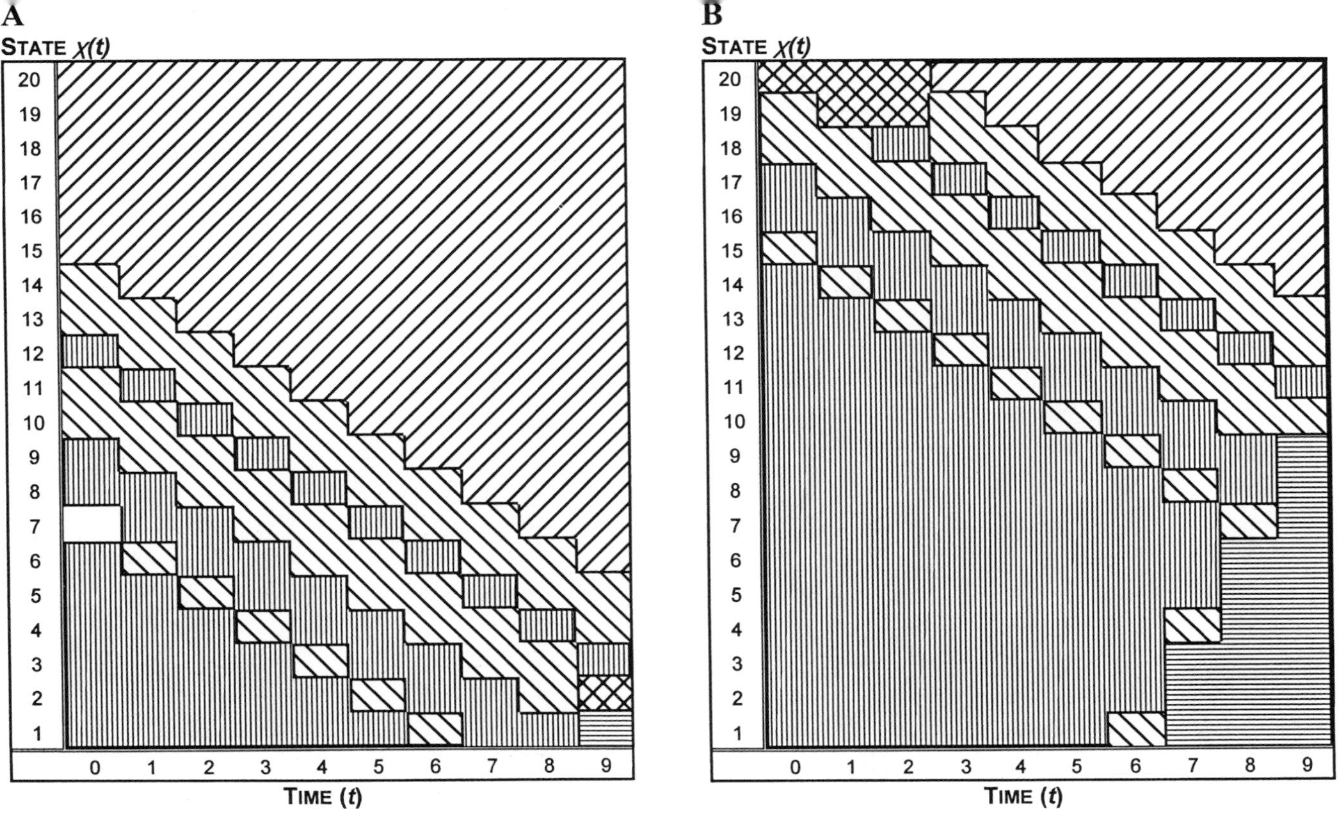

Figure 11.8 Social independence and incompatible skills; Y_2 more variable. In (A) and (B), $\alpha = 1.0$, and $\gamma = 0.3$. In (A), $R = 2$; in (B), $R = 6$. Left-hatched areas show when the skill pool and specialization are both optimal. Cross-hatched areas indicate when all four strategies have the same, positive survival probability. Other area patterns as in figure 11.6.

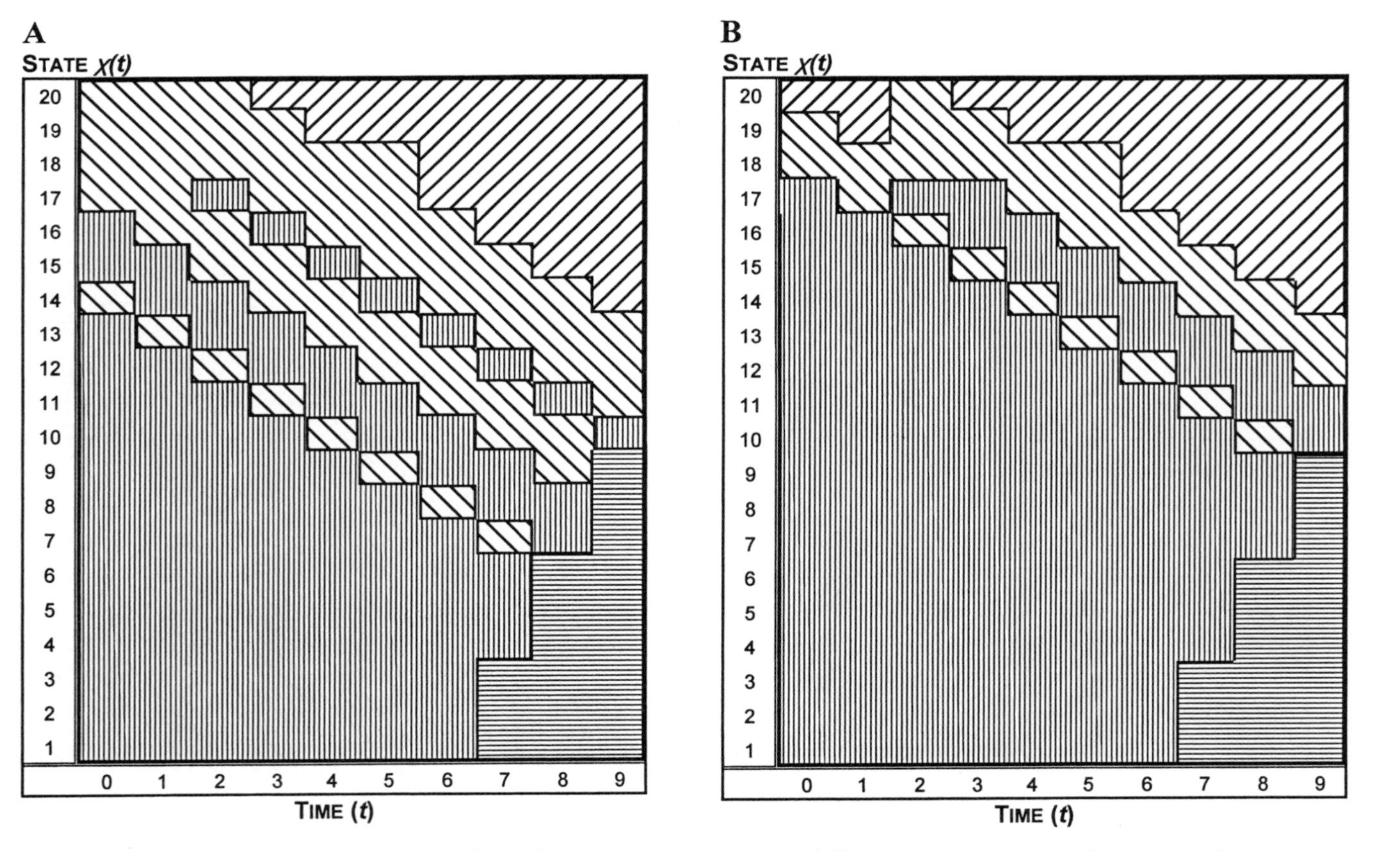

Figure 11.9 Social independence and incompatible skills; Y_2 less variable. In (A) and (B), $\alpha = 1.0$, $\gamma = 0.3$, and $R = 6$. In (A) $E[Y_2] = 3.0$; in

individual constraint (γ) did not affect the results under social interference ($\alpha = 0.5$).

Skill Enhancement without Social Interference

In the fourth and last series, we set $\alpha = 1.0$, and $\gamma = 1.3$. Under these conditions, i.e., social independence between foragers and compatible searching skills, the static model usually favored paired generalists. However, if type-2 clumps were insufficiently rewarding, the static model could favor specialists under these conditions.

Figure 11.10 (A and B) shows results of the fourth series. When $R = 2$ (fig. 11.10A), either matching generalists or marginal-rate generalists are favored across lower reserve levels. For sufficiently larger reserves, all strategies become equivalent. Increasing the physiological requirement to $R = 6$ (fig. 11.10B) makes marginal-rate generalists the sole optimal strategy at lower reserves and earlier times.

Figure 11.11 (A and B) offers two comparisons. First, comparing 11.11A with 11.10B shows that increasing the variance of type-2 clumps increases the likelihood that marginal-rate generalists will be favored over matching generalists. The former strategy relies more on type-1 clumps, which provide a greater mean and less variable reward. Second, comparing 11.11A with 11.11B indicates that decreasing the expected reward in type-2 clumps, not surprisingly, increases the frequency at which marginal-rate generalists are favored over matching generalists (and the other available alternatives).

In the fourth series of calculations, the skill pool never was the sole optimal strategy, and specialization was only rarely superior to generalization. Removing social interference between foragers (i.e., setting $\alpha = 1$) and assuming highly compatible searching skills ($\gamma > 1$) favors generalization, even more so than in the static model.

11.4 Conclusions

Our skill-pool economy assumes opportunistic foragers. Each searches for divisible resource clumps, and both feed whenever one finds food. The behavioral assumptions, including competitive symmetry, permit comparison of the various strategies. Our objective was demonstrating that within-individual constraints, as well as between-individual social effects, on a forager's ability to find two resources might influence the distribution of foraging skills among a group's members.

In the static model the best strategy minimizes the probability of an energetic failure at the end of the foraging period. Incompatible skills ($\gamma < 1$)

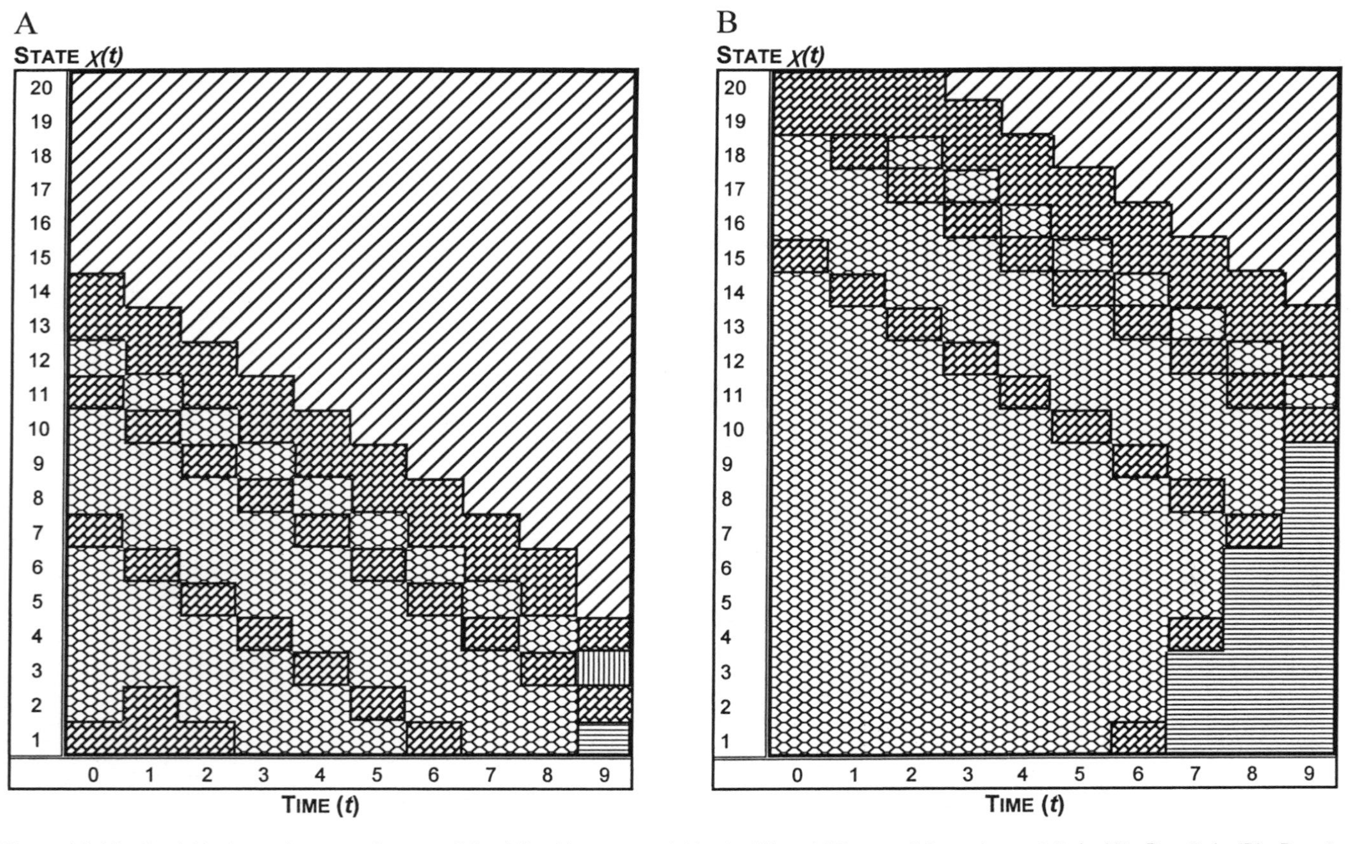

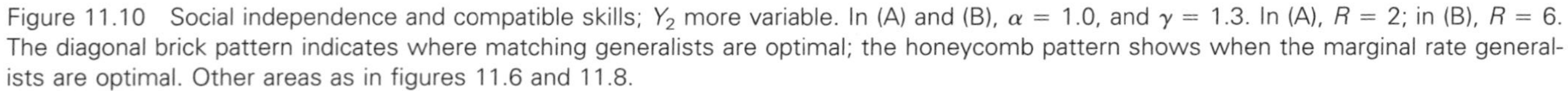

Figure 11.10 Social independence and compatible skills; Y_2 more variable. In (A) and (B), $\alpha = 1.0$, and $\gamma = 1.3$. In (A), $R = 2$; in (B), $R = 6$. The diagonal brick pattern indicates where matching generalists are optimal; the honeycomb pattern shows when the marginal rate generalists are optimal. Other areas as in figures 11.6 and 11.8.

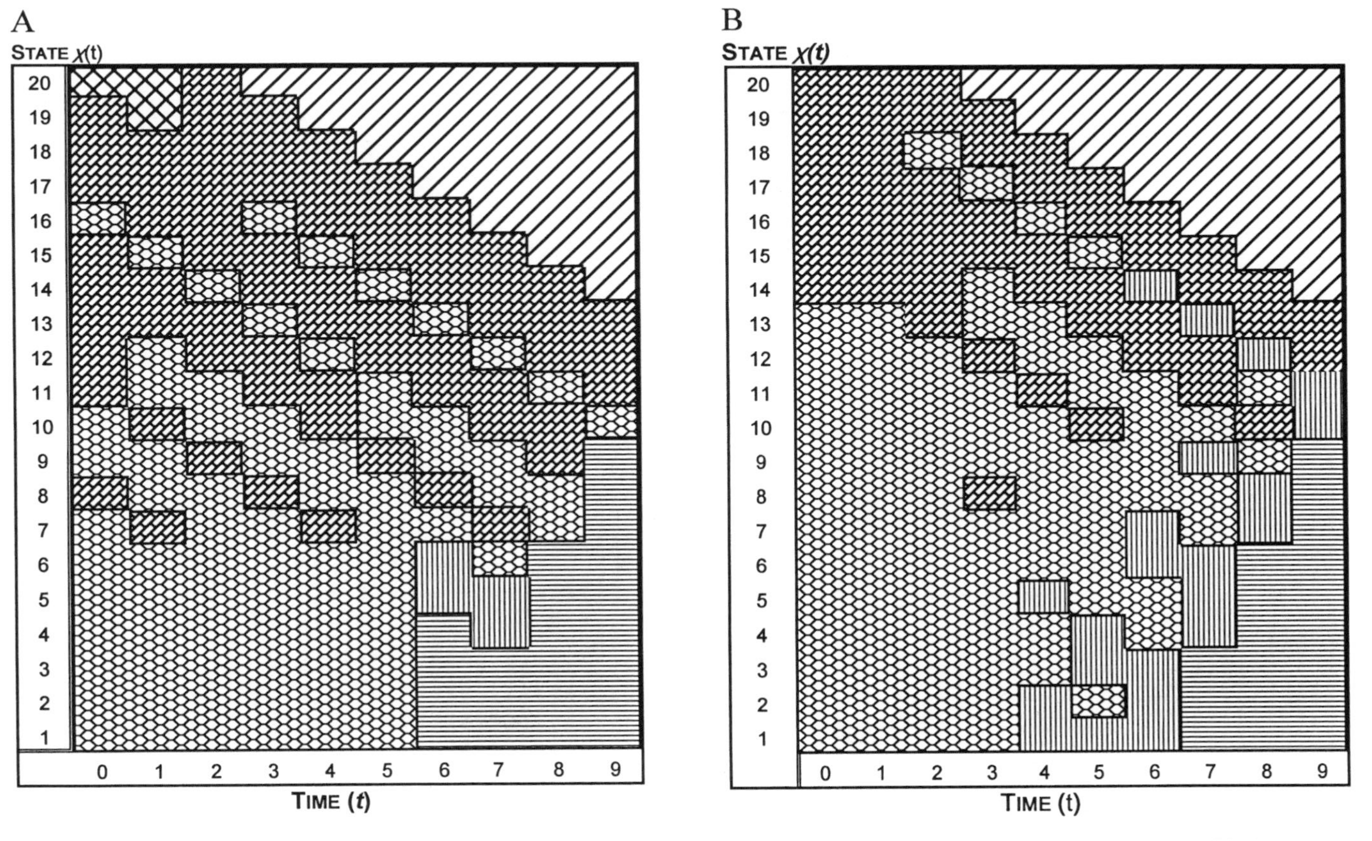

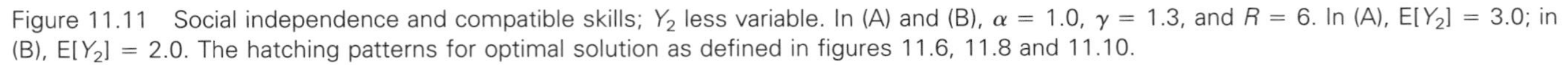

Figure 11.11 Social independence and compatible skills; Y_2 less variable. In (A) and (B), $\alpha = 1.0$, $\gamma = 1.3$, and $R = 6$. In (A), $E[Y_2] = 3.0$; in (B), $E[Y_2] = 2.0$. The hatching patterns for optimal solution as defined in figures 11.6, 11.8 and 11.10.

imply that generalists are necessarily inferior to one or both alternatives. For this case, the skill pool should be favored when

1. Social interference ($\alpha < 1$) inhibits specialists' clump discovery.
2. Type-2 (i.e., smaller) clumps yield sufficient energy.

Increasing the physiological requirement decreases the likelihood that the skill pool is favored, and removing social interference ($\alpha \geq 1$) renders the skill pool necessarily inferior to specialization.

Skill compatibility ($\gamma > 1$) can provide generalists an economic advantage in the static model. In this case, the skill pool may be favored by the combination of

1. Social interference
2. Relatively large type-2 clumps
3. γ not too large

Removing social interference ($\alpha \geq 1$) renders the skill pool necessarily inferior to one or both alternatives. Small type-2 clumps can favor specialization if γ is not too large. But as the within-individual constraint on clump production is relaxed (i.e., γ is increased), generalists become the most efficient foragers.

In the dynamic programming model, the best strategy maximizes the probability that the individual avoids starvation while accumulating enough energy to meet a postforaging physiological requirement. For the parameter values we examined, combining social interference ($\alpha = 0.5$) and skill incompatibility ($\gamma = 0.3$) strongly favors the skill pool. Increasing γ to 1.3, while retaining social interference, produces the same prediction; the skill pool is strongly favored. Removing social interference (letting $\alpha = 1$), while retaining skill incompatibility, favors specialists, at least when energy reserves are low. Finally, combining social independence ($\alpha = 1$) and skill enhancement ($\gamma = 1.3$) favors generalists. The dynamic programming model's predictions generally match predictions of the static model. Perhaps the most significant prediction added by the dynamic model is that lower energy reserves tend to favor specialization. When this occurs, an increase in reserve level (the state variable) may favor the skill pool.

Comparison to Solitary Foraging

Like several models in this book, our skill pool analysis fixes group size and predicts patterns in diversity among group members. We might suppose that a solitary's predation hazard is sufficient to promote group foraging. After groups of two (or more) form, individual survival might depend primarily on averting starvation, so that clump-discovery strategies govern survival probabilities. However, Giraldeau's (1984) original hypothesis suggests that we compare a solitary and a skill pool member strictly as alternate foraging options. The comparison requires a static analysis, where each forager retains the same strategy throughout the time available for feeding.

A solitary either specializes on type-1 clumps or generalizes. For $\gamma > 1$, we compare both a solitary specialist and solitary generalist with a skill pool member. For $\gamma < 1$, we exclude the generalist because of the within-individual inefficiency imposed by the strong skill incompatibility. We let $D = 1$ for simplicity. Between-individual effects of social interference (the level of α) neither apply to the skill pool nor, of course, to solitary foragers.

The skill pool encounters clumps as described in section 11.2. Then a member of a skill pool has expected intake $E[\chi(\tau)/2] = \tau(1 + f)/2$; see expression (11.18). The z-score associated with this individual's probability of an energetic failure must be identical to expression (11.20), with $D = 1$:

$$z(R) = [2R - \tau(1 + f)]/[\tau(1 + f^2)]^{1/2}. \tag{11.39}$$

The probability that a member of a skill pool obtains no food during the available foraging time is $\Pr[\chi(\tau) = 0] = e^{-2\tau}$.

A solitary specialist's expected intake is simply τ. The associated z-score is

$$z(R) = (R - \tau)/\tau^{1/2}. \tag{11.40}$$

The probability a solitary type-1 specialist finds no food is $e^{-\tau}$.

A solitary generalist searches for both types of clumps simultaneously, and the within-individual constraint governs the total discovery rate. We assume the generalist's clump-discovery rates are given by expressions (11.22) and (11.23), a "marginal rate generalist." Then the generalist's expected energy intake is $\tau(s_1^* + fs_2^*)$. The associated z-score is

$$z(R) = \frac{R - \tau\,(s_1^* + fs_2^*)}{[\tau\,(s_1^* + f^2 s_2^*)]^{1/2}}. \tag{11.41}$$

The generalist's probability of finding no food is $\exp\{-[\tau(s_1^* + s_2^*)]$, where $(s_1^*)^\gamma + (s_2^*)^\gamma = 1$.

A skill pool member cannot have an expected intake greater than the mean of one or both solitary strategies. The mean intake for a skill pool member is less than a solitary specialist's mean, since $\tau(1 + f) < 2\tau$ whenever $f < 1$. When $\gamma > 1$, the generalist is less constrained by skill incompatibility, and the mean intake of a skill pool member is also less than a solitary generalist's mean. The skill pool expectation is the average of two specialists (type 1 and type 2), each of which is inferior (when $\gamma > 1$) to the marginal rate generalist. Hence, one or both of the solitary strategies will always have an expected intake greater than the mean for a skill pool member.

A skill pool member's chance of an energetic failure can be lower than a solitary's failure probability. A skill pool member has a lower probability of failure than does a solitary specialist whenever

$$R - \tau f < (1 + f^2)^{1/2} \qquad (11.42)$$

from expressions (11.39) and (11.40). Not surprisingly, the skill pool does better if (R/τ) is small (i.e., when the physiological requirement can easily be met) and type-2 clumps are relatively large. If $R = \tau$, a solitary specialist has (approximately) equal probabilities of success and failure. But a skill pool member's failure probability must exceed 1/2, since $2R > \tau(1 + f)$ when $R = \tau$. That is, $z(R)$ must exceed 0 for the skill pool; see expression (11.39). Further analysis reveals that a skill pool member has a lower failure probability than a solitary specialist only when skill-pool membership provides a positive expected energy budget.

When $\gamma > 1$, a solitary generalist has a greater mean intake than a skill pool member since skills are more compatible. But each skill pool member can have a lower chance of an energetic failure than the generalist when f is sufficiently large, γ not too large, and (again) the skill pool provides a positive expected energy budget. Figure 11.12 shows how varying γ and f influences whether the skill pool or solitary foraging achieves the lower failure probability when the physiological requirement is not too severe.

A member of a skill pool always has a greater chance of obtaining some food than does a solitary. A skill pool fails to produce any food with probability $e^{-2\tau}$. A solitary specialist fails to discover any food with probability $e^{-\tau}$; the skill pool's advantage in this context is clear. The skill pool has a similar advantage over a solitary generalist if $e^{-2\tau} < \exp[-\tau(s_1^* + s_2^*)]$, hence when $(s_1^* + s_2^*) < 2$. This must always be true, since $[(s_1^*)^\gamma + (s_2^*)^\gamma] = 1$; the same result holds for any value of D.

Summarizing, a skill pool member's average energy intake will not exceed the better solitary's expected intake. However, a skill pool member, as a group forager, has a lesser chance of going without food. More importantly, a skill pool member's probability of energetic failure can be lower than a solitary's failure probability when the required intake level is not too severe, provided that type-2 clumps are sufficiently large and γ is sufficiently small.

Skill Pool Formation

The diversified specializations that compose a skill pool might be assembled in two ways. Individuals might acquire specializations as juveniles and later sort into groups with quite large resource-production differences among individuals (Krebs 1973). Alternatively, foraging groups might form first, and then individual specializations could be expressed in a pattern sufficiently diverse to avoid extensive social interference during resource acquisition (Giraldeau 1984). Clearly, stable group membership should promote

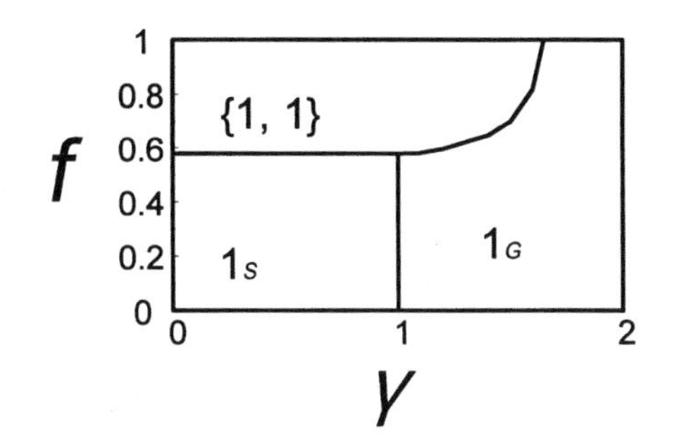

Figure 11.12 Lower *z*-score; skill pool versus solitary foraging. *f* is yield of type-2 clumps; γ determines shape of within-individual constraint. $R = 0.5$, $D = \tau = 1$. Skill pool has lower *z*-score in region indicated by {1, 1}. 1_S indicates where solitary specialist has lower *z*-score; 1_G does same for solitary generalist.

skill pool maintenance, and may be necessary for skill pool formation (Giraldeau and Lefebvre 1986).

The skill pool hypothesis (like several of our models) emphasizes that adaptive plasticity can generate behavioral diversity (e.g., Moran 1992; Thomas 1985). Individual production specializations might be acquired as "irreversible" developmental choices (Moran 1992); plasticity permits different individuals to adopt different searching skills. In this case skill pools would form after individuals specialize. However, specializations might exhibit reversible plasticity (e.g., Stephens 1987), so that group formation could precede skill diversification among group members. Giraldeau and Lefebvre (1986) report evidence for the latter process; generalities await further empirical work.

Math Box 11.1 Maximizing Generalists' Expected Energy Intake

E[$X(\tau)$], given by expression (11.15), is the expected energy intake for paired generalists. To maximize E[$X(\tau)$] with respect to the s_j, note that $\tau 2^{\alpha}$ is a constant. Then we can maximize the objective ($s_1 + fs_2$), subject to equality constraint (11.5). We use the objective and (11.5) to form the Lagrangian L:

$$L = s_1 + fs_2 + \xi(s_1{}^{\gamma} + s_2{}^{\gamma} - D^{\gamma}), \qquad (11.1.1)$$

where ξ is the Lagrange multiplier (e.g., Vincent and Grantham 1981). We can locate a solution (s_1*, s_2*) from the condition

$$\frac{\partial L}{\partial s_1} = \frac{\partial L}{\partial s_2} = \frac{\partial L}{\partial \xi} = 0. \qquad (11.1.2)$$

When $\gamma > 1$, the first two partial derivatives yield

$$\xi = -(1/\gamma)\,(s_1{}^*)^{1-\gamma}. \qquad (11.1.3)$$

Substituting (11.1.3) into $\partial L / \partial s_2$ and employing the within-individual equality constraint (11.5) together specify the solution,

$$s_1{}^* = \frac{D}{(1 + f^{\gamma/\gamma-1})^{1/\gamma}} \qquad (11.1.4)$$

$$s_2{}^* = f^{1/\gamma-1}\, s_1{}^* \qquad (11.1.5)$$

for each individual. (s_1*, s_2*) maximizes E[$X(\tau)$] for the paired generalists when $\gamma > 1$. When $\gamma \leq 1$, paired generalists cannot have a greater expected energy intake than two specialists searching for type-1 clumps.

PART FIVE Final Thoughts

12 Synthesis and Conclusions

12.1 Introduction

In preceding chapters we present a series of models that together provide a framework for a theory for social foraging. The models, which include both novel developments and results taken from the literature, fall into four broad families. Previous analyses of the evolutionary advantages of group foraging lead us to models of group-membership decisions in aggregation economies. The study of animals' spatial and temporal distributions across resource clumps leads to models of group membership in dispersion economies. Figure 12.1 summarizes this family of models. Topics as diverse as gamete dimorphism and food kleptoparasitism suggest a family of producer-scrounger models (Parker 1984b; see fig. 12.2). Conventional foraging theory directs us immediately to models of social foragers' patch-residence and prey-choice behavior; figure 12.3 summarizes these models. Finally, questions in community ecology lead us to models of behavioral diversity within and among group members (see fig. 12.4).

We think a coherent theory for social foraging can help unify research questions that have developed as distinct traditions. Quantitative models promote better theory and guide empirical research, often determining what experimentalists measure and how results are interpreted. In this chapter we try to synthesize the elements of the theory by reviewing our models' themes and emphasizing new directions for research they suggest.

12.2 Group Membership Models

Organizational logic and a respect for history led us to consider first a family of models dealing with conditions that may promote group foraging over a solitary existence. Questions about group membership had previously engendered two research directions that have remained separate (see Pulliam and Caraco 1984). One direction focuses principally on the survival value of foraging-group membership. Less inclined to predict group size, it has developed and tested a number of hypotheses linking advantages of energetic efficiency, avoiding predation or both to group foraging's adaptive significance. The other direction is concerned primarily with group size, and is less

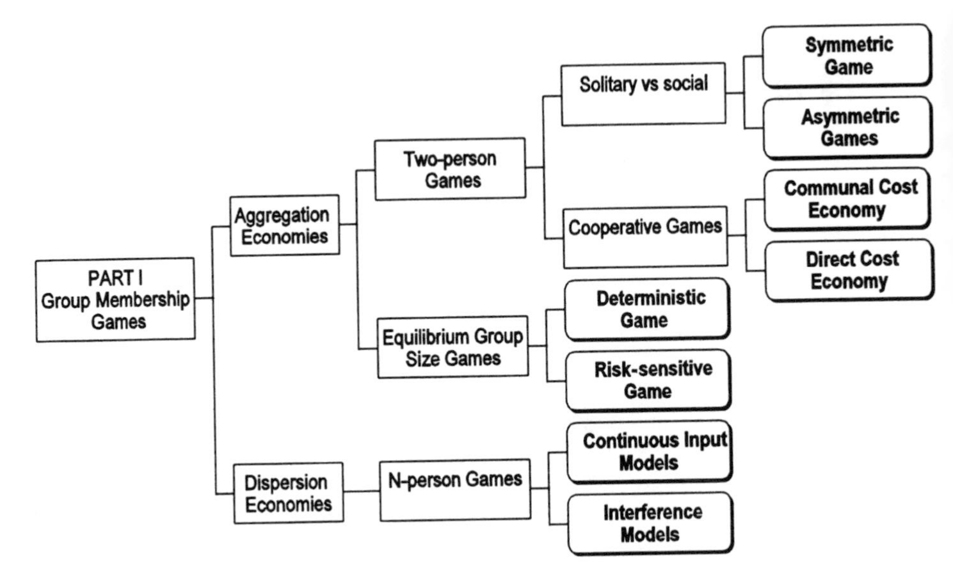

Figure 12.1 Social foraging theory: Family of group membership models. Note that each rounded box corresponds to a model specified in bold character.

concerned with social foraging as an adaptation. It has concentrated on exploring how spatial distributions of resource may influence patterns in consumer group sizes. Our first family of models (fig. 12.1) includes both types of analysis but retains the notable distinction between an aggregation economy and a dispersion economy. This simple unification lets us ask more effective questions about the ecological conditions that give rise to group foraging and about the processes governing group size. The goal of our two-person games is isolating ecological circumstances that might promote stable

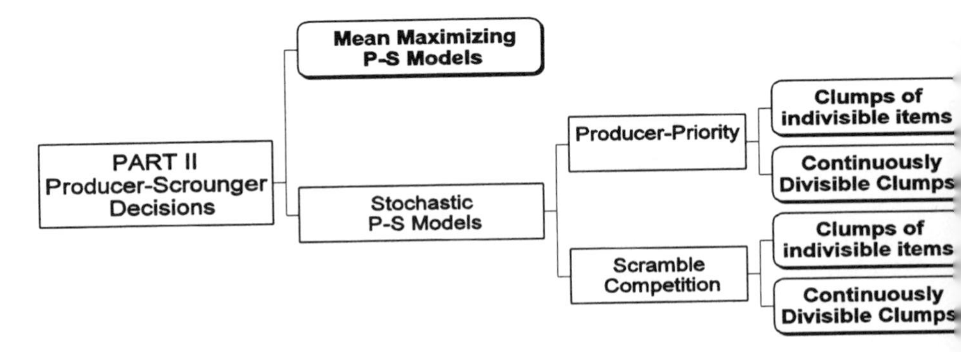

Figure 12.2 Social foraging theory: Family of searching within-group decisions. Note that each rounded box corresponds to a model in bold character.

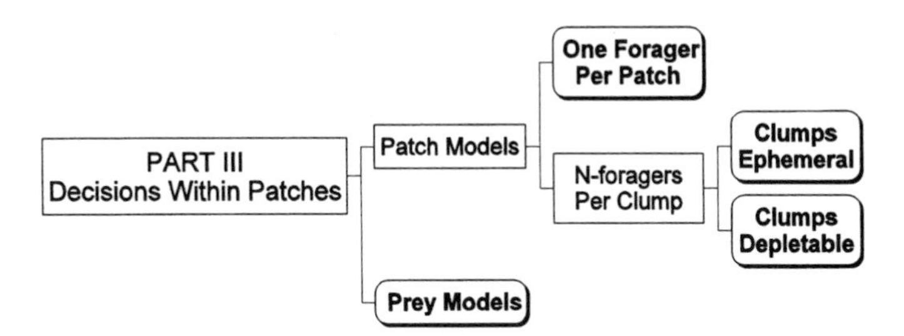

Figure 12.3 Social foraging theory: Family of within-patch models. Note that each rounded box corresponds to a model.

group foraging. Our N-person games go farther and predict the stable size of foraging groups, for both aggregation and dispersion economies. We review briefly what these models tell us.

AGGREGATION ECONOMIES

Our two-person games consider basic questions about the stability of group versus solitary foraging. We used a standard ESS analysis to examine how individuals might "decide" to forage with a partner or not. We found that stable group foraging requires that the group's patch encounter rate exceed some minimum threshold value that is greater than the clump-encounter rate

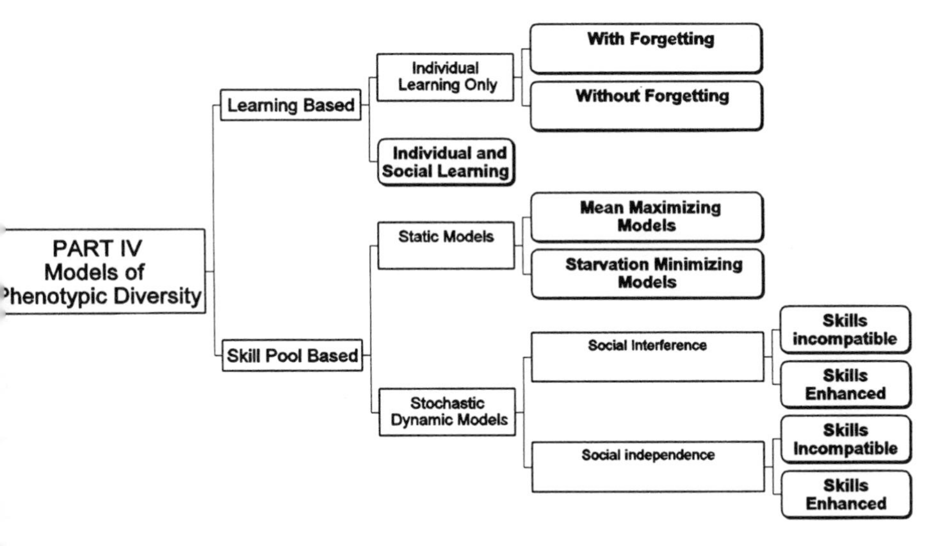

gure 12.4 Social foraging theory: Family of phenotypic diversity models. Note that each rounded x corresponds to a model.

of solitary foragers. This threshold is an important reference point for analyzing the effects of predation and competitor asymmetry on the likelihood of group formation. We showed that allowing group members a lower predation hazard reduces the required clump-encounter threshold, but increasing asymmetry of the division of food between group members increases the threshold.

We also used two-person games to model a forager deciding whether or not to inform its partner of a food discovery, and so share food. Based on the empirical literature, we think this question represents circumstances where the potential for conditionally cooperative foraging (in the sense of the Prisoner's Dilemma) is greatest. We categorized the analysis by discriminating cases with communal and direct costs. A communal cost means that a cost to either partner always implies that both foragers may be penalized. This constraint likely applies to foragers that breed communally or colonially, or are otherwise highly interactive members of a demographic group. For direct costs, each forager's survival is an independent process when neither shares food. Hence the different forager's fates are not as strongly linked as when there are communal costs.

Under communal cost, increasing food density makes sharing of food more likely. More interestingly, the likelihood of food-sharing depends on the probability that both foragers incur a cost when only one fails to find sufficient food. When there is no chance of this communal cost, food-sharing should not occur. For intermediate levels of the communal cost, not-sharing remains stable for both single and repeated play. But cooperators that share food each have a lower chance of a fitness penalty. Consequently, food-sharing, conditional upon mutual cooperation, might also be stable when the chance of repeated interaction is sufficiently large. Finally, when the communal cost exceeds a critical level, the only stable solution is sharing food. For a large enough communal cost, sharing does not require conditional cooperation. Rather, sharing becomes a simple unconditional mutualism, since failing to share will increase the individual's probability of a fitness penalty.

Under direct costs, a qualitatively similar pattern of solutions emerges. In this case, however, sharing of information and food is promoted not by a communal foraging cost, but by a sufficiently greater hazard of predation incurred when feeding alone, compared to exploiting the same food as a group member.

Our equilibrium group size models showed that both stable group size and the extent to which group foraging can provide benefits depend essentially on who controls group entry. A strictly ideal-free perspective allows intruders free entry to groups (Sibly 1983). Group members might profit little from group membership under free entry, because groups can increase in size until they are no longer attractive to solitary individuals. Under free entry, foraging groups may persist not because they are advantageous per se, but because they are the evolutionarily stable solution to aggregation economies.

The exact effect of genetic relatedness of group members on group size has been debated in a number of papers (Rodman 1981; Giraldeau 1988). Our analyses show that the effect of genetic relatedness on group size depends crucially on who controls group entry. Under free entry, increasing genetic relatedness leads to a reduction in group size simply because intruders, then, are less likely to be selected to join groups when this reduces the fitness of individuals to whom they are related. As a consequence, direct benefits of group membership under free entry should increase as relatedness increases, and group members should have fitness advantages over solitaries.

A more common approach to group membership has been to assume, often implicitly, that group members exert complete control over entry. Under group control, members stand to gain the most from group membership; group members will be selected to expel intruders that cause their benefits to decline. When the players are genetically related, kin-directed altruism implies that group members will accept intruder relatives into the group, increasing the size of groups and decreasing the average direct benefit per group member. It is worth pointing out that questions concerning group control are still in need of analysis. In recent studies, for instance, Heinsohn and Packer (1995) found that when lionesses defend their territory against intruding females, some are consistent "laggers" while others are "leaders."

The group size games we presented stress that any study attempting to document advantages of group foraging or predict group size should first establish whether free entry, group control, or a conflict of competing objectives between group members and intruders (Higashi and Yamamura 1993) applies. Moreover, those interested in predicting the effect of genetic relatedness on group size should specify a known, or assumed, rule of entry, since increasing genetic relatedness of group members reduces group size under free entry, but increases it under group control.

Dispersion Economies

N-person group membership models for dispersion economies are common and formulated in two general forms, continuous-input ideal free models and ideal interference models. All of the models assume grouping because there are more consumers than resource clumps. We presented these two types of models and reviewed a rather extensive list of modifications to the original assumptions, all of which address the tendency for consumers to be overrepresented in poorer resource clumps.

We noted a certain lack of parallelism between the amplifications and extensions of the two types of models. For instance, while perceptual constraints seem important in continuous input models, they are not dealt with in studies of the interference model. The same can be said for travel costs.

12.3 Searching Decisions within Groups

The models concerning searching decisions within groups followed from a long-standing disciplinary interest in kleptoparasitism. Our functional nomenclature based on whether a kleptoparasite uses stealth or aggression, and whether it exploits the food exclusively or in scramble competition, more readily reflects the distinguishing features of alternative models of kleptoparasitism than does the current profusion of terms for the phenomenon. So, assigning observed kleptoparasitism to one or the other proposed categories allows researchers to specify which group-foraging model more likely applies to the situation under study.

A common view of group foraging assumes that all individuals search equally and simultaneously for their food, and obtain a share of each group member's food discoveries. Several hypotheses for aggregation economies are based on this picture of group foraging. The hypothetical ability of group foragers to search for their food while simultaneously monitoring all other individual's searching is called the "compatibility assumption." More recent producer-scrounger games assume the opposite, that at any one moment foragers are either searching (producer) only or scrounging (looking for others' discoveries) only: complete incompatibility. The difference between these assumptions is critical; the degree of compatibility affects the frequency dependence of individual payoffs and, consequently, the predicted frequency of strategies within a group.

We argued that complete compatibility was more likely in small groups. Hence, our two-person games were based on complete compatibility. In large groups, on the other hand, monitoring all other foragers may become an onerous task, and so complete incompatibility may be a more appropriate assumption. We suggested that kleptoparasitism occurring in large groups can more realistically be analyzed as a "Producer-Scrounger" game. If this is so, then the way the producer and scrounger tactics (and the foraging payoffs obtained through each) are distributed among individuals will be a function of the relative competitive effectiveness of the strategies and the cost of scrounging. Producer-scrounger games emphasize the need to measure individual payoffs of both phenotypes, the importance of assessing whether these payoffs exhibit a frequency dependence assumed in the model, and, finally, the significance of estimating the distribution of role use across the population studied.

Assuming that the Producer-Scrounger game was a realistic depiction of group-feeding situations, we went on to develop several models. The first was based on mean rate maximization, and the others assumed risk sensitivity. The mean-maximizing model offers a useful first approximation. It predicts that the frequency of scrounger strategists within groups depends crucially on the way each clump is divided between its producer and the scroungers. It also predicts that scrounger should increase with increasing group size. The risk-sensitive models are more complete, and also more complicated, because they incorporate the effect that reward variance has on survival. Once again, the division of food clumps among the producer and scroungers is pivotally important in predicting the frequency of scrounger within groups. Our models showed that scrounging, despite its evolutionary stability, was often detrimental to all group members. It follows that factors governing the frequency of scrounger within groups will influence to no trivial extent the comparative economy of group feeding and solitary foraging. An important conclusion therefore is that more attention should be directed to describing the factors that govern how a food clump is divided among its discoverer and scroungers: the study of resource defense theory.

12.4 Models for Decisions within Patches

Most models in classical foraging theory address a solitary forager's patch exploitation or diet choice. Perhaps surprisingly, only a few models have asked the same questions about social foragers. We reviewed these models, some of which have been available for more than two decades, but somehow escaped testing. We categorized the patch-utilization models depending on whether they assumed individuals compete for access to patches, or compete within patches for access to food. The social patch exploitation models often made qualitatively similar predictions to their classical counterpart, the marginal value theorem, especially with regard to the effects of patch richness and travel time on patch exploitation time. In these cases, we gained relatively little from the added complexity of social foraging unless we went to the quantitative level (where empirical discrimination may prove difficult). In some cases, however, the social patch models made distinct predictions, especially regarding the effect of group size on patch emigration. For instance, when groups exploit depleting patches, they should leave patches sooner, yet exploit them more extensively, as group size increases.

Classical diet models made no direct attempt to deal with the complexities of diet choice under competition. We presented two social prey models. One was based on control theory, while the other was graphical. Their qualitative predictions were similar: the best policy of prey choice under competition was to start off as a specialist on the most profitable prey and then, at some critical prey density, to generalize and take all prey types. To our knowledge, no study has tested this prediction. We find the lack of empirical research on social patch and prey models remarkable, given the continued, strong interest in their counterparts for solitary foragers. Students of classical foraging theory searching for models to test should consider this section of the book closely.

12.5 Models of Phenotypic Diversity

This family of models was motivated by population and community ecologists' interest in the diversity of foraging behavior within feeding aggregations. Some models in this section differ from those usually relished in behavioral ecology because they do not deduce a preferred decision, but instead describe patterns attributable to ecological factors. We share the view that optimality models are not the sole class of tools capable of contributing to an economic theory for behavioral processes, especially those involving population-level phenomena.

We all appreciate the difference between a group of highly stereotyped individuals, each of which adopts a different behavior, and a group of individuals where each exhibits the same diverse set of behaviors. But how do we quantify the difference? Our first contribution in this section presented alternative methods for describing phenotypic variation by partitioning a group's behavioral diversity into its within-individual and among-individual components. Surprisingly few studies have attempted any sort of quantitative description of behavioral variation within foraging groups. The formulas we present offer behavioral ecologists accessible methods for analyzing a social group's behavioral diversity. We are confident that use of these methods will lead to identification of novel patterns in group members' use of resources.

The next chapter dealt with learning. Several of the models we developed in earlier chapters assume that foragers base their decisions on information concerning prevailing environmental conditions that influence survival. Learning is the mechanism through which this information is obtained. Furthermore, many animals' foraging behavior is shaped by learning during development. Hence, behavioral diversity can depend strongly on learning. We proposed a functional nomenclature, *learning about* and *learning how*, which is distinct from the more mechanistic categories favored in experimental psychology. Our terms correspond more closely to what we perceived as behavioral ecologists' interests in information gathering (*learning about*) and behavioral diversity (*learning how*).

Learning is sometimes assumed to homogenize populations behaviorally. Moreover, social groups have often been thought to foster conditions that enhance the acquisition of common behaviors through individual or social learning. We challenged this view by arguing that social foraging often created conditions that inhibit learning and therefore can maintain phenotypic diversity among group members. We first reviewed the types of influences, both positive and negative, that social foraging can exert on both individual learning and social learning. Then we developed models where naive group members' opportunities to practice a foraging skill are constrained by knowledgeable individuals' consistent production of food that is shared by the group. The results demonstrated that phenotypic diversity will persist within foraging groups despite individual learning. In fact, we found that fewer group members are expected to exhibit a new foraging skill than would be observed if the same individuals all foraged solitarily. Moreover, the proportion of group members exhibiting a foraging skill, when individual learning and forgetting equilibrate, decreases as group size increases.

The ability to use social learning—behavior copying in this context (Giraldeau 1997)—can reduce the total phenotypic diversity within the group. When social learning is possible, group size has an enhancing effect on learning so that more individuals within groups acquire the skill than would be expected of an equivalent population of solitary foragers. So, behavioral diversity among group members should decline with increasing group size. Our main point was that individual learning, by itself, will rarely be sufficient to homogenize behavior, even when animals forage at the same time in the same place. Social learning, i.e., any copying of a model's foraging behavior by a naive individual, will exert a more powerful homogenizing force on behavioral diversity. But even this relatively rare learning ability will seldom eliminate behavioral diversity among group members. The impact of learning on a group's behavioral diversity depends on how skills are learned, the effect of the skill on food-discovery rates, and group size.

One obvious source of phenotypic diversity in foraging behavior is the range of resource types exploited. Our final chapter developed a new risk-sensitive, information-sharing model where food occurs in two distinct types of clumps. We analyzed conditions under which we expect animals to form skill pools, where each forager specializes in searching for a different clump type but joins the discoveries of its partner. A skill pool maximizes between-forager diversity in searching behavior, but minimizes between-forager difference of diets. Both static and dynamic versions of the models made similar qualitative predictions. Skill pools should arise when searching for the same clump type produces interference between foragers, when the clump types yield similar energetic rewards, and when the individual's physiologically required food intake is small. The models confirmed that there exist ecological conditions capable of promoting skill pool formation, hence conditions that might economically diversify food searching behavior within a group. To date, however, very little effort has been devoted to estimating either diversity in the way group members search for food or diversity of group members' food consumption. The study of foraging groups as similar specialists, generalists, or skill pools might prove an innovative direction for research in behavioral ecology.

12.6 Conclusions

Our goal in writing this monograph has been to pull together elements of the four distinct research directions that have traditionally focused their attention on problems concerning social foraging. In doing so, we have highlighted a unity spanning what may have seemed as a diverse set of questions. A unified social foraging theory is more than the sum of its parts. Organized into a coherent body of questions and models, it provides an impetus for a renewed interest in foraging behavior. Moreover, social foraging theory provides the opportunity for behavioral ecologists to return to one of the field's initial interests: exploring the mechanisms underlying the structure of ecological communities. We suggested in several chapters how the models we proposed can be extended to the scale of ecological processes. Social foraging theory offers a surprisingly rich, unexploited patch of untested models and predictions. May many readers discover and exploit it.

References

Abrahams, M.V. 1986. Patch choice under perceptual constraints: A case for departures from an ideal free distribution. Behavioral Ecology and Sociobiology 19:409–415.

Abrahams, M.V., and L.M. Dill. 1989. A determination of the energetic equivalence of the risk of predation. Ecology 70:999–1007.

Adams, J.E., and E.D. McCune. 1979. Application of the generalized jackknife to Shannon's measure of information used as an index of diversity. *In* J.F. Grassle, G.P. Patil, W. Smith, and C. Taillie, eds., Ecological Diversity in Theory and Practice 117–132. Fairland, Md.: International Co-operative Publishing House.

Altmann, J. 1974. Observational study of behavior: Sampling methods. Behaviour 49:227–267.

Aoki, K. 1984. A quantitative genetic model of two-policy games between relatives. Journal of Theoretical Biology 109:111–126.

Axelrod, R., and D. Dion. 1988. The further evolution of cooperation. Science 242: 1385–1390.

Axelrod, R., and W.D. Hamilton. 1981.The evolution of cooperation. Science 211: 1390–1396.

Baczkowski, A.J., D.N. Joanes, and G.M. Shamia. 1997. Properties of a generalized diversity index. Journal of Theoretical Biology 188:207–213.

Baird, R.N., and L.M. Dill. 1996. Ecological and social determinants of group size in transient Killer whales. Behavioral Ecology 7:408–416.

Barnard, C.J. 1980. Factors affecting flock size mean and variance in a winter population of house sparrows (*Passer domesticus* L.). Behaviour 74:114–127.

Barnard, C.J., ed. 1984a. Producers and Scroungers: Strategies of Exploitation and Parasitism. New York: Chapman and Hall.

Barnard, C.J. 1984b. When cheats may prosper. In C.J. Barnard, ed., Producers and Scroungers: Strategies of Exploitation and Parasitism, 6–33. New York: Chapman and Hall.

Barnard, C.J., and R.M. Sibly. 1981. Producers and scroungers: A general model and its application to captive flocks of house sparrows. Animal Behaviour 29:543–555.

Barnard, C.J., and D.B.A. Thompson. 1985. Gulls and Plovers: The Ecology and Behaviour of Mixed-species Feeding Groups. London: Croom Helm.

Barta, Z., and L.-A. Giraldeau. 1998. The effect of dominance hierarchy on the use of alternative foraging tactics: A phenotype-limited producer-scrounger game. Behavioral Ecology and Sociobiology 42:217–223.

Barta, Z., and T. Szép. 1995. Frequency-dependent selection on information-transfer strategies at breeding colonies. Behavioral Ecology 6:308–310.

Barta, Z., R. Flynn, and L.-A. Giraldeau. 1997. Geometry for a selfish foraging group: A genetic algorithm approach. Proceedings of the Royal Society London B 264:1–6.

Bartholomew, D.J. 1983. Stochastic Models for Social Processes. New York: Wiley.

Barton, R.A., R.W. Byrne, and A. Whiten. 1996. Ecology, feeding competition and social structure in baboons. Behavioral Ecology and Sociobiology 38:321–329.

Beauchamp, G., and L.-A. Giraldeau 1996. Group foraging revisited: Information-sharing or producer-scrounger game? American Naturalist 148:738–743.

Beauchamp, G., and L.-A. Giraldeau. 1997. Patch exploitation in a producer-scrounger system: Test of a hypothesis using flocks of spice finches (*Lonchura punctulata*). Behavioral Ecology 8:54–59.

Beauchamp, G., and A. Kacelnik. 1991. Effects of the knowledge of partners on learning rates in zebra finches *Taenopygia guttata*. Animal Behaviour 41: 247–253.

Beauchamp, G., L.-A. Giraldeau, and N. Ennis. 1997. Experimental evidence for the maintenance of foraging specializations by frequency-dependent choice in flocks of spice finches. Ethology, Ecology and Evolution 9:105–117.

Beckerman, S. 1983. Optimal foraging group size for a human population: The case of Barí fishing. American Zoologist 23:283–290.

Beckerman, S. 1991. Barí spear fishing: Advantages to group formation? Human Ecology 19:529–554.

Begon, M., J.L. Harper, and C.R. Townsend. 1990. Ecology: Individuals, Populations and Communities. Boston: Blackwell Scientific.

Beissinger, S.R., T.J. Donnay, and R. Walton. 1994. Experimental analysis of diet specialization in the snail kite: The role of behavioral conservatism. Oecologia 100:54–65.

Bélisle, M. 1998. Foraging group size: Models and a test with jaegers kleptoparasitizing terns. Ecology 79:1922–1938.

Bélisle, M., and J.F. Giroux. 1995. Predation and kleptoparasitism by migrating parasitic Jaegers. Condor 97:771–781.

Bergelson, J.M., J.H. Willis, and P.E. Robakiewicz. 1986. Variance in search time: Do groups always reduce risk? Animal Behaviour 34:289–291.

Bernstein, C., A. Kacelnik, and J.R. Krebs. 1988. Individual decisions and the distribution of predators in a patchy environment. Journal of Animal Ecology 57:1007–1026.

Bernstein, C., A. Kacelnik, and J.R. Krebs. 1991.Individual decisions and the distribution of predators in a patchy environment, II. The influence of travel costs and structure of the environment. Journal of Animal Ecology 60:205–225.

Bertram, B.C.R. 1978. Living in groups: Predators and prey. In J.R. Krebs and N.B. Davies, eds., Behavioural Ecology: An Evolutionary Approach, 64–96. Sunderland, Mass.: Sinauer Associates.

Bhat, N. 1972. Elements of Applied Stochastic Processes. New York: Wiley.

Blanckenhorn, W.U. 1992. Group size and the cost of agonistic behavior in pumpkinseed sunfish. Ethology, Ecology and Evolution 4:255–271.

Blanckenhorn, W.U., and T. Caraco. 1992. Social subordinance and a resource queue. American Naturalist 139:442–449.

Blurton Jones, N.G. 1984. A selfish origin for human food sharing: Tolerated theft. Ethology and Sociobiology 5:1–3.

Boag, P. 1987. Adaptive variation in bill size of African seed-crackers. Nature 329: 669–670.

Boerlijst, M.C., M.A. Nowak, and K. Sigmund. 1997. The logic of contrition. Journal of Theoretical Biology 185:281–293.

Boswell, M.T., J.K. Ord, and G.P. Patil. 1979. Chance mechanisms underlying univariate distributions. In J.K. Ord, G.P. Patil, and C. Taillie, eds., Statistical Distributions in Ecological Work. 3–156. Burtonsville, Md.: International Cooperative Publishing House.

Bowman, K.O., K. Hutcheson, E.P. Odom, and L.R. Shenton. 1971. Comments on the distribution of indices of diversity. In G.P. Patil, E.C. Pielou, and W.E. Waters, eds., Statistical Ecology, vol. 3, 315–366. University Park, Penn.: Pennsylvania State University Press.

Boyd, R., and J.P. Lorberbaum. 1987. No pure strategy is evolutionarily stable in the repeated Prisoner's Dilemma game. Nature 327:58–59.

Boyd, R., and P. Richerson. 1988. The evolution of reciprocity in sizable groups. Journal of Theoretical Biology 132:337–356.

Bram, S.J., and W. Mattli. 1993. Theory of moves: Overview and examples. Conflict Management and Peace Science 12:1–39.

Brockmann, J.H., and C.J. Barnard. 1979. Kleptoparasitism in birds. Animal Behaviour 27:487–514.

Brockmann, H.J., A. Grafen, and R. Dawkins. 1979. Evolutionarily stable nesting strategy in a digger wasp. Journal of Theoretical Biology 77:473–496.

Brown, C.R. 1988. Social foraging in cliff swallows: Local enhancement, risk-sensitivity, competition and the avoidance of predators. Animal Behaviour 36: 780–792.

Brown C.R., and J.L. Hoogland. 1986. Risk in mobbing for solitary and colonial swallows. Animal Behaviour 34:1319–1323.

Brown, G.E., and J.A. Brown. 1993. Social dynamics in salmonid fishes: Do kin make better neighbours? Animal Behaviour 45:863–871.

Brown, J.L. 1983. Cooperation: A biologist's dilemma. Advances in the Study of Behavior 13:1–37.

Brown J.L. 1987. Helping and Communal Breeding in Birds. Princeton, N.J.: Princeton University Press.

Brown, J.L., and E.R. Brown. 1980. Reciprocal aid giving in a communal bird. Zeitschrift für Tierpsychologie 53:313–324.

Brown, J.L., and E.R. Brown. 1981. Kin selection and individual fitness in babblers. In R.D. Alexander and D.W. Tinkle, eds., Natural Selection and Social Behavior, 244–256. New York: Chiron Press.

Brown, J.S. 1988. Patch use as an indicator of habitat preference, predation risk, and competiton. Behavioral Ecology and Sociobiology 22:37–47.

Brown, J.S., and T.L. Vincent. 1987. Coevolution as an evolutionary game. Evolution 41:66–79.

Brown, K.M., and J.E. Alexander, Jr. 1994. Group foraging in a marine gastropod predator: Benefits and costs to individuals. Marine Ecology Progress Series 112: 97–105.

Campeiro, C.P. 1989. Cranial morphology and development: New light on the evolution of language. Human Evolution 4:9–32.

Caraco, T. 1979. Time budgeting and group size: A test of a theory. Ecology 60:618–627.

Caraco,T. 1980a. Stochastic dynamics of avian foraging flocks. American Naturalist 115:262–275.

Caraco, T. 1980b. On foraging time allocation in a stochastic environment. Ecology 61:119–128.

Caraco, T. 1981. Risk-sensitivity and foraging groups. Ecology 62:527–531.

Caraco, T. 1987. Foraging games in a random environment. In A.C. Kamil, J.R. Krebs, and H.R. Pulliam, eds., Foraging Behavior. New York: Plenum Press.

Caraco, T., and J.L. Brown. 1986. A game between communal breeders: When is food-sharing stable? Journal of Theoretical Biology 118:379–393.

Caraco, T., and L.-A. Giraldeau. 1991. Social foraging: Producing and scrounging in a stochastic environment. Journal of Theoretical Biology 153:559–583.

Caraco, T., and H.R. Pulliam. 1984. Sociality and survivorship in animals exposed to predation. In P.W. Price, C.N. Slobodchikoff, and W.S. Gaud, eds., A New Ecology: Novel Approaches to Interactive Systems, 279–309. New York: Wiley-Interscience.

Caraco, T., and L.L. Wolf. 1975. Ecological determinants of group sizes of foraging lions. American Naturalist 109:343–352.

Caraco, T., C. Barkan, J.L. Beacham, L. Brisbin, S. Lima, A. Mohan, J.A. Newman, W. Webb, and M.L. Withiam. 1989. Dominance and social foraging: A laboratory study. Animal Behaviour 38:41–58.

Caraco, T., W. Maniatty and B. Szymanski. 1997. Spatial effects and competitive coexistence. In F.L. Silva, J.C. Principe, and L.B. Almeida, eds., Spatiotemporal Models in Biological and Artificial Systems, 9–16. Amsterdam: IOS Press.

Caraco T., S. Martindale, and H.R. Pulliam. 1980a. Avian flocking in the presence of a predator. Nature 285:400–401.

Caraco, T., S. Martindale, and T.W. Whittam. 1980b. An empirical demonstration of risk-sensitive foraging preferences. Animal Behaviour 28:820–830.

Caraco, T., M. Duryea, G. Gardner, W. Maniatty, and B.K. Szymanski. 1998. Host spatial heterogeneity and extinction of an SIS epidemic. Journal of Theoretical Biology 192:351–361.

Caraco, T., G.W. Uetz, R.G. Gillespie, and L.-A. Giraldeau. 1995. Resource-consumption variance within and among individuals: on risk-sensitive coloniality in spiders. Ecology 76:196–205.

Cavalli-Sforza, L.L., and M.W. Feldman. 1981. Cultural Transmission and Evolution: A Quantitative Approach. N.J.: Princeton, Princeton University Press.

Chapman, C.A., and L. Lefebvre. 1990. Manipulating foraging group size: Spider monkey food calls at fruiting trees. Animal Behaviour 39:891–896.

Charnov, E.L., G.H. Orians, and K. Hyatt. 1976. The ecological implications of resource depression. American Naturalist 110:247–259.

Clark, C.W. 1987. The lazy, adaptable lion: A Markovian model of group foraging. Animal Behaviour 35:361–368.

Clark, C.W., and M. Mangel. 1984. Foraging and flocking strategies: Information in an uncertain environment. American Naturalist 123:626–641.

Clark, C.W., and M. Mangel. 1986. The evolutionary advantages of group foraging. Theoretical Population Biology 3:45–75.

Clements, K.C., and D.W. Stephens. 1995. Testing models of non-kin cooperation: Mutualism and the Prisoner's Dilemma. Animal Behaviour 50:527–549.

Clutton-Brock, T.H., and P.H. Harvey. 1984. Comparative approaches to investigating adaptation. In J.R. Krebs and N.B. Davies, eds., Behavioural Ecology: An Evolutionary Approach, 7–29. 2nd ed. Oxford: Blackwell.

Cohen, J.E. 1972. Markov population processes as models for primate social and population dynamics. Theoretical Population Biology 3:119–134.

Cohen, J. 1976. Irreproducible results and the breeding of pigs (or nondegenerate limit nonrandom variables in biology). BioScience 26:391–394.

Colwell, R.K. 1979. Toward a unified approach to the study of species diversity. In J.F. Grassle, G.P. Patil, W. Smith, and C. Taillie, eds., Ecological Diversity in Theory and Practice, 75–91. Fairland, Md.: International Co-operative Publishing House.

Commons, M.L. 1981. How reinforcement density is discriminated and scaled. In M.L. Commons and J.A. Nevins, eds., Quantitative Analyses of Behavior, vol. 1: Discriminative Properties of Reinforcement Schedules, 51–85. Cambridge, Mass.: Ballinger.

Cosmides, L., and J. Tooby. 1987. From evolution to behavior: Evolutionary psychology as the missing link. In J. Dupre, ed., The Latest on the Best: Essays on Evolution and Optimality, 277–306. Cambridge, Mass.: MIT Press.

Crook, J.H. 1965. The adaptive significance of avian social organizations. Symposia of the Zoological Society of London 14:181–218.

Crowley, P.H., and R.C. Sargent. 1996. Whence tit-for-tat? Evolutionary Ecology 10: 499–516.

Croy, M.I., and R.N. Hughes. 1990. The combined effects of learning and hunger in the feeding behaviour of the fifteen-spine stickleback (*Sinachia spinachia* L.). In R. N. Hughes, ed., Behavioural Mechanisms of Food Selection, 215–233. NATO ASI Series, vol. G 20, New York: Springer-Verlag.

Curio, E. 1976. The Ethology of Predation. New York: Springer-Verlag.

Czikeli, H. 1983. Agonistic interactions within a winter flock of slate-coloured juncos (*Junco hyemalis*) evidence for the dominants' strategy. Zeitschrift für Tierpsychologie 61:61–66.

Daly, M., L.F. Jacobs, M.I. Wilson, and P.R. Behrends. 1992. Scatter hoarding by kangaroo rats (*Dipodomys merriami*) and pilferage from their caches. Behavioral Ecology 3:102–111.

D'Amato, R.M. 1973. Delayed matching and short-term memory in monkeys. In G.H. Bower, ed., The Psychology of Learning and Motivation: Advances in Research and Theory, 27–48. New York: Academic Press.

Davies, N.B., and A.I. Houston. 1981. Owners and satellites: The economics of territory defence in the pied wagtail, *Motacilla alba*. Journal of Animal Ecology 50: 157–180.

Davies, N.B., and A.I. Houston. 1983. Time allocation between territories and flocks and owner-satellite conflict in foraging pied wagtails, *Motacilla alba*. Journal of Animal Ecology 52:621–634.

Davies, N.B., and A.I. Houston. 1984. Territory economics. In J.R. Krebs and N.B. Davies, eds., Behavioural Ecology: An Evolutionary Approach, Pages 148–169. 2nd ed. Sunderland, Mass.: Sinauer Associates.

Dawkins, R. 1980. Good strategy or evolutionarily stable strategy? In G.W. Barlow and S. Silverberg, eds., Sociobiology: Beyond Nature/Nurture?, 331–367. Boulder, Col.: Westview Press.

Dawson, B.V., and B.M. Foss. 1965. Observational learning in budgerigars. Animal Behaviour 13:470–474.

Dennis, B., and G.P. Patil. 1979. Species abundance, diversity, and environmental predictability. In J.F. Grassle, G.P. Patil, W. Smith and C. Taillie, eds., Ecological Diversity in Theory and Practice, 93–116. Fairland, Md.: International Co-operative Publishing House.

Diamond, J.M. 1987. Learned specializations of birds. Nature 330:16–17.

Dickinson, A. 1980. Contemporary Animal Learning Theory. Cambridge, U.K.: Cambridge University Press.

Dittus, W.P.G. 1977. The social regulation of population density and age-sex distribution of the Toque monkey. Behaviour 63:281–322.

Dugatkin, L.A. 1990. N-Person games and the evolution of cooperation: A model based on predator inspection in fish. Journal of Theoretical Biology 142:123–135.

Dugatkin, L.A. 1997. Cooperation among Animals: An Evolutionary Perspective. New York: Oxford University Press.

Dugatkin L.A., Mesterton-Gibbons, M., and Houston, A.I. 1992. Beyond the Prisoner's Dilemma: Towards models to discriminate among mechanisms of cooperation in nature. Trends in Ecology and Evolution 7:202–205.

Dunbrack, R. L. 1979. A re-examination of robbing behavior in foraging egrets. Ecology 60: 644–645.

Durrett, R., and S.A. Levin. 1994. Stochastic spatial models: A user's guide. Philosophical Transactions of the Royal Society, London, Series B 343:329–350.

Edwards, J. 1983. Diet shifts in moose due to predator avoidance. Oecologia 60:185–189.

Ekman, J.B. 1986. Tree use and predator vulnerability of wintering passerines. Ornis Scandinavica 17:261–267.

Ekman, J.B., and C.E.H. Askenmo. 1984. Social rank and habitat use in willow tit groups. Animal Behaviour 32:508–514.

Ekman, J., and M. Hake. 1988. Avian flocking reduces starvation risk: An experimental demonstration. Behavioral Ecology and Sociobiology 22:91–94.

Ekman, J., and B. Rosander. 1987. Starvation risk and flock size of the social forager: When there is a flocking cost. Theoretical Population Biology 31:166–177.

Ekman, J., G. Cederholm, and C. Askenmo. 1981. Spacing and survival in winter groups of the willow tit *Parus montanus* and the crested tit *P. cristatus*—a removal study. Journal of Animal Ecology 50:1–9.

Elgar, M. 1986a. House sparrows establish foraging flocks by giving chirrup calls if resources are divisible. Animal Behavior 34:169–174.

Elgar, M. 1986b. The establishment of foraging flocks of house sparrows: Risk of predation and daily temperature. Behavioral Ecology and Sociobiology 19:433–438.

Elgar, M., and C.P. Catterall. 1981. Flocking and predator surveillance in house sparrows: A test of a hypothesis. Animal Behaviour 29:868–872.

Elgar, M.A., H. McKay, and P. Woon. 1986. Scanning, pecking and alarm flights in house sparrows. Animal Behaviour 34:1892–1894.

Elliott, L. 1978. Social behavior and foraging ecology of the eastern chipmunk (*Tamias striatus*) in the Adirondack Mountains. Smithsonian Contributions to Zoology 265.

Emlen, S.T. 1984. Cooperative breeding in birds and mammals. In J.R. Krebs and N.B. Davies, eds., Behavioural Ecology: An Evolutionary Approach, 305–339. 2nd ed., Sunderland, Mass.: Sinauer Associates.

Emlen, S.T. 1995. An evolutionary theory of the family. Proceedings of the National Academy of Sciences USA 92:8092–8099.

Engen, S. 1979. Some basic concepts of ecological equitability. In J.F. Grassle, G.P. Patil, W. Smith and C. Taillie, eds., Ecological Diversity in Theory and Practice, 37–50. Fairland, Md.: International Co-operative Publishing House.

Ens, B.J., P. Esselink, and L. Zwarts. 1990. Kleptoparasitism as a problem of prey choice: A study of mudflat-feeding curlews, *Numenius arquata*. Animal Behaviour 39:219–230.

Erlandsson, A. 1988. Food-sharing vs. monopolising prey: A form of kleptoparasitism in *Velia caprai* (Heteroptera). Oikos 53: 23–26.

Fagen, R. 1987a. A generalized habitat matching rule. Evolutionary Ecology 1:5–10.

Fagen, R. 1987b. Phenotypic plasticity and social environment. Evolutionary Ecology 1:263–271.

Farrell, J., and R. Ware. 1989 Evolutionary stability in the repeated prisoner's dilemma. Theoretical Population Biology 36:161–166.

Feldman, M.W., and E.A.C. Thomas. 1987. Behavior-dependent contexts for repeated plays of the Prisoner's Dilemma, II: Dynamical aspects of the evolution of cooperation. Journal of Theoretical Biology 128:297–315.

Ferrière, R., and R.E. Michod. 1996. The evolution of cooperation in spatially heterogeneous populations. American Naturalist 147:692–717.

Field, J. 1989. Intraspecific parasitism and nesting success in the solitary wasp *Ammophila sabulosa*. Behaviour 110:23–46.

Field, J. 1992. Intraspecific parasitism as an alternative reproductive tactic in nest building wasps and bees. Biological Review 67:79–126.

Fisher, J., and R.A. Hinde. 1949. The opening of milk bottles by birds. British Birds 42:347–357.

Foster, S.A. 1985. Group foraging by a coral reef fish: Mechanism for gaining access to defended resources. Animal Behaviour 33:782–792.

Fragaszy, D.M., and E. Visalberghi. 1989. Social influences on the acquisition of tool-using behaviors in tufted capuchin monkeys (*Cebus apella*). Journal of Comparative Psychology 103:159–170.

Fragaszy, D.M., and E. Visalberghi. 1996. Social learning in monkeys: Primate "primacy" reconsidered. In C.M. Heyes and B.G. Galef Jr., eds., Social Learning in Animals, 65–84. Toronto: Academic Press.

Fretwell, S.D. 1972. Populations in a Seasonal Environment. Princeton, N.J.: Princeton University Press.

Fretwell, S.D., and H.L. Lucas. 1970. On territorial behavior and other factors influencing habitat distribution in birds. Acta Biotheoretica 19:16–36.

Galef, B.G., Jr. 1988. Imitation in animals: History, definition, and interpretation of data from the psychological laboratory. In T.R. Zentall and B.G. Galef Jr., eds., Social Learning: Psychological and Biological Perspectives, 3–28. Hillsdale, N.J.: Lawrence-Erlbaum Associates.

Galef, B.G., Jr. 1991. Information centres of Norway rats: Sites for information exchange and information parasitism. Animal Behaviour 41:295–301.

Galef, B.G., Jr. 1996. Social enhancement of food preferences in Norway rats: A brief review. In C.M. Heyes and B.G. Galef Jr., eds., Social Learning in Animals. Toronto: Academic Press.

Galef, B.G., Jr., L.A. Manzig, and R.M. Field. 1986. Imitation learning in budgerigars: Dawson and Foss (1965) revisited. Behavioural Processes 13:191–202.

Gese, E.M., O.J. Rongstad, and W.R. Mytton. 1988. Relationship between coyote group size and diet in southeastern Colorado. Journal of Wildlife Management 52:647–653.

Ghent, A.W. 1960. A study of the group-feeding behaviour of the jack pine sawfly, *Neodiprion pratti banksianae* Roh. Behaviour 16:110–148.

Giffin, W.C. 1978. Queueing: Basic Theory and Applications. Columbus, Ohio: Grid Inc.

Gilliam, J.F., and D.F. Fraser. 1987. Habitat selection under predation hazard: Test of a model with foraging minnows. Ecology 68:1856–1862.

Giraldeau, L.-A. 1984. Group foraging: The skill pool effect and frequency-dependent learning. American Naturalist 124:72–79.

Giraldeau L.-A. 1988. The stable group and the determinants of foraging group size. In C.N. Slobodchikoff, ed., The Ecology of Social Behavior, 33–53. New York: Academic Press.

Giraldeau, L.-A. 1997. The ecology of information use. In J.R. Krebs and N.B. Davies, eds., Behavioural Ecology: An Evolutionary Approach, 42–68. 4th ed. Oxford: Blackwell Scientific.

Giraldeau, L.-A., and T. Caraco. 1993. Genetic relatedness and group size in an aggregation economy. Evolutionary Ecology 7:429–438.

Giraldeau, L.-A., and D. Gillis. 1985. Optimal group size can be stable: A reply to Sibly. Animal Behaviour 33:666–667.

Giraldeau, L.-A., and D. Gillis. 1988. Do lions hunt in group sizes that maximize hunters' daily food returns? Animal Behaviour 36:611–613.

Giraldeau, L.-A., and L. Lefebvre. 1985. Individual feeding preferences in feral groups of rock doves. Canadian Journal of Zoology 63:189–191.

Giraldeau, L.-A., and L. Lefebvre. 1986. Exchangeable producer and scrounger roles in a captive flock of feral pigeons: A case for the skill pool effect. Animal Behaviour 34: 797–783.

Giraldeau, L.-A., and L. Lefebvre. 1987. Scrounging prevents cultural transmission of food-finding behaviour in pigeons. Animal Behaviour 35: 387–394.

Giraldeau, L.-A., and B. Livoreil. 1998. Game theory and social foraging. In L. A. Dugatkin and H.K. Reeve, eds., Game Theory and Animal Behavior, 16–37. New York: Oxford University Press.

Giraldeau, L.-A., and J.J. Templeton. 1991. Food scrounging and diffusion of foraging skills in pigeons, *Columba livia*: The importance of tutor and observer rewards. Ethology 89:63–72.

Giraldeau, L.-A., T. Caraco, and T. Valone. 1994c. Social foraging: Individual learning and cultural transmission of innovations. Behavioral Ecology 5:35–43.

Giraldeau, L.-A., J.A. Hogan, and M.J. Clinchy. 1990. The payoffs to producing and scrounging: What happens when patches are divisible? Ethology 85:132–146.

Giraldeau, L.-A., D.L. Kramer, I. Deslandes, and H. Lair. 1994a. The effect of competitors and distance on central place foraging in eastern chipmunks, *Tamias striatus*. Animal Behaviour 47:621–632.

Giraldeau, L.-A., C. Soos, and G. Beauchamp. 1994b. A test of the producer-scrounger foraging game in captive flocks of spice finches, *Lonchura punctulata*. Behavioral Ecology and Sociobiology 34:251–256.

Godin, J.G.J. 1986. Risk of predation and foraging behaviour in shoaling banded killifish (*Fundulus diaphanus*). Canadian Journal of Zoology 64:1675–1678.

Goodwin, D. 1954. Notes on feral pigeons. Avicultural Magazine 60:190–213.

Goss-Custard, J.D. 1976. Variation in the dispersion of redshank *Tringa totanus* on their winter feeding grounds. Ibis 118:257–263.

Goss-Custard, J.D., and S.E.A. le V. dit Durell. 1987a. Age-related effects in oystercatchers, *Haematopus ostralegus*, feeding on mussels, *Mytilus edulis*. I: Foraging efficiency and interference. Journal of Animal Ecology 56:521–536.

Goss-Custard, J.D., and S.E.A. le V. dit Durell. 1987b. Age-related effects in oystercatchers, *Haematopus ostralegus*, feeding on mussels, *Mytilus edulis*. II: Aggression. Journal of Animal Ecology 56:537–548.

Goss-Custard, J.D., and S.E.A. le V. dit Durell. 1987c. Age-related effects in oystercatchers, *Haematopus ostralegus*, feeding on mussels, *Mytilus edulis*. III: The effect of interference on overall foraging efficiency. Journal of Animal Ecology 56: 549–558.

Goss-Custard, J.D., R.W.G. Caldow, R.T. Clarke, S.E.A. le V. dit Durell, and W.J. Sutherland. 1995. Deriving population parameters from individual variations in foraging behaviour. I: Empirical game-theory distribution model of oystercatchers *Haematopus ostralegus* feeding on mussels *Mytilus edulis*. Journal of Animal Ecology 64:265–276.

Götmark, F.A. 1990. A test of the information-centre hypothesis in a colony of sandwich terns *Sterna sandvicensis*. Animal Behaviour 39:487–495.

Gould, S.J., and R.C. Lewontin. 1979. The spandrels of San Marco and the Panglossian paradigm: A critique of the adaptationist programme. Proceedings of the Royal Society of London, Series B 205:581–598.

Grafen, A. 1982. How not to measure inclusive fitness. Nature 298:425–426.

Grafen, A. 1984. Natural selection, kin selection and group selection. In J.R. Krebs and N.B. Davies, eds., Behavioural Ecology: An Evolutionary Approach, 62–84. 2nd ed. Sunderland, Mass.: Sinauer Associates.

Grafen, A. 1991. Modelling in behavioural ecology. In J.R. Krebs and N.B. Davies, eds., Behavioural Ecology: An Evolutionary Approach, 5–31. 3rd ed. Oxford: Blackwell.

Grand, T.C., and J.W.A. Grant. 1994. Spatial predictability of resources and the ideal free distribution in convict cichlids, *Cichlasoma nigrofasciatum*. Animal Behaviour 48:909–919.

Grant, D.S., and W.A. Roberts. 1973. Trace interaction in pigeon short-term memory. Journal of Experimental Psychology 101:21–29.

Grant, J.W.A., and D.L. Kramer. 1990. Temporal clumping of food arrival reduces its monopolization and defence by zebrafish *Brachydanio rerio*. Animal Behaviour 44:101–110.

Grassle, J.F., G.P. Patil, W. Smith, and C. Taillie, eds. 1979. Ecological Diversity in Theory and Practice. Fairland, Md.: International Co-operative Publishing House.

Gray, R. 1987. Faith and foraging: A critique of the "Paradigm Argument from Design." In A.C. Kamil, J.R. Krebs, and H.R. Pulliam, eds., Foraging Behavior, 69–138. New York: Plenum Press.

Green, R.F. 1984. Stopping rules for optimal foragers. American Naturalist 123:80–90.

Green, R.F. 1987. Stochastic models of optimal foraging. In A.C. Kamil, J.R. Krebs, and H.R. Pulliam, eds., Foraging Behavior, 273–302. New York: Plenum Press.

Green, R.F. 1989. Putting ecology back into optimal foraging theory. Mathematics and Statistics Technical Report 89–11, University of Minnesota, Duluth.

Greene, E. 1987. Individuals in an osprey colony discriminate between high and low quality information. Nature 329:239–241.

Gross, M.R. 1996. Alternative reproductive strategies and tactics: Diversity within the sexes. Trends in Ecology and Evolution 11:92–97.

Haigh, J. 1975. Game theory and evolution. Advances in Applied Probability 7:8–11.

Hake, M., and J. Ekman. 1988. Finding and sharing depletable patches: When group foraging decreases intake rates. Ornis Scandinavica 19:275–279

Hamilton, W.D. 1964. The genetical evolution of social behavior. Journal of Theoretical Biology 7:1–52.

Hammerstein, P. 1996. Darwinian adaptation, population genetics and the streetcar theory of evolution. Journal of Mathematical Biology 34:511–532.

Hansen, A. 1986. Fighting behavior in bald eagles: A test of game theory. Ecology 67: 787–797.

Hanski, I. 1980. Movement patterns in dung beetles and in the dung fly. Animal Behaviour 28:953–964.

Harley, C.B. 1981. Learning the evolutionarily stable strategy. Journal of Theoretical Biology 89:611–633.

Hart A., and D.W. Lendrem. 1984. Vigilance and scanning patterns in birds. Animal Behaviour 32:1216–1224.

Harvey, P.H., and M.D. Pagel. 1991. The Comparative Method in Evolutionary Biology. Oxford: Oxford University Press.

Hassel, M.P., and R.M. May. 1985. From individual behaviour to population dynamics. In R.M. Sibly and R.H. Smith, eds., Behavioural Ecology: Ecological Consequences of Adaptive Behaviour, 3–32. Oxford: Blackwell Scientific.

Hatch, J.J. 1973. Predation and piracy by gulls at a ternery in Maine. Auk 87:244–254.

Heinsohn, R., and C. Packer. 1995. Complex cooperative strategies in group-territorial African lions. Science 269:1260–1262.

Heller, R. 1980. On optimal diet in a patchy environment. Theoretical Population Biology 17:201–214.

Herbers, J.M. 1981. Time resources and laziness in animals. Oecologia 49:252–262.

Heyes, C. 1993. Imitation, culture and cognition. Animal Behaviour 46:999–1010.

Heyes, C.M., and B.G. Galef, Jr. 1996. Social Learning in Animals: The Roots of Culture. Toronto: Academic Press.

Heyes, C.M., and G.R. Dawson. 1990. A demonstration of observational learning using a bidirectional control. Quarterly Journal of Experimental Psychology 42B: 59–71.

Higashi, M., and N. Yamamura. 1993. What determines the animal group size? Insider-outsider conflict and its resolution. American Naturalist 142:553–563.

Hines, W.G.S. 1980. Three characterizations of population strategy stability. Journal of Applied Probability 17:333–340.

Hines, W.G.S. 1987. Evolutionarily stable strategies: A review of basic theory. Theoretical Population Biology 79:19–30.

Hirshleifer, D., and E. Rasmussen. 1988. Cooperation in a repeated prisoner's dilemma with ostracism. Technical Report 18–86, Anderson Graduate School of Management, University of California at Los Angeles.

Hockey, P.A.R., P.G. Ryan, and A.L. Bosman. 1989. Age-related intraspecific kleptoparasitism and foraging success of kelp gulls, *Larus domenicanus*. Ardea 77: 205–210.

Holling, C.S. 1959. Some characteristics of simple types of predation and parasitism. Canadian Entomologist 91:385–398.

Hoogland, J.L. 1979. Aggression, ectoparasitism and other possible costs of prairie dog (Sciuridae:*Cynomys* spp.) coloniality. Behaviour 69:1–35.

Hoogland J.L., and P.W. Sherman. 1976. The advantages and disadvantages of bank swallow (*Riparia riparia*) coloniality. Ecological Monographs 46:33–58.

Houston, A.I. 1987. The control of foraging decisions. In M.L. Commons, A. Kacelnik, and S.J. Shettleworth, eds., Quantitative Analyses of Behavior, 41–61. Hillsdale, N.J.: Lawrence Erlbaum Associates.

Houston, A.I. 1993. Mobility limits cooperation. Trends in Ecology and Evolution 8:194–196.

Houston, A.I. 1996. Estimating oystercatcher mortality from foraging behaviour. Trends in Ecology and Evolution 11:108–110.

Houston, A.I., and J.M. McNamara. 1981. How to maximize reward rate on two variable-interval paradigms. Journal of the Experimental Analysis of Behavior 35: 367–396.

Houston, A.I., and J.M. McNamara. 1982. A sequential approach to risk-taking. Animal Behaviour 30:1260–1261.

Houston, A.I., and J.M. McNamara. 1985. The choice of two prey types that minimises the probability of starvation. Behavioral Ecology and Sociobiology 17:135–141.

Houston, A.I., and J.M. McNamara. 1986. Evaluating the selection pressures on foraging decisions. In R. Campan and R. Zayan, eds., Relevance of Models and Theories in Ethology, 61–75. Toulouse: Privat.

Houston, A.I., and J.M. McNamara. 1987. Singing to attract a mate: A stochastic dynamic game. Journal of Theoretical Biology 129:57–68.

Houston A.I., and J.M. McNamara. 1988. Fighting for food: A dynamic version of the Hawk-Dove game. Evolutionary Ecology 2:51–64.

Houston A.I., and J.M. McNamara. 1992. Phenotypic plasticity as a state-dependent life-history decision. Evolutionary Ecology 6:243–253.

Houston A.I., and J.M. McNamara. 1993. A theoretical investigation of the fat reserves and mortality levels of small birds in winter. Ornis Scandanavica 24:205–219.

Houston, A.I., C.W. Clark, J.M. McNamara, and M. Mangel. 1988. Dynamic models in behavioural and evolutionary ecology. Nature 332:29–34.

Houston A.I., J.M. McNamara, and J.M. Hutchinson. 1993. General results concerning the trade-off between gaining energy and avoiding predation. Philosophical Transactions of the Royal Society of London, Series B 341:375–397.

Houston A.I., J.M. McNamara, and M. Milinski. 1995. The distribution of animals between resources: A compromise between equal numbers and equal intake rates. Animal Behaviour 49:248–251.

Huey, R.B. 1991. Physiological consequences of habitat selection. American Naturalist 137:S91–S115.

Hughes, R.N. 1979. Optimal diets under the energy maximisation premise: The effects of recognition time and learning. American Naturalist 113:209–221.

Huntingford, F.A. 1993. Can cost-benefit analysis explain fish distribution patterns? Journal of Fish Biology 43:289–308.

Inman, A., L. Lefebvre, and L.-A. Giraldeau. 1987. Individual diet differences in feral pigeons: Evidence for resource partitioning. Animal Behaviour 35:1902–1903.

Ives, A.R. 1995. Spatial heterogeneity and host-parasitoid population dynamics: Do we need to study behavior? Oikos 74:366–376.

Izawa, K. 1976. Group sizes and compositions of monkeys in the upper Amazon basin. Primates 17:367–399.

Janson, C.H. 1984. Female choice and mating system of the brown-capuchin monkey *Cebus apella* (Primates: Cebidae). Zeitschrift für Tierpsychologie 65:177–200.

Janson, C.H. 1985. Aggressive competition and individual food consumption in wild brown capuchin monkeys (*Cebus apella*). Behavioral Ecology and Sociobiology 18:125–138.

Janson, C.H. 1988. Food competition in brown capuchin monkeys (*Cebus apella*): Quantitative effects of group size and tree productivity. Behaviour 105:53–76.

Jarman, P.J. 1982. Prospects for interspecific comparison in sociobiology. In King's College Sociobiology Group, ed., Current Problems in Sociobiology, 323–342. Cambridge, U.K.: Cambridge University Press.

Kacelnik, A., J.R. Krebs, and C. Bernstein. 1992. The ideal free distribution and predator-prey populations. Trends in Ecology and Evolution 7:50–54.

Kamil, A.C. 1983. Optimal foraging theory and the psychology of learning. American Zoologist 23:291–302.

Karmeshu, and R.K. Pathria. 1979. Cooperative behavior in a non-linear model of diffusion of information. Canadian Journal of Physics 57:1572–1578.

Keeny, R.L., and H. Raiffa. 1976. Decisions with Multiple Objectives: Preferences and Value Tradeoffs. New York: Wiley.

Kelly, F.P. 1979. Reversibility and Stochastic Networks. New York: Columbia University Press.

Kennedy, M., and R.D. Gray. 1993. Can ecological theory predict the distribution of foraging animals? A critical evaluation of experiments on the ideal free distribution. Oikos 68:158–166.

Koops, M., and L.-A. Giraldeau. 1996. Producer-scrounger foraging games in starlings: A test of mean-maximizing and risk-minimizing foraging models. Animal Behaviour 51: 773–783.

Kramer, D.L. 1985. Are colonies supraoptimal groups? Animal Behavior 33:1031–1032.

Krebs, J.R. 1973. Social learning and the significance of mixed-species flocks of chickadees. Canadian Journal of Zoology 51:1275–1288.

Krebs, J.R., and N.B. Davies. 1984. Behavioural Ecology: An Evolutionary Approach. 2nd ed. Oxford: Blackwell.

Krebs, J.R., and J.A. Inman. 1992. Learning and foraging: Individuals, groups and populations. American Naturalist 140:S63–S84.

Krebs, J.R., A. Kacelnik, and P. Taylor. 1978. Test of optimal sampling by foraging great tits. Nature 275:539–542.

Krebs, J.R., M.H. MacRoberts, and J.M. Cullen. 1972. Flocking and feeding in the great tit *Parus major*—an experimental study. Ibis 114:507–530.

Krebs, J.R., D.W. Stephens, and W.J. Sutherland. 1983. Perspectives in foraging theory. In A.H. Brush and G.A. Clark Jr., eds., Perspective in Ornithology, 165–221. Cambridge, U.K.: Cambridge University Press.

Krivan, C. 1996. Optimal foraging and predator-prey dynamics. Theoretical Population Dynamics 49:265–290.

Kurland, J.A., and S.J. Beckerman. 1985. Optimal foraging and hominid evolution: Labor and reciprocity. American Anthropologist 87:73–93.

Kushlan, J. A. 1978. Nonrigorous foraging by robbing egrets. Ecology 59: 649–653.

Laland, K., P.J. Richerson, and R. Boyd. 1996. Developing a theory of animal social learning. In C.M. Heyes and B.G. Galef Jr., eds., Social Learning in Animals, 129–154. Toronto: Academic Press.

Lande, R. 1982. A quantitative genetic theory of life history evolution. Ecology 63: 607–615.

Langen, T.A., and K.N. Rabenold. 1994. Dominance and diet selection in Juncos. Behavioral Ecology 5:334–338.

Laverty, T. M. 1994. Costs to foraging bumble bees of switching plant species. Canadian Journal of Zoology 72:43–47.

Lazarus, J. 1972. Natural selection and the function of flocking in birds: A reply to Murton. Ibis 114:556–558.

Lefebvre, L. 1983. Equilibrium distribution of feral pigeons at multiple food sources. Behavioral Ecology and Sociobiology 12:11–17.

Lefebvre, L. 1986. Cultural diffusion of a novel food-finding behaviour in urban pigeons: An experimental field test. Ethology 71:295–34.

Lefebvre, L. 1995. The opening of milk bottles by birds: Evidence for accelerating learning rates, but against the wave-of-advance model of cultural transmission. Behavioural Processes 34:43–54.

Lefebvre, L., and L.-A. Giraldeau. 1984. Daily feeding site use of urban pigeons. Canadian Journal of Zoology 62:1425–1428.

Lefebvre, L., and L.-A. Giraldeau. 1994. Cultural transmission in pigeons is affected by the number of tutors and bystanders present during demonstrations. Animal Behaviour 47:331–337.

Leigh, E.G. 1991. Genes, bees and ecosystems: The evolution of a common interest among individuals. Trends in Ecology and Evolution 6:257–262.

Lendrem, D. 1986. Modelling in Behavioural Ecology. Portland, Ore.: Timber Press.

León, J.A. 1993. Plasticity in fluctuating environments. In J. Yoshimura and C.W. Clark, eds., Adaptation in Stochastic Environments. Lecture Notes in Biomathematics 98. New York: Springer-Verlag.

Lessells, C.M. 1991. The evolution of life histories. In J.R. Krebs and N.B. Davies, eds., Behavioural Ecology: An Evolutionary Approach, 32–68. 3rd ed. Oxford: Blackwell.

Lessells, C.M. 1995. Putting resource dynamics into continuous input ideal free distribution models. Animal Behaviour 49:487–494.

Levin, S.A., B. Grenfell, A. Hastings, and A.S. Perelson. 1997. Mathematical and computational challenges in population biology and ecosystems science. Science 275:334–343.

Lima, S.L., and L.M. Dill. 1990. Behavioural decisions made under the risk of predation: A review and prospectus. Canadian Journal of Zoology 68:619–640.

Lima, S.L., T.J. Valone, and T. Caraco. 1985. Foraging efficiency-predator risk tradeoff in the grey squirrel. Animal Behaviour 33:155–165.

Lindström, Å. 1989. Finch flock size and risk of hawk predation at a migratory stopover site. Auk 106:225–232.

Lively, C.M. 1986. Canalization versus developmental conversion in a spatially variable environment. American Naturalist 128:561–572.

Livoreil, B., and L.-A. Giraldeau. 1997. Departure decisions by spice finches foraging singly or in groups. Animal Behaviour 54:967–977.

Lorberbaum, J. 1994. No strategy is evolutionarily stable in the repeated prisoner's dilemma. Journal of Theoretical Biology 168:117–130.

Łomnicki, A. 1988. Population Ecology of Individuals. Princeton, N.J.: Princeton University Press.

Lovegrove, B.G., and C. Wissel. 1988. Sociality in molerats: Metabolic scaling and the role of risk sensitivity. Oecologia 74:600–606.

Mangel, M. 1990. Resource divisibility, predation and group formation. Animal Behaviour 39:1163–1172.

Mangel, M., and C.W. Clark. 1983. Uncertainty, search and information in fisheries. Journal of the International Council for Exploring the Seas 41:93–103.

Mangel, M., and C.W. Clark. 1986. Towards a unified foraging theory. Ecology 67: 1127–1138.

Mangel, M., and C.W. Clark. 1988. Dynamic Modelling in Behavioral Ecology. Princeton, N.J.: Princeton University Press.

Mangel, M., and R.E. Plant. 1985. Regulatory mechanisms and information processing in uncertain fisheries. Marine Resource Economics 1:389–418.

Maniatty, W.A., B.K. Szymanski, and T. Caraco. 1998. Parallel computing with generalized cellular automata. Parallel and Distributed Computing Practices 1:31–50.

Martindale, S. 1982. Nest defense and central place foraging: A model and experiment. Behavioral Ecology and Sociobiology 10:85–89.

May, R.M. 1974. Stability and Complexity in Model Ecosystems. Princeton, N.J.: Princeton University Press.

May, R.M. 1975. Patterns of species abundance and diversity. In M.L. Cody and J. Diamond, eds., Ecology and Evolution of Communities, 81–120. Cambridge, Mass.: Belknap Press of Harvard University Press.

May, R.M. 1987. More evolution of cooperation. Nature 327:15–17.

Maynard Smith, J. 1978. Optimization theory in evolution. Annual Review of Ecology and Systematics 9:31–56.

Maynard Smith, J. 1982. Evolution and the Theory of Games. Cambridge, U.K.: Cambridge University Press.

Maynard Smith, J. 1984. Game theory and the evolution of behaviour. Behavior and Brain Sciences 7:95–125.

Maynard Smith, J. 1988. Can a mixed strategy be stable in a finite population? Journal of Theoretical Biology 130:247–251.
Maynard Smith, J., and G.A. Parker. 1976. The logic of asymmetric contests. Animal Behaviour 24:159–175.
Mayr, E. 1983. How to carry out the adaptationist program? American Naturalist 121:324–334.
McNamara, J.M. 1982. Optimal patch use in a stochastic environment. Theoretical Population Biology 21:269–285.
McNamara, J.M. 1990. The policy which maximizes long-term survival of an animal faced with the risks of starvation and predation. Advances in Applied Probability 22:295–308.
McNamara, J. M., and A.I. Houston. 1980. The application of statistical decision theory to animal behaviour. Journal of Theoretical Biology 85:673–690.
McNamara, J.M., and A.I. Houston. 1985. A simple model of information use in the exploitation of patchily-distributed food. Animal Behaviour 33:553–560.
McNamara, J.M., and A.I. Houston. 1986. The common currency for behavioral decisions. American Naturalist 127:358–378.
McNamara, J.M., and A.I. Houston. 1989. State-dependent contests for food. Journal of Theoretical Biology 137:457–479.
McNamara, J.M., and A.I. Houston. 1990. State-dependent ideal free distributions. Evolutionary Ecology 4:298–311.
McNamara, J.M., and A.I. Houston. 1992. Risk-sensitive foraging: A review of the theory. Bulletin of Mathematical Biology 54:355–378.
McNamara, J.M., and A.I. Houston. 1994.The effect of a change in foraging options on intake rate and predation rate. American Naturalist 144:978–1000.
McNamara, J.M., A.I. Houston, and J.N. Webb. 1994. Dynamic kin selection. Proceedings of the Royal Society of London, Series B 258:23–28.
McNamara, J.M., J.N. Webb, E.J. Collins, T. Székeli, and A.I. Houston. 1997. A general technique for computing evolutionarily stable strategies based on errors in decision-making. Journal of Theoretical Biology 189:211–225.
McQuoid, L.M., and B.G. Galef, Jr. 1992. Social influences of feeding site selection by Burmese fowl *Gallus gallus*. Journal of Comparative Psychology 106:137–141.
Mesterton-Gibbons, M. 1991. An escape from "the prisoner's dilemma." Journal of Mathematical Biology 29:251–269.
Mesterton-Gibbons, M. 1992. An Introduction to Game-theoretic Modelling. New York: Addison-Wesley.
Mesterton-Gibbons, M., and L.A. Dugatkin. 1992. Cooperation among unrelated individuals: Evolutionary factors. Quarterly Review of Biology 67:267–281.
Mesterton-Gibbons, M., and L.A. Dugatkin. 1997. Cooperation and the prisoner's dilemma: Towards testable models of mutualism versus reciprocity. Animal Behaviour 54:551–557.
Michod, R.E. 1984. Constraints on adaptation, with special reference to social behavior. In P.W. Price, C.N. Slobodchikoff and W.S. Gaud, eds., A New Ecology: Novel Approaches to Interactive Systems, 253–278. New York: Wiley-Interscience.
Milinski, M. 1979. An evolutionarily stable feeding strategy in sticklebacks. Zeitschrift für Tierpsychologie 51:36–40.

Milinski, M. 1984. Competitive resource sharing: An experimental test of a learning rule for ESSs. Animal Behaviour 32:233–242.
Milinski, M. 1988. Games fish play: Making decisions as a social forager. Trends in Ecology and Evolution 3:325–330.
Milinski, M., and G.A. Parker. 1991. Competition for resources. In J.R. Krebs and N.B. Davies, eds., Behavioural Ecology: An Evolutionary Approach, 137–168. 3rd ed. Oxford: Blackwell Scientific.
Mitchell, W.A. 1990. An optimal control theory of diet selection: The effects of resource depletion and exploitative competition. Oikos 58:16–24.
Moon, R.D., and H.P. Zeigler. 1979. Food preferences in the pigeon (*Columba livia*). Physiology and Behavior 22:1171–1182.
Moran, N.A. 1992. The evolutionary maintenance of alternative phenotypes. American Naturalist 139:971–989.
Morse, D.H. 1980. Behavioral Mechanisms in Ecology. Cambridge, Mass.: Harvard University Press.
Morse, D.H. and R.S. Fritz. 1987. The consequences of foraging for reproductive success. In A.C. Kamil, J.R. Krebs, and H.R. Pulliam, eds., Foraging Behaviour, 443–456. New York: Plenum Press.
Motro, U. 1991. Co-operation and defection: Playing the field and the ESS. Journal of Theoretical Biology 151:145–154.
Mottley, K., and L.-A. Giraldeau. In press. Experimental evidence that group foragers can reach stable equilibria when engaged in a producer-scrounger game. Animal Behaviour.
Mowbray, M. 1997. Evolutionarily stable strategy distributions for the repeated prisoner's dilemma. Journal of Theoretical Biology 187:223–229.
Murdoch, W.W., and A. Oaten. 1975. Predation and population stability. Advances in Ecological Research 9:2–131.
Murton, R.K. 1971. The significance of a specific search image in the feeding behaviour of the wood-pigeon. Behaviour 40:10–42.
Newman, J.A. 1991. Patch use under predation hazard: Foraging behavior in a simple stochastic environment. Oikos 61:29–44.
Newman, J.A., and T. Caraco. 1989. Co-operative and non-co-operative bases of food-calling. Journal of Theoretical Biology 141:197–209.
Nicol, C.J., and S.J. Pope. 1994. Social learning in small flocks of laying hens. Animal Behaviour 47:1289–1296.
Nishimura, K. 1992. Foraging in an uncertain environment: Patch exploitation. Journal of Theoretical Biology 156:91–111.
Nishimura, K., and D.W. Stephens. 1997. Iterated prisoner's dilemma: Pay-off variance. Journal of Theoretical Biology 141:197–209.
Noë, R.A. 1990. A veto game played by baboons: A challenge to the use of the prisoner's dilemma as a paradigm for reciprocity and cooperation. Animal Behavior 39:78–90.
Norton-Griffiths, M. 1967. Some ecological aspects of the feeding behaviour of the oystercatcher *Haematopus ostralegus* on the edible mussel *Mytilus edulis*. Ibis 109:412–424.
Nowak, M.A., and R.M. May. 1992. Evolutionary games and spatial chaos. Nature 359:826–829.

Nowak, M.A., and K. Sigmund. 1992. Tit for tat in heteregeneous populations. Nature 355:250–252.

Nowak, M.A., and K. Sigmund. 1993. Win-stay, lose-shift out-performs tit-for-tat. Nature 364:56–58.

Nowak, M.A., and K. Sigmund. 1994. The alternative prisoner's dilemma. Journal of Theoretical Biology 168:219–226.

Nudds, T.D. 1978. Convergence of group size strategies by mammalian social carnivores. American Naturalist 112:957–960.

Ollason, J.G. 1987. Learning to forage in a regenerating patchy environment: Can it fail to be optimal? Theoretical Population Biology 31:13–32.

Olsén, H.K., T. Järvi, and A.-C. Löf. 1996. Aggressiveness and kinship in brown trout (*Salmo trutta*) parr. Behavioral Ecology 7:445–450.

Ord, J.K. 1972. Families of Frequency Distributions. New York: Hafner Publishing.

Orians, G.H., and J.F. Wittenberger. 1991. Spatial and temporal scales in habitat selection. American Naturalist 137:S29–S49.

Oster, G.F., and E.O. Wilson. 1978. Caste and Ecology in the Social Insects. Princeton, N.J.: Princeton University Press.

Owen, G. 1968. Game Theory. Philadelphia: Saunders.

Packer, C. 1986. The ecology of sociality in felids. In D.I. Rubenstein and W. Wrangham, eds., Ecological Aspects of Social Evolution: Birds and Mammals, 429–451. Princeton, N.J.: Princeton University Press.

Packer, C., and L. Ruttan. 1988. The evolution of cooperative hunting. American Naturalist 132:159–198.

Packer, C., D. Sheel, and A.E. Pusey. 1990. Why lions form groups: Food is not enough. American Naturalist 136:1–19.

Palameta, B. 1989. The importance of socially transmitted information in the acquisition of novel foraging skills by pigeons and canaries. Ph.D. thesis, King's College, Cambridge University, England.

Palameta, B., and L. Lefebvre. 1985. The social transmission of a food-finding technique in pigeons: What is learned? Animal Behaviour 33:892–896.

Parker, G.A. 1970. Sperm competition and its evolutionary effect on copulation duration in the fly *Scatophaga stercoraria*. Journal of Insect Physiology 16:1301–1328.

Parker, G.A. 1978. Searching for mates. In J.R. Krebs and N. B. Davies, eds., Behavioural Ecology: An Evolutionary Approach, 3–61. Sunderland, Mass.: Sinauer Associates.

Parker, G.A. 1982. Phenotype-limited evolutionarily stable strategies. In King's College Sociobiology Group, eds., Current Problems in Sociobiology, 173–201. Cambridge, U.K.: Cambridge University Press.

Parker, G. A. 1984a. Evolutionarily stable strategies. In J.R. Krebs and N. B. Davies, eds., Behavioural Ecology: An Evolutionary Approach, 3–61. 2nd ed. Sunderalnd, Mass.: Sinauer Associates.

Parker, G.A. 1984b. The producer/scrounger model and its relevance to sexuality. In C.J. Barnard, ed., Producers and Scroungers: Strategies of Exploitation and Parasitism, 127–153. New York: Chapman and Hall.

Parker G.A. 1984c. Sperm competition and the evolution of animal mating strategies. In R.L. Smith, ed., Sperm Competition and Evolution of Animal Mating Systems, 1–60. New York: Academic Press.

Parker, G.A. 1985. Population consequences of evolutionarily stable strategies. In R.M. Sibly and H.R. Smith, eds., Behavioural Ecology: Ecological Consequences of Adaptive Behaviour. Oxford: Blackwell Scientific.

Parker, G.A., and R.A. Stuart. 1976. Animal behaviour as a strategy optimizer: Evolution of resource assessment strategies and optimal emigration thresholds. American Naturalist 110:1055–1076.

Parker, G.A., and W.J. Sutherland. 1986. Ideal free distributions when individuals differ in competitive ability: Phenotype-limited ideal free models. Animal Behaviour 34:1222–1242.

Parker, G.A., L.W. Simmons, and P.I. Ward. 1993. Optimal copula duration in dungflies: effects of frequency dependence and female mating status. Behavioral Ecology and Sociobiology 32:157–166.

Partridge, L., and P. Green. 1985. Intraspecific feeding specializations and population dynamics. In R.M. Sibly and R.H. Smith, eds., Behavioural Ecology: Ecological Consequences of Adaptive Behaviour, 207–226. Oxford: Blackwell Scientific.

Partridge, L., and P. Green. 1987. An advantage for specialist feeding in jackdaws *Corvus monedula*. Animal Behaviour 35:982–990.

Patil, G.P., and M.L. Rosenzweig. 1979. Contemporary Quantitative Ecology and Related Ecometrics. Fairland, Md.: International Co-operative Publishing House.

Patil, G.P., and C. Taillie. 1979. An overview of diversity. In J.F. Grassle, G.P. Patil, W. Smith, and C. Taillie, eds., Ecological Diversity in Theory and Practice, 3–27. Fairland, Md.: International Co-operative Publishing House.

Pielou, E.C. 1975. Ecological Diversity. New York: Wiley.

Pietriwicz, A.T., and A.C. Kamil. 1979. Search image formation in the blue jay (*Cyanocitta cristata*). Science 204:1332–1333.

Pitcher, T. J. 1986. Functions of schooling behaviour in teleosts. In T.J. Pitcher, ed., The Behaviour of Teleost Fishes. London: Croom Helm.

Pitcher, T.J., A.E. Magurran, and I.J. Winfield. 1982. Fish in larger shoals find food faster. Behavioral Ecology and Sociobiology 10:149–151.

Polis, G.A. 1984. Age structure component of niche width and intraspecific resource partitioning: Can age groups function as ecological species? American Naturalist 123:541–564.

Post, D.G., G. Hausfater, and S.A. McClusky. 1980. Feeding behavior of yellow baboons (*Papio cynocephalis*): Relationship to age, gender, and dominance rank. Folia Primatologica 34:170–195.

Price, T.D. 1984. The evolution of sexual size dimorphism in Darwin's finches. American Naturalist 123:500–518.

Pulliam, H. R. 1980. Learning to forage optimally. In A.C. Kamil and T.S. Sargent, eds., Foraging Behavior. New York: Garland Press.

Pulliam, H.R. 1983. Ecological community theory and the coexistence of sparrows. Ecology 64:45–52.

Pulliam, H.R. 1988. Sources, sinks and population regulation. American Naturalist 132: 652–661.

Pulliam, H.R., and T. Caraco. 1984. Living in groups: Is there an optimal group size? In J.R. Krebs and N.B. Davies, eds., Behavioural Ecology: An Evolutionary Approach, 122–147. 2nd ed. Sunderland, Mass.: Sinauer Associates.

Pulliam, H.R., and B.J. Danielson. 1991. Sources, sinks and habitat selection: A landscape perspective on population dynamics. American Naturalist 137:S50–S66.

Pulliam, H.R., and G.C. Millikan. 1982. Social organization in the non-reproductive season. In D.S. Farmer and J.R. King, eds., Avian Biology, vol. 6, 117–132. New York: Academic Press.

Pulliam, H.R., G.H. Pyke, and T. Caraco. 1982. The scanning behavior of juncos: A game-theoretical approach. Journal of Theoretical Biology 95:89–103.

Pyke, G.H. 1984. Optimal foraging theory: A critical review. Annual Review of Ecology and Systematics 15:523–575.

Rachlin, H., L. Green, J.H. Kagel, and R.C. Battalio. 1976. Economic demand theory and psychological studies of choice. In G. Bower, ed., The Psychology of Learning and Motivation, vol. 10, 129–154. New York: Academic Press.

Rachlin, H., J.H. Kagel, and R.C. Battalio. 1980. Substitutability in time allocation. Psychological Review 87:355–374.

Ranta, E., N. Peukhuri, A. Laurila, H. Rita, and N.B. Metcalfe. 1996. Producers, scroungers and foraging group structure. Animal Behaviour 51:171–175.

Ranta, E., H. Rita, and K. Lindström. 1993. Competition versus cooperation: Success of individuals foraging alone and in groups. American Naturalist 142:42–58.

Rapport, D.J. 1980. Optimal foraging for complementary resources. American Naturalist 116:324–346.

Rayor, S. L., and G. W. Uetz. 1990. Trade-offs in foraging success and predation risk with spatial position in colonial spiders. Behavioral Ecology and Sociobiology 27:77–85.

Recer, G.M., and T. Caraco. 1989a. Food sharing, survival and the habitat-matching rule. Animal Behaviour 37:153–168.

Recer, G.M., and T. Caraco. 1989b. Sequential-encounter prey choice and effects of spatial resource variability. Journal of Theoretical Biology 139:239–249.

Recer, G.M., W.U. Blanckenhorn, J.A. Newman, E.M. Tuttle, M.L. Withiam, and T. Caraco. 1987. Temporal resource variability and the habitat-matching rule. Evolutionary Ecology 1:363–378.

Reeve, H.K., and P. Nonacs. 1997. Within-group aggression and the value of group membership: Theory and field test with social wasps. Behavioral Ecology 8:75–82.

Regelmann, K. 1984. Competitive resource sharing: A simulation model. Animal Behaviour 32:226–232.

Riechert S.E., R. Roeloffs, and A.C. Echternacht. 1986. The ecology of the cooperative spider *Agelena consociata* in equatorial Africa (Aranae, Agelenidae). Journal of Arachnology 14:175–192.

Riley, J.G. 1979. Evolutionary equilibrium strategies. Journal of Theoretical Biology 76:109–123.

Rita, H., E. Ranta, and N. Peuhkuri. 1997. Group foraging, patch exploitation time and the finder's advantage. Behavioral Ecology and Sociobiology 40:35–39.

Robichaud, D., L. Lefebvre, and L. Robidoux. 1996. Dominance affects resource partitioning in pigeons, but pair bonds do not. Canadian Journal of Zoology 74: 833–840.

Rodman, P.S. 1981. Inclusive fitness and group size with a reconsideration of group sizes in lions and wolves. American Naturalist 118:275–283.

Rohwer, S., and P.W. Ewald. 1981. The cost of dominance and advantage of subordination in a badge signaling system. Evolution 35:441–454.

Rosenzweig, M.L. 1981. A theory of habitat selection. Ecology 62:327–335.

Rosenzweig, M.L. 1991. Habitat selection and population interactions: The search for mechanism. American Naturalist 137:S5–S28.

Roughgarden, J. 1974. Niche width: Biogeographic patterns among *Anolis* lizard populations. American Naturalist 128:561–572.

Roughgarden, J. 1976. Resource partitioning among competing species—a coevolutionary approach. Journal of Theoretical Biology 9:338–424.

Roughgarden, J. 1979. Theory of Population Genetics and Evolutionary Ecology: An Introduction. Upper Saddle River, N.J.: Prentice Hall.

Rubenstein, D.I., R.J. Barnett, R.S. Ridgeley, and P.H. Klopfer. 1977. Adaptive advantages of mixed-species feeding flocks among seed-eating finches in Costa Rica. Ibis 119:10–21.

Ruxton, G.D., S.J. Hall, and Gurney, W.S.C. 1995. Attraction toward feeding conspecifics when food patches are exhaustible. American Naturalist 145:653–660.

Schaffer, M.E. 1988. Evolutionarily stable strategies for a finite population and a variable contest size. Journal of Theoretical Biology 132:469–478.

Schaffer, W.M. 1978. A note on the theory of reciprocal altruism. American Naturalist 112:250–253.

Schaller, G.B. 1972. The Serengeti Lion. Chicago: University of Chicago Press.

Schoener, T.W. 1986. Mechanistic approaches to community ecology: A new reductionism? American Zoologist 26:81–106.

Schoener, T.W. 1987. A brief history of optimal foraging ecology. In A.C. Kamil, J.R. Krebs, and H.R. Pulliam, eds., Foraging Behavior. New York: Plenum Press.

Schulman, S.E., and B. Chapais.1980. Reproductive value and rank relations among macaque sisters. American Naturalist 115:589–593.

Seeley, T.D., and P.K. Visscher. 1988. Assessing the benefits of cooperation in honeybee foraging: Search costs, forage quality, and competitive ability. Behavioral Ecology and Sociobiology 22:229–237.

Selten, R., and P. Hammerstein. 1984. Gaps in Harley's argument on evolutionarily stable learning rules and the logic of "tit-for-tat." Behavior and Brain Sciences 7:115.

Sharma, C.L., R.K. Pathria, and Karmeshu. 1982. Critical behavior of a class of nonlinear stochastic models of diffusion of information. Physical Review A 26: 3567–3674.

Sharma, C.L., R.K. Patria, and Karmeshu. 1983. Diffusion of information in a social group. Journal of Mathematical Sociology 9:211–226.

Sherry, D.F., and B.G. Galef, Jr. 1984. Cultural transmission without imitation: Milk bottle opening by birds. Animal Behaviour 32:937–938.

Shettleworth, S. J. 1984. Learning and behavioural ecology. In J.R. Krebs and N.B. Davies, eds., Behavioural Ecology: An Evolutionary Approach, 170–194. 2nd ed. Sunderland, Mass.: Sinauer Associates.

Shettleworth, S.J. 1998. Cognition, Evolution, and Behavior. New York: Oxford University Press.

Shettleworth, S.J., J.R. Krebs, D.W. Stephens, and J. Gibbon. 1988. Tracking a fluctuating environment: A study of sampling. Animal Behaviour 36:87–105.

Shumway, R., and J. Gurland. 1960. Fitting the Poisson binomial distribution. Biometrics 16:522–533.

Sibly, R.M. 1983. Optimal group size is unstable. Animal Behaviour 31:947–948.

Sjerps, M., and P. Haccou. 1994. Effects of competition on optimal patch leaving: A War of Attrition. Theoretical Population Biology 46:300–318.

Slatkin, M., and R. Lande. 1976. Niche width in a fluctuating environment–density independent model. American Naturalist 110:31–55.

Smith, B.H. 1987. Effects of genealogical relationship and colony age on the dominance hierarchy in the primitively eusocial bee, *Lasioglossum zephyrum.* Animal Behaviour 35:211–217.

Smith, E.A. 1985. Inuit foraging groups: Some simple models incorporating conflicts of interest, relatedness, and central place sharing. Ethology and Sociobiology 6:27–47.

Stacy, P.B. 1986. Group size and foraging efficiency in yellow baboons. Behavioral Ecology and Sociobiology 18:175–187.

Staddon, J.E.R., and R.H. Ettinger. 1989. Learning: An Introduction to the Principles of Adaptive Behavior. San Diego: Harcourt, Brace and Jovanovich.

Stephens, D.W. 1981. The logic of risk-sensitive foraging preferences. Animal Behaviour 29:628–629.

Stephens, D.W. 1982. Stochasticity in foraging theory: risk and information. D.Phil. thesis, Oxford University, Oxford, England.

Stephens, D.W. 1987. On economically tracking a variable environment. Theoretical Population Biology 32:15–25.

Stephens, D.W., and E.L. Charnov. 1982. Optimal foraging: Some simple stochastic models. Behavioral Ecology and Sociobiology 10:251–263.

Stephens, D.W., and J.R. Krebs. 1986. Foraging Theory. Princeton, N.J.: Princeton University Press.

Stephens, D.W., K. Nishimura, and K.B. Toyer. 1995. Error and discounting in the Iterated Prisoner's Dilemma. Journal of Theoretical Biology 176:457–469.

Stokes, A.W., and H.W. Williams. 1972. Courtship feeding calls in gallinaceous birds. Auk 89:177–180.

Sullivan, K. A. 1988. Ontogeny of time budgets in yellow-eyed juncos: Adaptation to ecological constraints. Ecology 69:118–124.

Sutherland, W.J. 1983. Aggregation and the "ideal free" distribution. Journal of Animal Ecology 52:821–828.

Sutherland, W.J. 1996. From Individual Behaviour to Population Ecology. Oxford: Oxford University Press.

Sutherland, W.J., and G.A. Parker. 1985. Distribution of unequal competitors. In R.M. Sibly and R.H. Smith, eds., Behavioural Ecology: Ecological Consequences of Adaptive Behaviour, 255–273. Oxford: Blackwell Scientific.

Székely, T., P.D. Sozou, and A.I. Houston.1991. Flocking behaviour of passerines: A dynamic model for the non-reproductive season. Behavioral Ecology and Sociobiology 28:203–213.

Tamm, S. 1987. Tracking varying environments: Sampling by hummingbirds. Animal Behaviour 35:1725–1734.

Taylor, P.D., and L.B. Jonker. 1978. Evolutionarily stable strategies and game dynamics. Mathematical Biosciences 40:145–156.

Templeton, J.J. 1994. The use of personal and public information in foraging flocks of European starlings. Ph.D. thesis, Concordia University, Montréal, Canada.

Templeton, J.J., and L.-A. Giraldeau. 1995a. Patch assessment in foraging flocks of European starlings: Evidence for public information use. Behavioral Ecology 6:65–72.

Templeton, J.J., and L.-A. Giraldeau. 1995b. Public information and personal decisions: The costs and benefits of information use in foraging flocks of European Starlings (*Sturnus vulgaris*). Animal Behaviour 49:1617–1626.

Templeton, J.J., and L.-A. Giraldeau. 1996. Vicarious sampling: The use of personal and public information by starlings foraging in a simple patchy environment. Behavioural Ecology and Sociobiology 38:105–113.

Terborgh, J. 1983. Five New World Primates. Princeton, N.J.: Princeton University Press.

Thomas, B. 1985. On evolutionarily stable sets. Journal of Mathematical Biology 22:105–115.

Thompson, D.B.A. 1986. The economics of kleptoparasitism: Optimal foraging, host and prey selection by gulls. Animal Behaviour 34:1189–1205

Thompson, D.B.A., and D.W. Lendrem. 1985. Gulls and plovers: Host vigilance, kleptoparasite success and a model of kleptoparasitic detection. Animal Behaviour 33:1318–1324.

Thorpe, W. H. 1956. Learning and Instinct in Animals. London: Methuen.

Tinbergen, N. 1964. Social Behaviour in Animals with Special Reference to Vertebrates. London: Chapman and Hall.

Tregenza, T. 1995. Building on the ideal free distribution. Advances in Ecological Research 26:253–307.

Tregenza, T., G.A. Parker, and D.J. Thompson. 1996a. Interference and the ideal free distribution: Models and tests. Behavioral Ecology 7: 379–386.

Tregenza, T., D.J. Thompson, and G.A. Parker. 1996b. Interference and the ideal free distribution: Oviposition in a parasitoid wasp. Behavioral Ecology 7:387–394.

Uetz, G.W. 1989. The "ricochet effect" and prey capture in colonial spiders. Oecologia 81:154–159.

Uetz, G.W., and M.A. Hodge. 1990. Influence of habitat and prey availability on spatial organization and behavior of colonial web-building spiders. National Geographic Research 6:22–40.

Valone, T.J. 1989. Group foraging, public information and patch estimation. Oikos 56:357–363.

Valone, T.J. 1993. Patch information and estimation: A cost of group foraging. Oikos 68:258–266.

Valone, T.J., and J.S. Brown. 1989. Measuring patch assessment abilities of desert granivores. Ecology 70:1800–1810.

Valone, T., and L.-A. Giraldeau. 1993. Patch estimation in group foragers: What information is used? Animal Behaviour 45:721–728.

Vehrencamp, S.L. 1983. A model for the evolution of despotic versus egalitarian societies. Animal Behaviour 31:667–682.

Via, S., and R. Lande. 1985. Genotype-environment interaction and the evolution of phenotypic plasticity. Evolution 39:505–522.

Vickery, W.L. 1987. How to cheat against a simple mixed strategy ESS. Journal of Theoretical Biology 127:133–139.

Vickery, W.L., L.-A. Giraldeau, J.J. Templeton, D.L. Kramer, and C.A. Chapman. 1991. Producers, scroungers and group foraging. American Naturalist 137:847–863.

Vincent, T.L., and J.S. Brown. 1984. Stability in an evolutionary game. Theoretical Population Biology 26:408–427.

Vincent, T.L., and W.J. Grantham. 1981. Optimality in Parametric Systems. New York: Wiley.

Vollrath, F. 1984. Kleptobiotic interactions in invertebrates. In C.J. Barnard, ed., Producers and Scroungers: Strategies of Exploitation and Parasitism. New York: Chapman and Hall.

Waltz, E.C. 1983. On tolerating followers in information-centers, with comments on testing the adaptive significance of coloniality. Colonial Waterbirds 6:31–36.

Ward, P. I. 1986. Prey availability increases less quickly than nest size in the social spider *Stegodyphus mimosarum.* Behaviour 94:213–225.

Ward, P.I., and M.M. Enders. 1985. Conflict and cooperation in the group feeding social spider *Stegodyphus mimosarum.* Behaviour 94:167–182.

Weathers, W.W., and K.A. Sullivan. 1989. Juvenile foraging proficiency, parental effort and avian reproductive success. Ecological Monographs 59:223–246.

Weissing, F.J. 1996. Genetic versus phenotypic models of selection: Can genetics be neglected in a long-term perspective? Journal of Mathematical Biology 34:533–555.

Werner T.K., and T.W. Sherry. 1987. Behavioral feeding specialization in *Pinaroloxias inornata*, the "Darwin's finch" of Cocos Island, Costa Rica. Proceedings of the National Academy of Sciences USA 84:5506–5510.

West-Eberhard, M.J. 1989. Phenotypic plasticity and the origins of diversity. Annual Review of Ecology and Systematics 20:249–278.

Whiten, A., and R. Ham. 1992. On the nature and evolution of imitation in the animal kingdom: Reappraisal of a century of research. Advances in the Study of Behaviour 21:239–283.

Whitten, P.L. 1983. Diet and dominance among female vervet monkeys (*Cercopithecus aethiops*). American Journal of Primatology 5:139–159.

Wilkinson, G.S. 1984. Reciprocal food sharing in vampire bats. Nature 309:181–184.

Wilkinson, G.S. 1985. The social organization of the common vampire bat. II: Mating systems, genetic structure and relatedness. Behavioral Ecology and Sociobiology 17:123–134.

Wilson, D.S. 1980. The Natural Selection of Populations and Communities. Menlo Park, Calif.: Benjamin/Cummings.

Wilson, D.S., and M. Turelli. 1986. Stable underdominance and the evolutionary invasion of empty niches. American Naturalist 127:835–850.

Winterhalder, B. 1996. Social foraging and the behavioral ecology of intra-group resource transfers. Evolutionary Anthropology. 5:46–57.

Winterhalder, B. 1997. Gifts given, gifts taken: The behavioral ecology of non-market, intra-group exchange. Journal of Archaeological Research. 5:121–169.

Wolpoff, M.H. 1980. Paleoanthropology. New York: Knopf.

Woodward, G.L., and T.M. Laverty. 1992. Recall of flower handling skills by bumble bees: A test of Darwin's interference hypothesis. Animal Behaviour 44:1045–1051.

Wrangham, R.W. 1977. Feeding behaviour of chimpanzees in Gombe National Park, Tanzania. In T.H. Clutton-Brock, ed., Primate Ecology: Studies of Feeding and Ranging Behaviour in Lemurs, Monkeys and Apes, 504–538. London: Academic Press.

Wrangham, R.W. 1986. The evolution of social structure. In B.B. Smuts, D.L. Cheyney, R.M. Seyfarth, R.W. Wrangham, and T.T. Struhsaker, eds., Primate Societies. Chicago: University of Chicago Press.

Yamamura, N., and M. Higashi. 1992. An evolutionary theory of conflict resolution between relatives: Altruism, manipulation, compromise. Evolution 46:1236–1239.

Yamamura, N., and N. Tsuji. 1987. Optimal patch time under exploitative competition. American Naturalist 129:553–567.

Ydenberg, R.C., and L.M. Dill. 1986. The economics of fleeing from predators. Advances in the Study of Behavior 16:229–249.

Zentall, T.R., and B.G. Galef, Jr. 1988. Social Learning: Psychological and Biological Perspectives. Hillsdale, N.J.: Lawrence-Erlbaum Associates.

Subject Index

Species Index